Conceptual Mathematics, 2nd Edition

In the last 60 years, the use of the notion of category has led to a remarkable unification and simplification of mathematics. *Conceptual Mathematics* introduces this tool for the learning, development, and use of mathematics, to beginning students and general readers, but also to practicing mathematical scientists. This book provides a skeleton key, making explicit some concepts and procedures that are common to all branches of pure and applied mathematics.

The treatment does not presuppose knowledge of specific fields, but rather develops, from basic definitions, such elementary categories as discrete dynamical systems and directed graphs; the fundamental ideas are then illuminated by examples in these categories.

This second edition provides links with more advanced topics of possible study. In the new appendices and annotated bibliography the reader will find concise introductions to adjoint functors and geometrical structures, as well as sketches of relevant historical developments.

Conceptual Mathematics, 2nd Edition

A first introduction to categories

F. WILLIAM LAWVERE
SUNY at Buffalo

STEPHEN H. SCHANUEL
SUNY at Buffalo

CAMBRIDGE
UNIVERSITY PRESS

University Printing House, Cambridge CB2 8BS, United Kingdom

One Liberty Plaza, 20th Floor, New York, NY 10006, USA

477 Williamstown Road, Port Melbourne, VIC 3207, Australia

314-321, 3rd Floor, Plot 3, Splendor Forum, Jasola District Centre, New Delhi - 110025, India

79 Anson Road, #06-04/06, Singapore 079906

Cambridge University Press is part of the University of Cambridge.

It furthers the University's mission by disseminating knowledge in the pursuit of education, learning and research at the highest international levels of excellence.

www.cambridge.org
Information on this title: www.cambridge.org/9780521719162

First published 1997
Second edition 2009
9th printing 2019

A catalogue record for this publication is available from the British Library

Library of Congress Cataloging in Publication data
Lawvere, F.W.
Conceptual mathematics : a first introduction to categories / F.William Lawvere, Stephen H. Schanuel. – 2nd ed.
p. cm.
Includes index.
ISBN 978-0-521-71916-2 (pbk.) – ISBN 978-0-521-89485-2 (hardback)
1. Categories (Mathematics) I. Schanuel, S. H. (Stephen Hoel), 1933– II. Title.
QA169.L355 2008
512′.62–dc22 2007043671

ISBN 978-0-521-89485-2 Hardback
ISBN 978-0-521-71916-2 Paperback

to Fatima

Contents

	Preface	xiii
	Organisation of the book	xv
	Acknowledgements	xvii
	Preview	
Session 1	Galileo and multiplication of objects	3
	1 Introduction	3
	2 Galileo and the flight of a bird	3
	3 Other examples of multiplication of objects	7
	Part I The category of sets	
Article I	**Sets, maps, composition**	13
	1 Guide	20
Summary:	Definition of category	21
Session 2	Sets, maps, and composition	22
	1 Review of Article I	22
	2 An example of different rules for a map	27
	3 External diagrams	28
	4 Problems on the number of maps from one set to another	29
Session 3	Composing maps and counting maps	31
	Part II The algebra of composition	
Article II	**Isomorphisms**	39
	1 Isomorphisms	39
	2 General division problems: Determination and choice	45
	3 Retractions, sections, and idempotents	49
	4 Isomorphisms and automorphisms	54
	5 Guide	58
Summary:	Special properties a map may have	59

Session 4	Division of maps: Isomorphisms	60
	1 Division of maps versus division of numbers	60
	2 Inverses versus reciprocals	61
	3 Isomorphisms as 'divisors'	63
	4 A small zoo of isomorphisms in other categories	64
Session 5	Division of maps: Sections and retractions	68
	1 Determination problems	68
	2 A special case: Constant maps	70
	3 Choice problems	71
	4 Two special cases of division: Sections and retractions	72
	5 Stacking or sorting	74
	6 Stacking in a Chinese restaurant	76
Session 6	Two general aspects or uses of maps	81
	1 Sorting of the domain by a property	81
	2 Naming or sampling of the codomain	82
	3 Philosophical explanation of the two aspects	84
Session 7	Isomorphisms and coordinates	86
	1 One use of isomorphisms: Coordinate systems	86
	2 Two abuses of isomorphisms	89
Session 8	Pictures of a map making its features evident	91
Session 9	Retracts and idempotents	99
	1 Retracts and comparisons	99
	2 Idempotents as records of retracts	100
	3 A puzzle	102
	4 Three kinds of retract problems	103
	5 Comparing infinite sets	106
Quiz		108
How to solve the quiz problems		109
Composition of opposed maps		114
Summary/quiz on pairs of 'opposed' maps		116
Summary:	On the equation $p \circ j = 1_A$	117
Review of 'I-words'		118
Test 1		119
Session 10	Brouwer's theorems	120
	1 Balls, spheres, fixed points, and retractions	120
	2 Digression on the contrapositive rule	124
	3 Brouwer's proof	124

4 Relation between fixed point and retraction theorems 126
5 How to understand a proof:
The objectification and 'mapification' of concepts 127
6 The eye of the storm 130
7 Using maps to formulate guesses 131

Part III Categories of structured sets

Article III **Examples of categories** 135
1 The category $\mathcal{S}^{\circlearrowright}$ of endomaps of sets 136
2 Typical applications of $\mathcal{S}^{\circlearrowright}$ 137
3 Two subcategories of $\mathcal{S}^{\circlearrowright}$ 138
4 Categories of endomaps 138
5 Irreflexive graphs 141
6 Endomaps as special graphs 143
7 The simpler category $\mathcal{S}^{\downarrow}$: Objects are just maps of sets 144
8 Reflexive graphs 145
9 Summary of the examples and their general significance 146
10 Retractions and injectivity 146
11 Types of structure 149
12 Guide 151

Session 11 Ascending to categories of richer structures 152
1 A category of richer structures: Endomaps of sets 152
2 Two subcategories: Idempotents and automorphisms 155
3 The category of graphs 156

Session 12 Categories of diagrams 161
1 Dynamical systems or automata 161
2 Family trees 162
3 Dynamical systems revisited 163

Session 13 Monoids 166

Session 14 Maps preserve positive properties 170
1 Positive properties versus negative properties 173

Session 15 Objectification of properties in dynamical systems 175
1 Structure-preserving maps from a cycle to another endomap 175
2 Naming elements that have a given period by maps 176
3 Naming arbitrary elements 177
4 The philosophical role of N 180
5 Presentations of dynamical systems 182

Session 16	Idempotents, involutions, and graphs	187
	1 Solving exercises on idempotents and involutions	187
	2 Solving exercises on maps of graphs	189
Session 17	Some uses of graphs	196
	1 Paths	196
	2 Graphs as diagram shapes	200
	3 Commuting diagrams	201
	4 Is a diagram a map?	203
Test 2		204
Session 18	Review of Test 2	205

Part IV Elementary universal mapping properties

Article IV	**Universal mapping properties**	213
	1 Terminal objects	213
	2 Separating	215
	3 Initial object	215
	4 Products	216
	5 Commutative, associative, and identity laws for multiplication of objects	220
	6 Sums	222
	7 Distributive laws	222
	8 Guide	223
Session 19	Terminal objects	225
Session 20	Points of an object	230
Session 21	Products in categories	236
Session 22	Universal mapping properties and incidence relations	245
	1 A special property of the category of sets	245
	2 A similar property in the category of endomaps of sets	246
	3 Incidence relations	249
	4 Basic figure-types, singular figures, and incidence, in the category of graphs	250
Session 23	More on universal mapping properties	254
	1 A category of pairs of maps	255
	2 How to calculate products	256

Session 24	Uniqueness of products and definition of sum	261
	1 The terminal object as an identity for multiplication	261
	2 The uniqueness theorem for products	263
	3 Sum of two objects in a category	265
Session 25	Labelings and products of graphs	269
	1 Detecting the structure of a graph by means of labelings	270
	2 Calculating the graphs $A \times Y$	273
	3 The distributive law	275
Session 26	Distributive categories and linear categories	276
	1 The standard map $A \times B_1 + A \times B_2 \longrightarrow A \times (B_1 + B_2)$	276
	2 Matrix multiplication in linear categories	279
	3 Sum of maps in a linear category	279
	4 The associative law for sums and products	281
Session 27	Examples of universal constructions	284
	1 Universal constructions	284
	2 Can objects have negatives?	287
	3 Idempotent objects	289
	4 Solving equations and picturing maps	292
Session 28	The category of pointed sets	295
	1 An example of a non-distributive category	295
Test 3		299
Test 4		300
Test 5		301
Session 29	Binary operations and diagonal arguments	302
	1 Binary operations and actions	302
	2 Cantor's diagonal argument	303
	Part V Higher universal mapping properties	
Article V	**Map objects**	313
	1 Definition of map object	313
	2 Distributivity	315
	3 Map objects and the Diagonal Argument	316
	4 Universal properties and 'observables'	316
	5 Guide	319
Session 30	Exponentiation	320
	1 Map objects, or function spaces	320

	2 A fundamental example of the transformation of map objects	323
	3 Laws of exponents	324
	4 The distributive law in cartesian closed categories	327
Session 31	Map object versus product	328
	1 Definition of map object versus definition of product	329
	2 Calculating map objects	331
Article VI	The contravariant parts functor	335
	1 Parts and stable conditions	335
	2 Inverse Images and Truth	336
Session 32	Subobject, logic, and truth	339
	1 Subobjects	339
	2 Truth	342
	3 The truth value object	344
Session 33	Parts of an object: Toposes	348
	1 Parts and inclusions	348
	2 Toposes and logic	352
Article VII	The Connected Components Functor	358
	1 Connectedness versus discreteness	358
	2 The points functor parallel to the components functor	359
	3 The topos of right actions of a monoid	360
Session 34	Group theory and the number of types of connected objects	362
Session 35	Constants, codiscrete objects, and many connected objects	366
	1 Constants and codiscrete objects	366
	2 Monoids with at least two constants	367
Appendices		368
Appendix I	Geometery of figures and algebra of functions	369
	1 Functors	369
	2 Geometry of figures and algebra of functions as categories themselves	370
Appendix II	Adjoint functors with examples from graphs and dynamical systems	372
Appendix III	The emergence of category theory within mathematics	378
Appendix IV	Annotated Bibliography	381
Index		385

Preface

Since its first introduction over 60 years ago, the concept of category has been increasingly employed in all branches of mathematics, especially in studies where the relationship between different branches is of importance. The categorical ideas arose originally from the study of a relationship between geometry and algebra; the fundamental simplicity of these ideas soon made possible their broader application.

The categorical concepts are latent in elementary mathematics; making them more explicit helps us to go beyond elementary algebra into more advanced mathematical sciences. Before the appearance of the first edition of this book, their simplicity was accessible only through graduate-level textbooks, because the available examples involved topics such as modules and topological spaces.

Our solution to that dilemma was to develop from the basics the concepts of directed graph and of discrete dynamical system, which are mathematical structures of wide importance that are nevertheless accessible to any interested high-school student. As the book progresses, the relationships between those structures exemplify the elementary ideas of category. Rather remarkably, even some detailed features of graphs and of discrete dynamical systems turn out to be shared by other categories that are more continuous, e.g. those whose maps are described by partial differential equations.

Many readers of the first edition have expressed their wish for more detailed indication of the links between the elementary categorical material and more advanced applications. This second edition addresses that request by providing two new articles and four appendices. A new article introduces the notion of connected component, which is fundamental to the qualitative leaps studied in elementary graph theory and in advanced topology; the introduction of this notion forces the recognition of the role of functors.

The appendices use examples from the text to sketch the role of adjoint functors in guiding mathematical constructions. Although these condensed appendices cannot substitute for a more detailed study of advanced topics, they will enable the student, armed with what has been learned from the text, to approach such study with greater understanding.

Buffalo, January 8, 2009

F. William Lawvere
Stephen H. Schanuel

Organisation of the book

The reader needs to be aware that this book has two very different kinds of 'chapters':

The **Articles** form the backbone of the book; they roughly correspond to the written material given to our students the first time we taught the course.

The **Sessions**, reflecting the informal classroom discussions, provide additional examples and exercises. Students who had difficulties with some of the exercises in the Articles could often solve them after the ensuing Sessions. We have tried in the Sessions to preserve the atmosphere (and even the names of the students) of that first class. The more experienced reader could gain an overview by reading only the Articles, but would miss out on many illuminating examples and perspectives.

Session 1 is introductory. Exceptionally, Session 10 is intended to give the reader a taste of more sophisticated applications; mastery of it is not essential for the rest of the book.

Each Article is further discussed and elaborated in the specific subsequent Sessions indicated below:

Article I	Sessions 2 and 3
Article II	Sessions 4 through 9
Article III	Sessions 11 through 17
Article IV	Sessions 19 through 29
Article V	Sessions 30 and 31
Article VI	Sessions 32 and 33
Article VII	Sessions 34 and 35

The **Appendices**, written in a less leisurely manner, are intended to provide a rapid summary of some of the main possible links of the basic material of the course with various more advanced developments of modern mathematics.

Acknowledgements

First Edition
This book would not have come about without the invaluable assistance of many people:

Emilio Faro, whose idea it was to include the dialogues with the students in his masterful record of the lectures, his transcriptions of which grew into the Sessions;
Danilo Lawvere, whose imaginative and efficient work played a key role in bringing this book to its current form;
our students (some of whom still make their appearance in the book), whose efforts and questions contributed to shaping it;
John Thorpe, who accepted our proposal that a foundation for discrete mathematics *and* continuous mathematics could constitute an appropriate course for beginners.

Special thanks go to Alberto Peruzzi, who provided invaluable expert criticism and much encouragement. Many helpful comments were contributed by John Bell, David Benson, Andreas Blass, Aurelio Carboni, John Corcoran, Bill Faris, Emilio Faro, Elaine Landry, Fred Linton, Saunders Mac Lane, Kazem Mahdavi, Mara Mondolfo, Koji Nakatogawa, Ivonne Pallares, Norm Severo, and Don Schack, as well as by many other friends and colleagues. We are grateful also to Cambridge University Press, in particular to Roger Astley and Maureen Storey, for all their work in producing this book.

Above all, we can never adequately acknowledge the ever-encouraging generous and graceful spirit of Fatima Fenaroli, who conceived the idea that this book should exist, and whose many creative contributions have been irreplaceable in the process of perfecting it.

Thank you all,

Buffalo, New York
2009

F. William Lawvere
Stephen H. Schanuel

Second Edition
Thanks to the readers who encouraged us to expand to this second edition, and thanks to Roger Astley and his group at Cambridge University Press for their help in bringing it about.

2009

F. William Lawvere
Stephen H. Schanuel

Preview

SESSION 1

Galileo and multiplication of objects

1. Introduction

Our goal in this book is to explore the consequences of a new and fundamental insight about the nature of mathematics which has led to better methods for understanding and using mathematical concepts. While the insight and methods are simple, they are not as familiar as they should be; they will require some effort to master, but you will be rewarded with a clarity of understanding that will be helpful in unravelling the mathematical aspect of any subject matter.

The basic notion which underlies all the others is that of a *category*, a 'mathematical universe'. There are many categories, each appropriate to a particular subject matter, and there are ways to pass from one category to another. We will begin with an informal introduction to the notion and with some examples. The ingredients will be objects, maps, and composition of maps, as we will see.

While this idea, that mathematics involves different categories and their relationships, has been implicit for centuries, it was not until 1945 that Eilenberg and Mac Lane gave *explicit* definitions of the basic notions in their ground-breaking paper 'A general theory of natural equivalences', synthesizing many decades of analysis of the workings of mathematics and the relationships of its parts.

2. Galileo and the flight of a bird

Let's begin with Galileo, four centuries ago, puzzling over the problem of motion. He wished to understand the precise motion of a thrown rock, or of a water jet from a fountain. Everyone has observed the graceful parabolic arcs these follow; but the motion of a rock means more than its track. The motion involves, for each instant, the position of the rock at that instant; to record it requires a motion picture rather than a time exposure. We say the motion is a 'map' (or 'function') from time to space.

The flight of a bird as a map from time to space

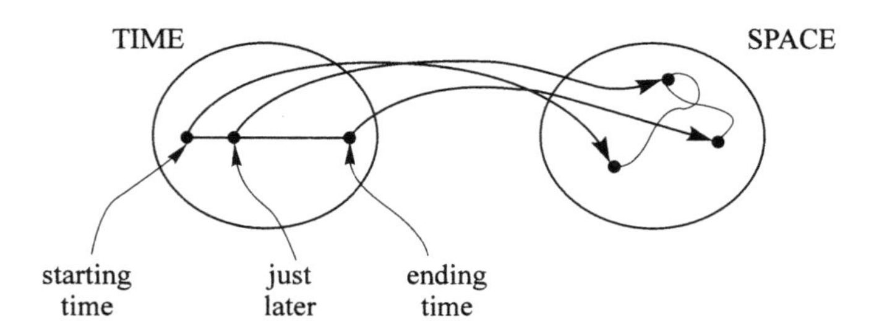

Schematically:

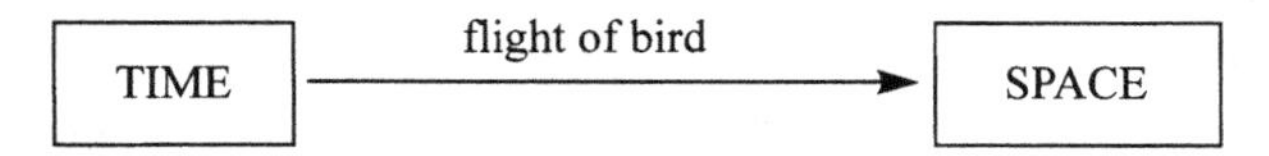

You have no doubt heard the legend; Galileo dropped a heavy weight and a light weight from the leaning tower of Pisa, surprising the onlookers when the weights hit the ground simultaneously. The study of vertical motion, of objects thrown straight up, thrown straight down, or simply dropped, seems too special to shed much light on general motion; the track of a dropped rock is straight, as any child knows. However, the motion of a dropped rock is not quite so simple; it accelerates as it falls, so that the last few feet of its fall takes less time than the first few. Why had Galileo decided to concentrate his attention on this special question of vertical motion? The answer lies in a simple equation:

$$\text{SPACE} = \text{PLANE} \times \text{LINE}$$

but it requires some explanation!

Two new maps enter the picture. Imagine the sun directly overhead, and for each point in space you'll get a shadow point on the horizontal plane:

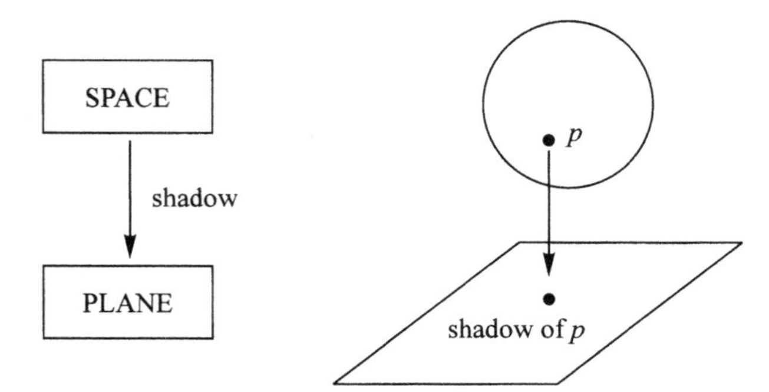

This is one of our two maps: the 'shadow' map from space to the plane. The second map we need is best imagined by thinking of a vertical line, perhaps a pole stuck into the ground. For each point in space there is a corresponding point on the line, the one at the same level as our point in space. Let's call this map 'level':

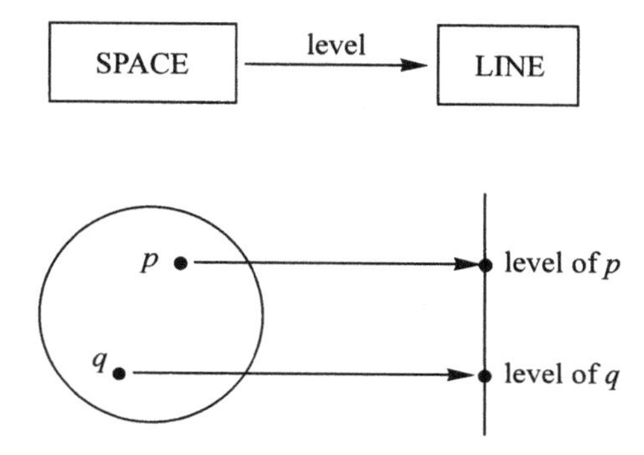

Together, we have:

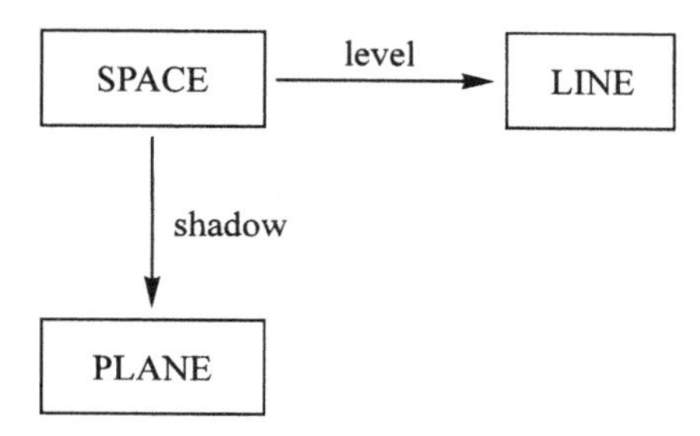

These two maps, 'shadow' and 'level', seem to reduce each problem about space to two simpler problems, one for the plane and one for the line. For instance, if a bird is in our space, and you know only the shadow of the bird and the level of the bird, then you can reconstruct the position of the bird. There is more, though. Suppose you have a motion picture of the bird's shadow as it flies, and a motion picture of its level – perhaps there was a bird-watcher climbing on our line, keeping always level with the bird, and you filmed the watcher. From these two motion pictures you can reconstruct the entire flight of the bird! So not only is a position in space reduced to a position in the plane and one on the line, but also a motion in space is reduced to a motion in the plane and one on the line.

Let's assemble the pieces. From a motion, or flight, of a bird

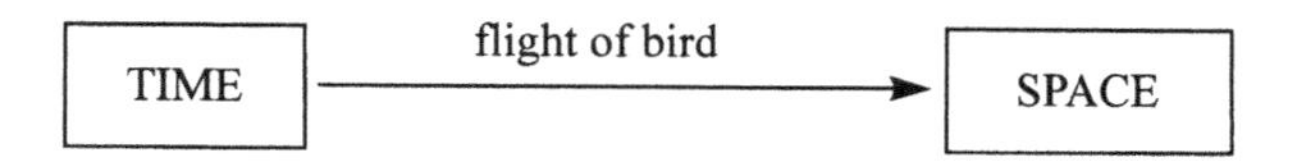

we get two simpler motions by 'composing' the flight map with the shadow and level maps. From these three maps,

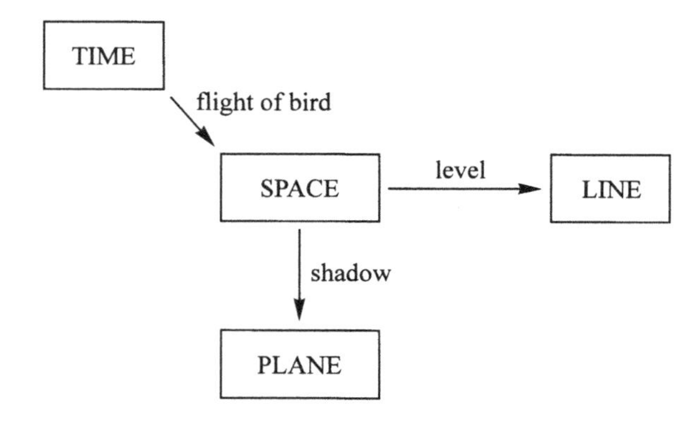

we get these two maps:

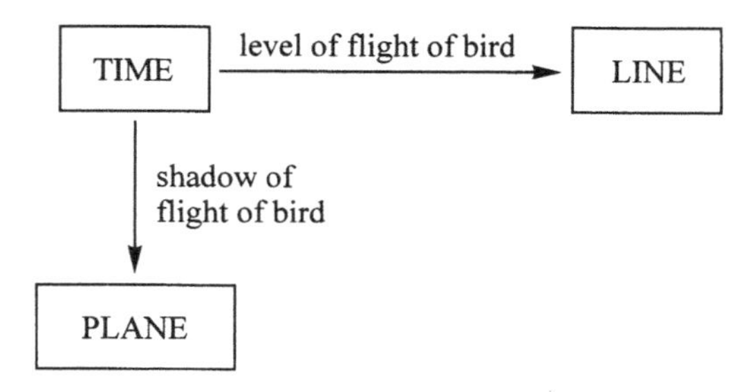

and now space has disappeared from the picture.

Galileo's discovery is that from these two simpler motions, in the plane and on the line, he could completely recapture the complicated motion in space. In fact, if the motions of the shadow and the level are 'continuous', so that the shadow does not suddenly disappear from one place and instantaneously reappear in another, the motion of the bird will be continuous too. This discovery enabled Galileo to reduce the study of motion to the special cases of horizontal and vertical motion. It would take us too far from our main point to describe here the beautiful experiments he designed to study these, and what he discovered, but I urge you to read about them.

Does it seem reasonable to express this relationship of space to the plane and the line, given by two maps,

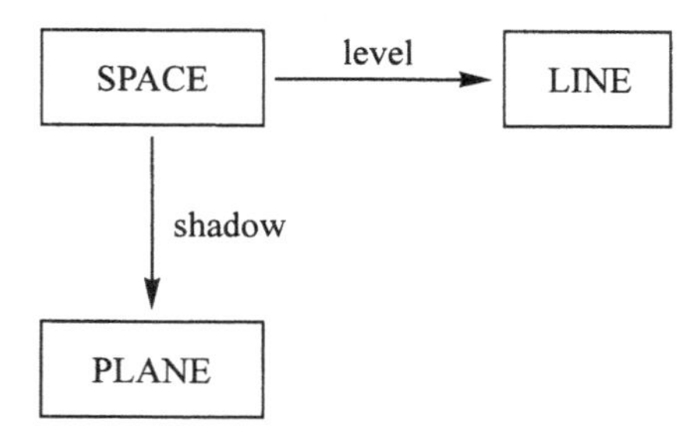

by the equation SPACE = PLANE × LINE? What do these maps have to do with multiplication? It may be helpful to look at some other examples.

3. Other examples of multiplication of objects

Multiplication often appears in the guise of *independent choices*. Here is an example. Some restaurants have a list of options for the first course and another list for the second course; a 'meal' involves one item from each list. First courses: soup, pasta, salad. Second courses: steak, veal, chicken, fish.

So, one possible 'meal' is: 'soup, then chicken'; but 'veal, then steak' is not allowed. Here is a diagram of the possible meals:

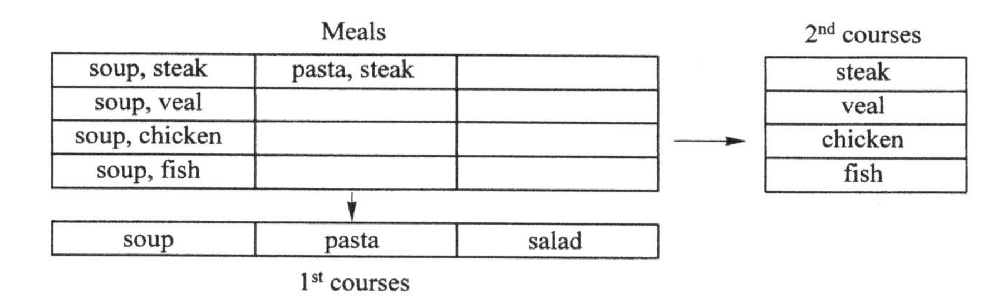

(Fill in the other meals yourself.) Notice the analogy with Galileo's diagram:

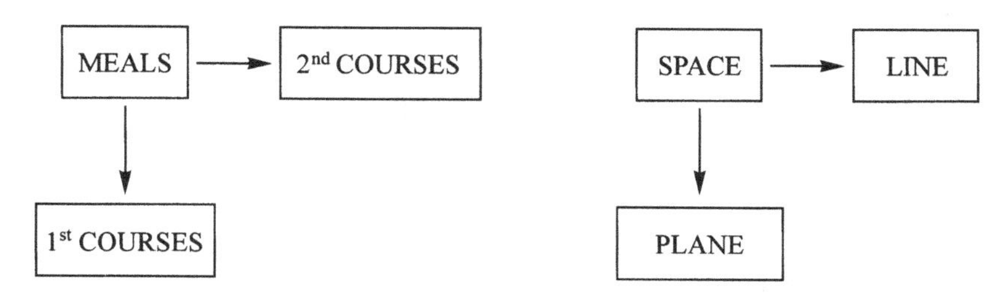

This scheme with three 'objects' and two 'maps' or 'processes' is the right picture of multiplication of objects, and it applies to a surprising variety of situations. The idea of multiplication is the same in all cases. Take for example a segment and a disk from geometry. We can multiply these too, and the result is a cylinder. I am not referring to the fact that the *volume* of the cylinder is obtained by multiplying the area of the disk by the length of the segment. The cylinder *itself* is the product, segment times disk, because again there are two processes or projections that take us from the cylinder to the segment and to the disk, in complete analogy with the previous examples.

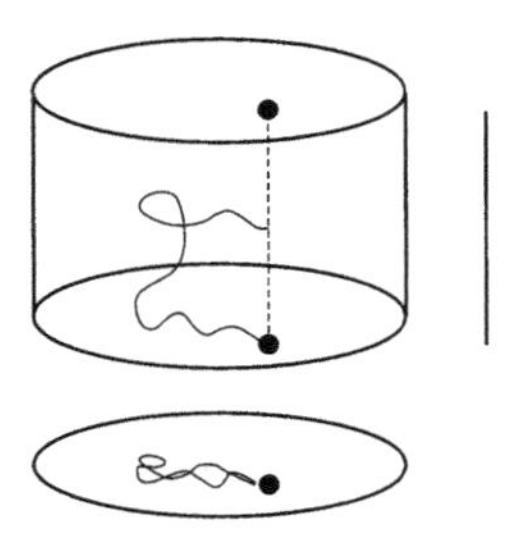

Every point in the cylinder has a corresponding 'level' point on the segment and a corresponding 'shadow' point in the disk, and if you know the shadow and level points, you can find the point in the cylinder to which they correspond. As before, the motion of a fly trapped in the cylinder is determined by the motion of its level point in the segment and the motion of its shadow point in the disk.

An example from logic will suggest a connection between multiplication and the word '*and*'. From a sentence of the form '*A and B*' (for example, 'John is sick *and* Mary is sick') we can deduce *A* and we can deduce *B*:

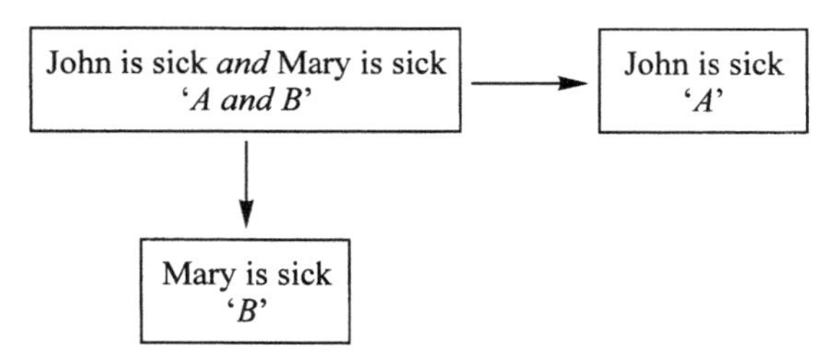

But more than that: to deduce the single sentence 'John is sick and Mary is sick' from some other sentence *C* is the same as deducing each of the two sentences from *C*. In other words, the two deductions

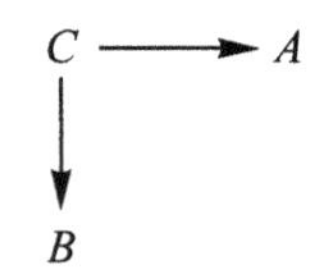

amount to one deduction $C \longrightarrow (A \text{ and } B)$. Compare this diagram

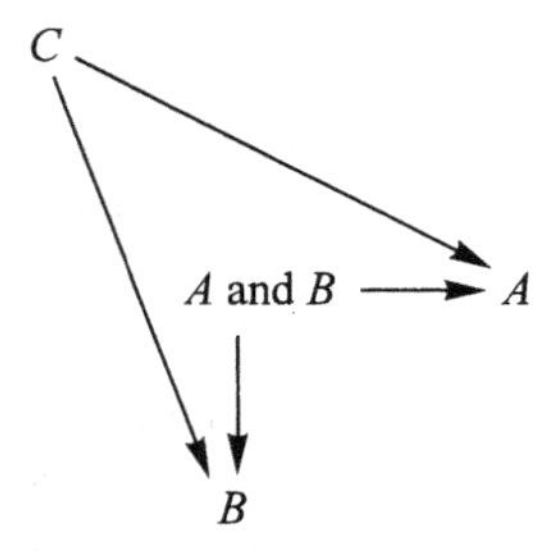

with the diagram of Galileo's idea.

One last picture, perhaps the simplest of all, hints at the relation to multiplication of numbers:

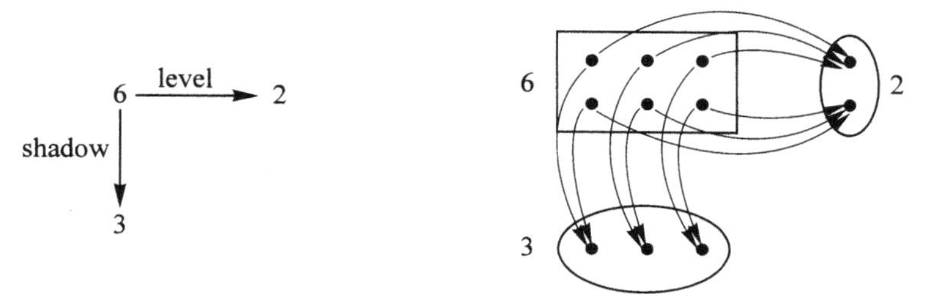

Why does $3 \times 2 = 6$?

I hope these pictures seem suggestive to you. Our goal is to learn to use them as precise instruments of understanding and reasoning, not merely as intuitive guides.

Exercise 1:
Find other examples of combining two objects to get a third. Which of them seem to fit our pattern? That is, for which of them does the third object seem to have 'maps' to the two you began with? It may be helpful to start by thinking of real-life problems for which multiplication of numbers is needed to calculate the solution, but not all examples are related to multiplication of numbers.

Exercise 2:
The part of Galileo's work which we discussed is really concerned with only a small portion of space, say the immediate neighbourhood of the tower of Pisa. Since the ground might be uneven, what could be meant by saying that two points are at the same level? Try to describe an experiment for deciding whether two nearby points are at the same level, without using 'height' (distance from an imaginary plane of reference.) Try to use the most elementary tools possible.

PART I

The category of sets

A *map* of sets is a process for getting from one set to another. We investigate the *composition* of maps (following one process by a second process), and find that the algebra of composition of maps resembles the algebra of multiplication of numbers, but its interpretation is much richer.

ARTICLE I

Sets, maps, composition

A first example of a category

Before giving a precise definition of 'category', we should become familiar with one example, the **category of finite sets and maps**.

An [object] in this category is a finite *set* or *collection*. Here are some examples:

(the set of all students in the class) is one object,
(the set of all desks in the classroom) is another,
(the set of all the twenty-six letters in our alphabet) is another.

You are probably familiar with some notations for finite sets:

$$\{John,\ Mary,\ Sam\}$$

is a name for the set whose three elements are, of course, John, Mary, and Sam. (You know some infinite sets also, e.g. the set of all natural numbers: $\{0, 1, 2, 3, \ldots\}$.) Usually, since the order in which the elements are listed is irrelevant, it is more helpful to picture them as scattered about:

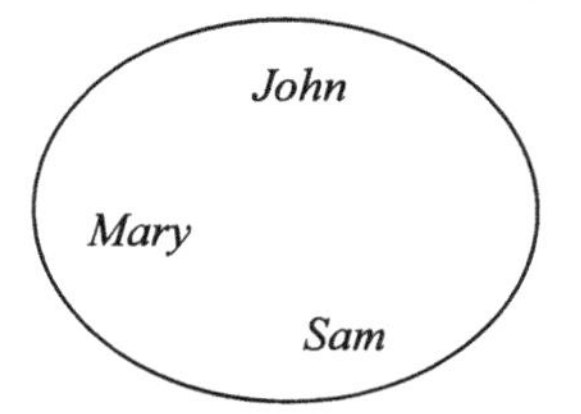

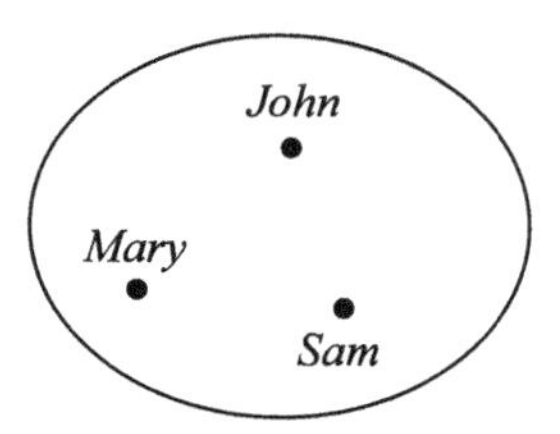

where a dot represents each element, and we are then free to leave off the labels when for one reason or another they are temporarily irrelevant to the discussion, and picture this set as:

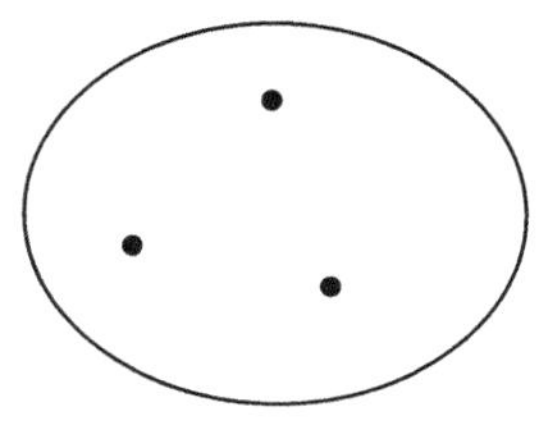

Such a picture, labeled or not, is called an *internal diagram* of the set.

A map f in this category consists of three things:

1. a set A, called the *domain* of the map,
2. a set B, called the *codomain* of the map,
3. a rule assigning to each element a in the domain, an element b in the codomain. This b is denoted by $f \circ a$ (or sometimes '$f(a)$'), read 'f of a'.

(Other words for map are 'function', 'transformation', 'operator', 'arrow', and 'morphism'.)

An example will probably make it clearer: Let $A = \{John, Mary, Sam\}$, and let $B = \{eggs, oatmeal, toast, coffee\}$, and let f assign to each person his or her favorite breakfast. Here is a picture of the situation, called the *internal diagram* of the map:

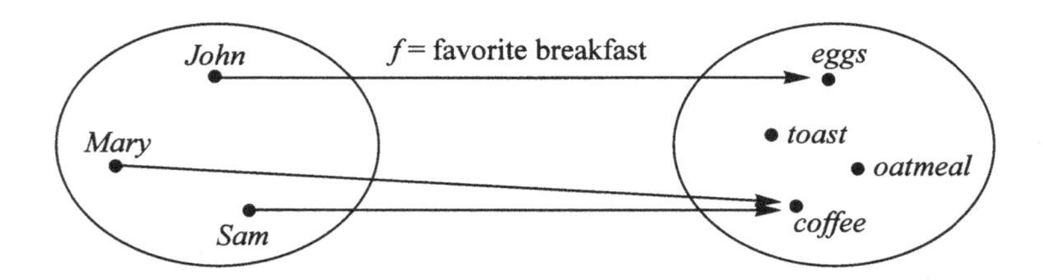

This indicates that the favorite breakfast of John is eggs, written $f(John) = eggs$, while Mary and Sam prefer coffee. Note some pecularities of the situation, because these are features of the internal diagram of any map:

(a) From each dot in the *domain* (here $\{John, Mary, Sam\}$), there is exactly one arrow leaving.
(b) To a dot in the *codomain* (here $\{eggs, oatmeal, toast, coffee\}$), there may be any number of arrows arriving: zero or one or more.

The important thing is: For each dot in the domain, we have exactly one arrow leaving, and the arrow arrives at some dot in the codomain.

Nothing in the discussion above is intended to exclude the possibility that A and B, the domain and codomain of the map, could be the *same* set. Here is an internal diagram of such a map g:

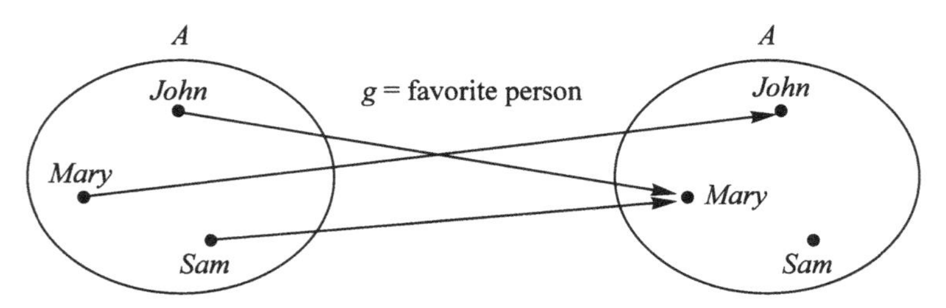

(Many 1950s movie plots are based on this diagram.)

A map in which the domain and codomain are the same object is called an *endomap*. (Why? What does the prefix 'endo' mean?) For endomaps *only*, an alternative form of internal diagram is available. Here it is, for the endomap above:

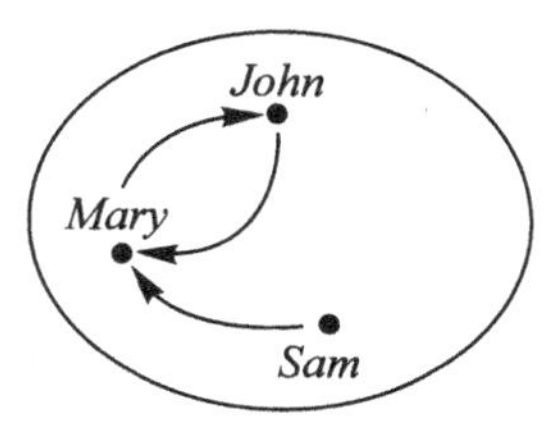

For each object A, there is a special, especially simple, endomap which has domain and codomain both A. Here it is for our example:

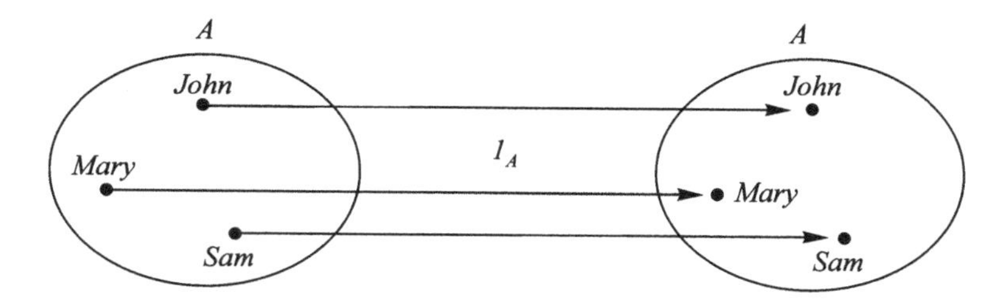

Here is the corresponding special internal diagram, available because the map is an endomap:

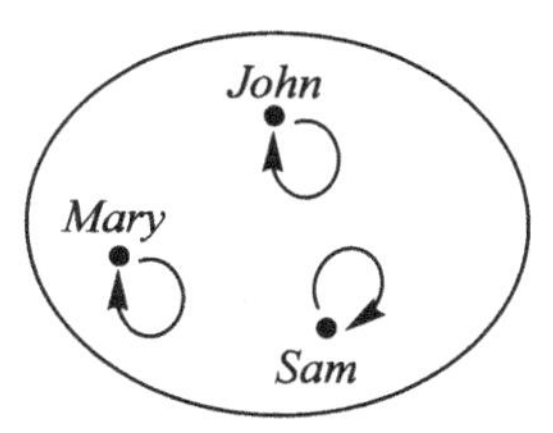

A map like this, in which the domain and the codomain are the same set A, *and* for each a in $A, f(a) = a$, is called an identity map. To state it more precisely, this map is 'the identity map from {*John*, *Mary*, *Sam*} to {*John*, *Mary*, *Sam*},' or 'the identity map on the object {*John*, *Mary*, *Sam*}.' (Simpler still is to give that object a short name, A = {*John*, *Mary*, *Sam*}; and then call our map 'the identity map on A', or simply '1_A'.)

Sometimes we need a scheme to keep track of the domain and codomain, without indicating in the picture all the details of the map. Then we can use just a letter to stand for each object, and a single arrow for each map. Here are the *external diagrams* corresponding to the last five internal diagrams:

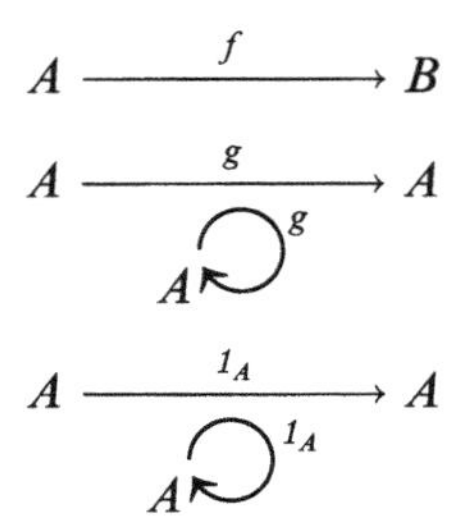

External diagrams are especially helpful when there are *several* objects and maps to be discussed, or when some of the exact details of the maps are temporarily irrelevant.

The final basic ingredient, which is what lends all the dynamics to the notion of category, is **composition of maps**, by which two maps are combined to obtain a third map. Here is an example:

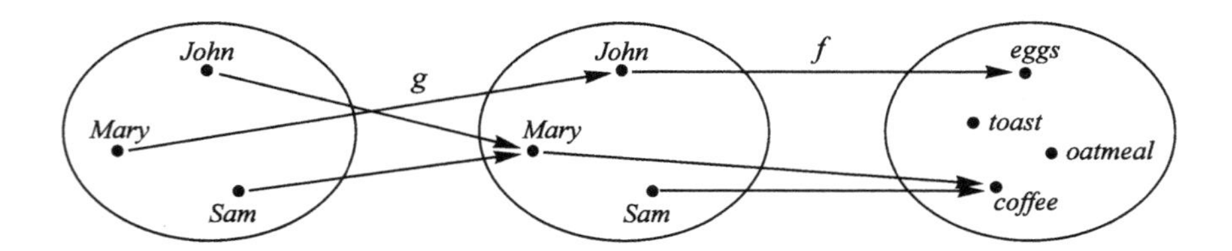

Or, in the *external* diagram:

$$A \xrightarrow{g} A \xrightarrow{f} B$$

If we ask: 'What should each person serve for breakfast to his or her favorite person?' we are led to answers like this: 'John likes Mary, and Mary prefers coffee, so John should serve coffee.' Working out the other two cases as well, we get: 'Mary likes John, and John likes eggs, so Mary should serve eggs; Sam likes Mary, and Mary likes coffee, so Sam should serve coffee.' Pictorially:

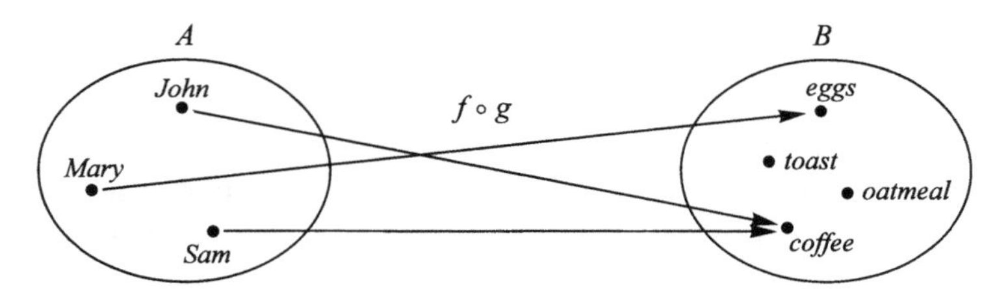

Or in the external diagram:

$$A \xrightarrow{f \circ g} B$$

'$f \circ g$' is read '*f following g*', or sometimes '*f* of *g*', as in: 'The favorite breakfast of the favorite person of John is coffee,' for '$f \circ g \circ John = coffee$.' Let's sum up: If we have two maps f and g, and if the domain of f is the same object as the codomain of g, pictorially

$$X \xrightarrow{g} Y \xrightarrow{f} Z$$

then we can build from them a single map

$$X \xrightarrow{f \circ g} Z$$

We will soon be considering an analogy between *composition of maps* and multiplication of numbers. This analogy should not be confused with the analogy in Session 1, between *multiplication of objects* and multiplication of numbers.

That's all! These are all the basic ingredients we need, to have a CATEGORY, or 'mathematical universe':

Data for a category:

Objects: $A, B, C \dots$

Maps: $A \xrightarrow{f} B, \dots$

Identity maps: (one per object): $A \xrightarrow{1_A} A, \dots$

Composition of maps: assigns to each pair of maps of type $A \xrightarrow{g} B \xrightarrow{f} C$, another map called '$f$ following g', $A \xrightarrow{f \circ g} C$

Now comes an important, even crucial, aspect. These data must fit together nicely, as follows.

Rules for a category:

1. The **identity laws:**

(a) If $A \xrightarrow{1_A} A \xrightarrow{g} B$

then $A \xrightarrow{g \circ 1_A = g} B$

(b) If $A \xrightarrow{f} B \xrightarrow{1_B} B$

then $A \xrightarrow{1_B \circ f = f} B$

2. The **associative law:**

If $A \xrightarrow{f} B \xrightarrow{g} C \xrightarrow{h} D$

then $A \xrightarrow{h \circ (g \circ f) = (h \circ g) \circ f} D$

Here are some pictures to illustrate these properties in the category of sets:

1. The identity laws:

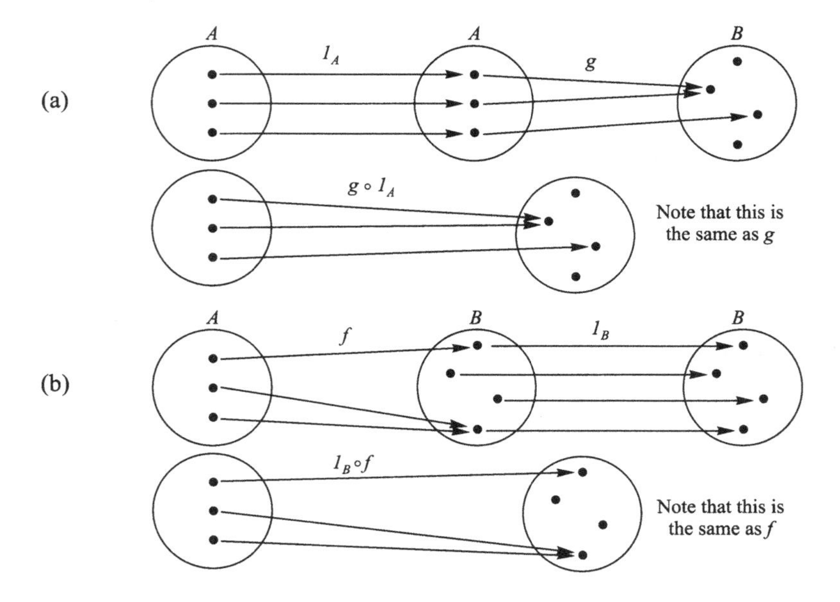

2. The associative law:

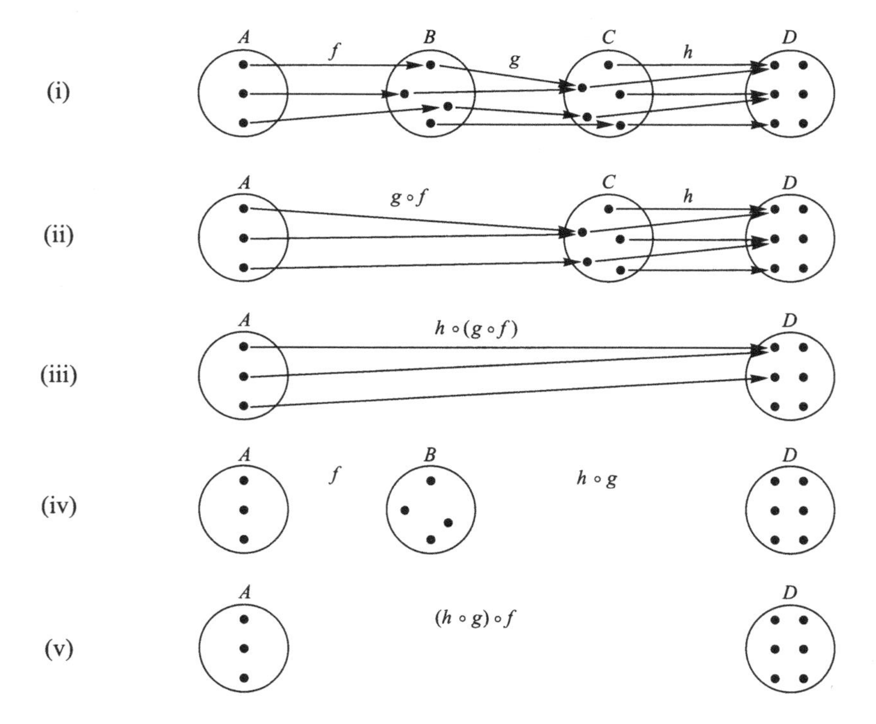

Exercise 1:
Check to be sure you understand how we got diagrams (ii) and (iii) from the given diagram (i). Then fill in (iv) and (v) yourself, starting over from (i). Then check to see that (v) and (iii) are the same.

Is this an accident, or will this happen for any three maps in a row? Can you give a simple explanation why the results

$$h \circ (g \circ f) \text{ and } (h \circ g) \circ f$$

will always come out the same, whenever we have three maps in a row

$$X \xrightarrow{f} Y \xrightarrow{g} Z \xrightarrow{h} W?$$

What can you say about *four* maps in a row?

One very useful sort of set is a 'singleton' set, a set with exactly one element. Fix one of these, say (*me*), and call this set '**1**'. Look at what the maps from **1** to {*John, Mary, Sam*} are. There are exactly three of them:

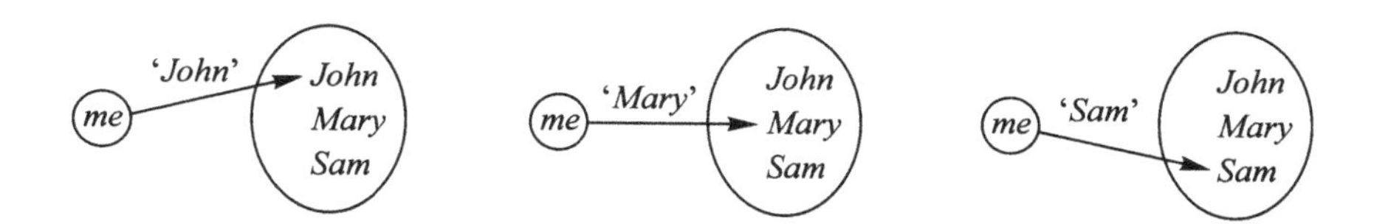

Definition: *A* **point** *of a set X is a map* $\mathbf{1} \longrightarrow X$.

(If A is some familiar set, a map from A to X is called an 'A-element' of X; thus '**1**-elements' are points.) Since a point is a map, we can compose it with another map, and get a point again. Here is an example:

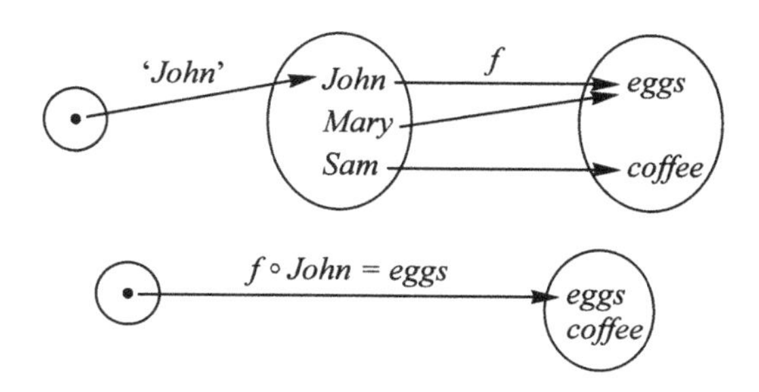

The equation $f \circ John = eggs$ is read 'f following *John* is *eggs*' or more briefly, 'f of *John* is *eggs*' (or sometimes 'f sends *John* to *eggs*').

To help familiarize yourself with the category of finite sets, here are some exercises. *Take* $A = \{John, Mary, Sam\}$, $B = \{eggs, coffee\}$ *in all of these.*

Exercise 2:
How many different maps f are there with domain A and codomain B? One example is

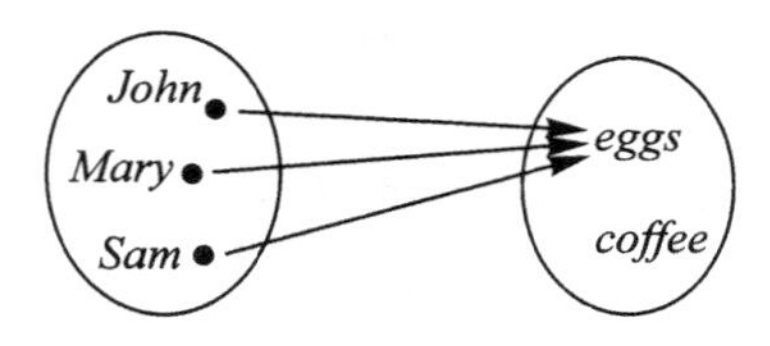

but there are lots of others: How many in all?

Exercise 3:
Same, but for maps $A \xrightarrow{f} A$

Exercise 4:
Same, but for maps $B \xrightarrow{f} A$

Exercise 5:
Same, but for maps $B \xrightarrow{f} B$

Exercise 6:
How many maps $A \xrightarrow{f} A$ satisfy $f \circ f = f$?

Exercise 7:
How many maps $B \xrightarrow{g} B$ satisfy $g \circ g = g$?

Exercise 8:
Can you find a pair of maps $A \xrightarrow{f} B \xrightarrow{g} A$ for which $g \circ f = 1_A$?
If so, how many such pairs?

Exercise 9:
Can you find a pair of maps $B \xrightarrow{h} A \xrightarrow{k} B$ for which $k \circ h = 1_B$?
If so, how many such pairs?

1. Guide

Our discussion of maps of sets has led us to the general definition of *category*, presented for reference on the next page. This material is reviewed in Sessions 2 and 3.

Definition of CATEGORY

A category consists of the DATA: ***. . . with corresponding notation***

(1) *OBJECTS* — $A, B, C, \ldots$

(2) *MAPS* — $f, g, h, \ldots$

(3) *For each map f, one object as DOMAIN of f and one object as CODOMAIN of f*

To indicate that f is a map, with domain A and codomain B, we write $A \xrightarrow{f} B$ (or $f\colon A \longrightarrow B$) and we say '$f$ is a map from A to B.'

(4) *For each object A an IDENTITY MAP, which has domain A and codomain A*

We denote this map by 1_A, so

$$A \xrightarrow{1_A} A$$

is one of the maps from A to A.

(5) *For each pair of maps*

$$A \xrightarrow{f} B \xrightarrow{g} C,$$

a COMPOSITE MAP

$$A \xrightarrow{g \text{ following } f} C$$

We denote this map by

$$A \xrightarrow{g \circ f} C$$

(and sometimes say 'g of f').

satisfying the following RULES:

These notations are used in the following external diagrams illustrating the rules:

(i) *IDENTITY LAWS: If* $A \xrightarrow{f} B$, *then* $1_B \circ f = f$ *and* $f \circ 1_A = f$

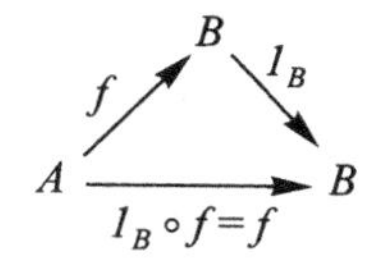

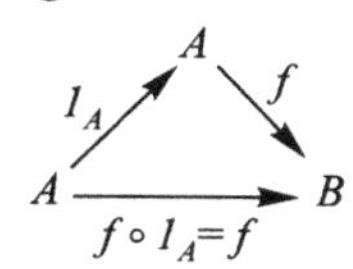

(ii) *ASSOCIATIVE LAW:*
If $A \xrightarrow{f} B \xrightarrow{g} C \xrightarrow{h} D$,
then $(h \circ g) \circ f = h \circ (g \circ f)$

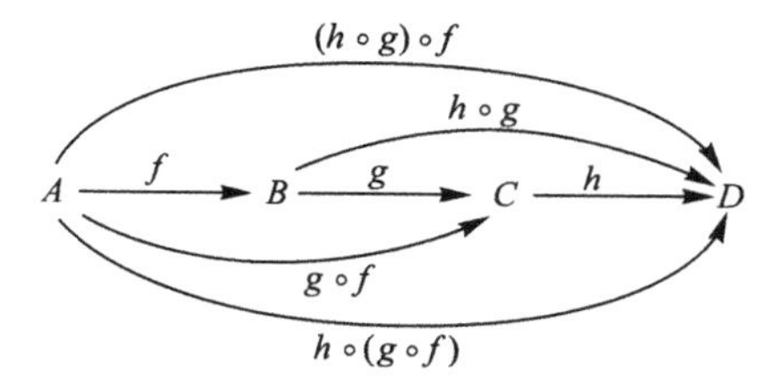

The associative law allows us to leave out the parentheses and just write '$h \circ g \circ f$', which we read as 'h following g following f'. A longer composite like $h \circ g \circ f \circ e \circ d$ is also unambiguous; all ways of building it by composition of pairs give the same result.

Hidden in items (4) and (5) above are the BOOKKEEPING rules. Explicitly these are:

the domain and codomain of 1_A are both A;

$g \circ f$ is only defined if the domain of g is the codomain of f;

the domain of $g \circ f$ is the domain of f and the codomain of $g \circ f$ is the codomain of g.

SESSION 2

Sets, maps and composition

1. Review of Article I

Before discussing some of the exercises in Article I, let's have a quick review. A **set** is any collection of things. You know examples of infinite sets, like the set of all natural numbers, $\{0, 1, 2, 3, \ldots\}$, but we'll take most of our examples from finite sets. Here is a typical **internal diagram** of a **function**, or **map**:

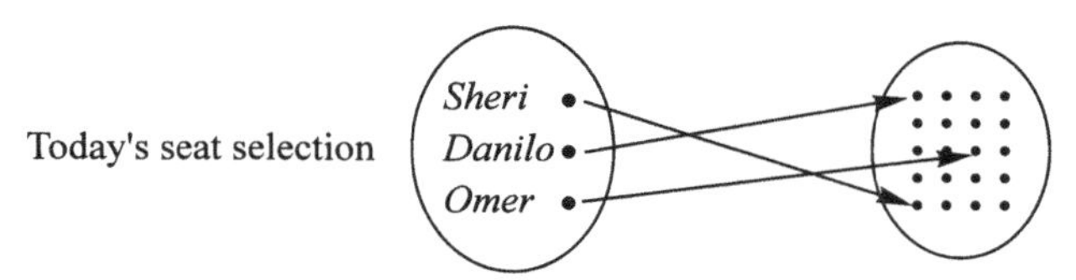

Other words that mean the same as *function* and *map* are **transformation**, **operator**, **morphism**, and **functional**; the idea is so important that it has been rediscovered and renamed in many different contexts.

As the internal diagram suggests, to have a **map** f **of sets** involves three things:

1. a set A, called the **domain** of the map f;
2. a set B, called the **codomain** of the map f; and then the main ingredient:
3. a **rule** (or process) for f, assigning to each element of the domain A exactly one element of the codomain B.

That is a fairly accurate description of what a map is, but we also need a means to tell when two different rules give the same map. Here is an example. The first map will be called f and has as domain and as codomain the set of all natural numbers. The rule for f will be: '*add* 1 *and then square*'. (This can be written in mathematical shorthand as $f(x) = (x + 1)^2$, but that is not important for our discussion.) Part of the internal picture of f is:

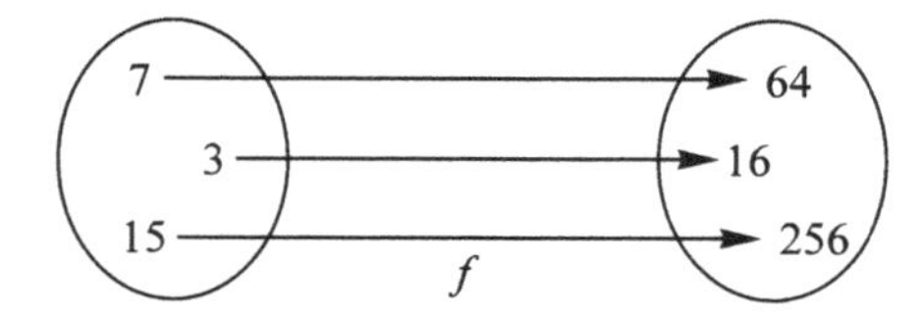

The second map will be called g. As domain and codomain of g we take again the set of all natural numbers, but the rule for g will be '*square the input*, *double the input*, *add the two results, and then add* 1', a very different rule indeed. Still, part of the internal diagram of g is:

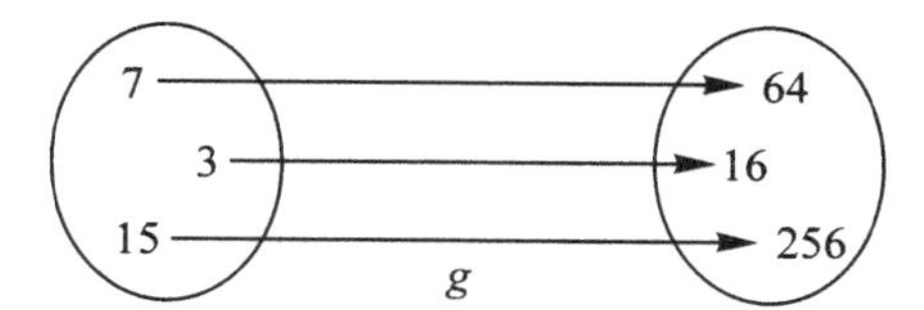

the same as for f. Not only that, you can check with any number you like and you will always get the same thing with the rule for f as with the rule for g. So, because the two rules produce the same result for each input (and the domains are the same and the codomains are the same), we say that f and g are the *same map*, and we write this as $f = g$. (Do you know how the encoded formula for the rule g looks? Right, $g(x) = x^2 + 2x + 1$.) What the equation $(x + 1)^2 = x^2 + 2x + 1$ says is precisely that $f = g$, not that the two rules are the same rule (which they obviously are not; in particular, one of them takes more steps than the other.) The idea is that a function, or map of sets, is not the rule itself, but what the rule accomplishes. This aspect is nicely captured by the pictures, or internal diagrams.

In categories other than the category of sets, 'a map from A to B' is typically some sort of 'process for getting from A to B,' so that in any category, maps f and g are not considered the same unless they have *at least* the properties:

1. f and g have the same domain, say A, and
2. f and g have the same codomain, say B.

Of course, there may be many different maps from A to B, so that these two properties alone do not guarantee that f and g are the same map. If we recall that a point of a set A is a map from a singleton set $\mathbf{1}$ to A, we see that there is a simple **test for equality of maps of sets** $A \xrightarrow{f} B$ and $A \xrightarrow{g} B$:

$$\text{If for each } point\ \mathbf{1} \xrightarrow{a} A,\ f \circ a = g \circ a, \text{ then } f = g.$$

(Notice that $f \circ a$ and $g \circ a$ are points of B.) Briefly, 'if maps of sets agree at points they are the same map.'

In doing the exercises you should remember that the two maps

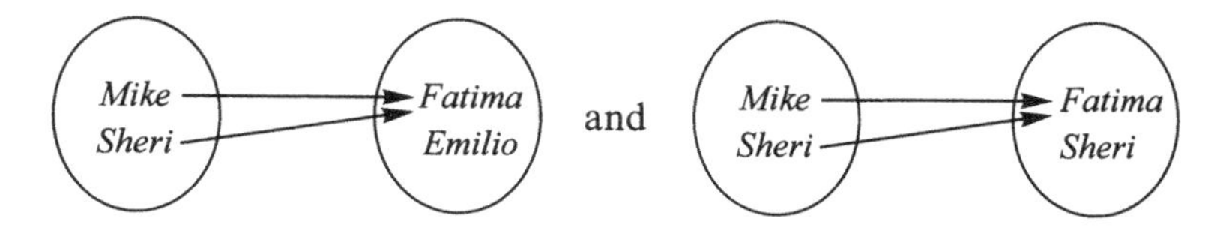

are not the same even though they have the same rule ('Mike likes Fatima and Sheri likes Fatima'), because they have different codomains. On the other hand the two maps

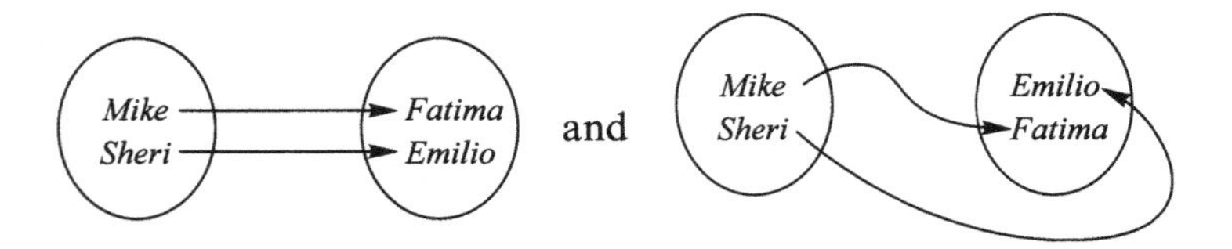

are the same, even though their pictures don't look quite the same.

You should also remember that the composite of two maps like this:

$$A \xrightarrow{g} B \xrightarrow{f} C$$

is called '$f \circ g$', in the opposite order! This is because of a choice that was made by our great-grandparents. To say 'Mike is sent by the map f to Fatima', they wrote:

$$f(Mike) = Fatima$$

(read: 'f **of** Mike is Fatima'). A better choice might have been:

$$Mike\, f = Fatima$$

Let me show you how the notation '$f(Mike) = Fatima$' gave rise to the convention of writing '$f \circ g$' for the composite, g followed by f. Imagine we write the composite gf. Then we would get

$$(gf)(John) = f(g(John))$$

which is too complicated. With the present convention, we get

$$(f \circ g)(John) = f(g(John))$$

which is easier to remember. So, in order not to get confused between the order in '$f \circ g$' and the order in the diagram (which is the order in which the rules are applied), you should get used to reading '$f \circ g$' as 'f **following** g'.

The first exercise in Article I was to use internal diagrams to check the associative law for the composition of the maps

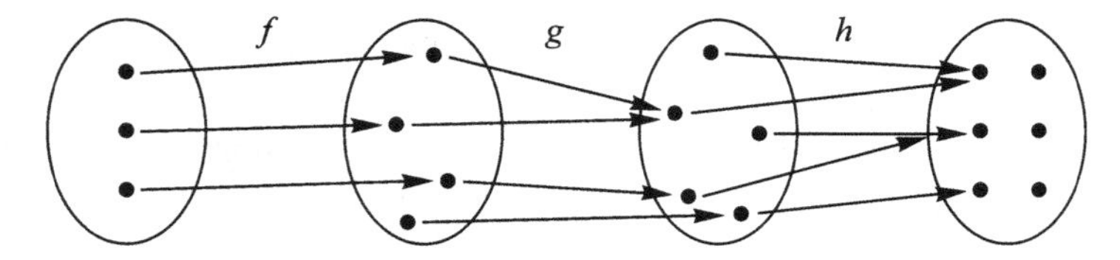

A first step is to fill in the figure

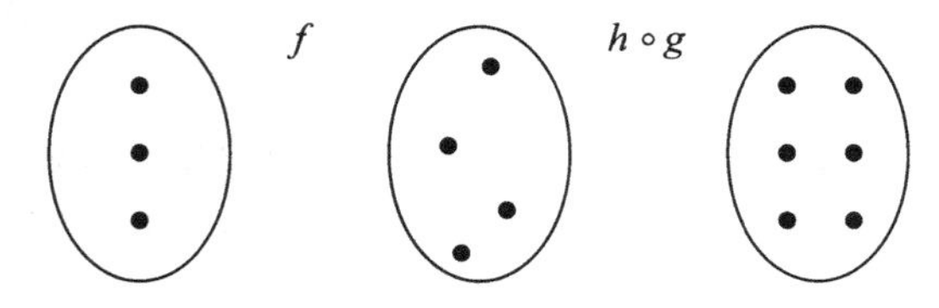

which Chad has done like this:

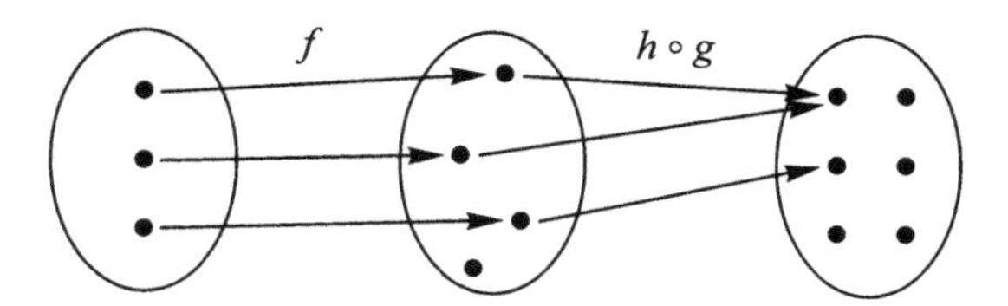

Is this correct? Not quite, because we are supposed to draw two maps, and the thing drawn for $h \circ g$ is not a map; one of the points of the domain of $h \circ g$ has been left without an assigned output. This deficiency won't matter for the next step, because that information is going to get lost anyhow, but it belongs in this step and it is incorrect to omit it. Chad's trouble was that in drawing $h \circ g$, he noticed that the last arrow would be irrelevant to the composite $(h \circ g) \circ f$, so he left it out.

CHAD: It seems the principle is like in multiplication, where the order in which you do things doesn't matter; you get the same answer.

I am glad you mention order. Let me give you an example to show that the order *does* matter. Consider the two maps

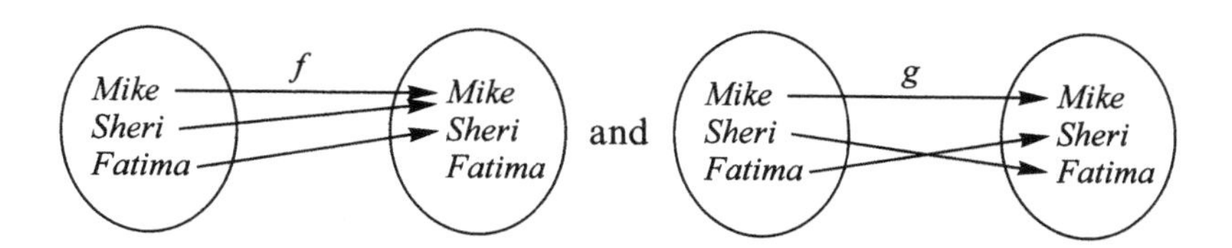

Work out the composite $f \circ g$, and see what you get:

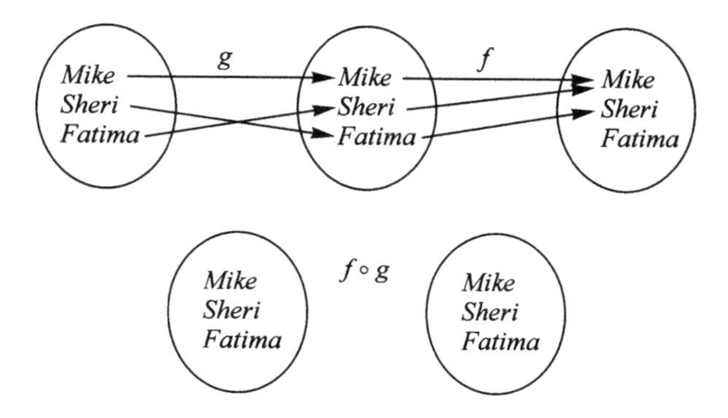

Now work out the composite in the opposite order:

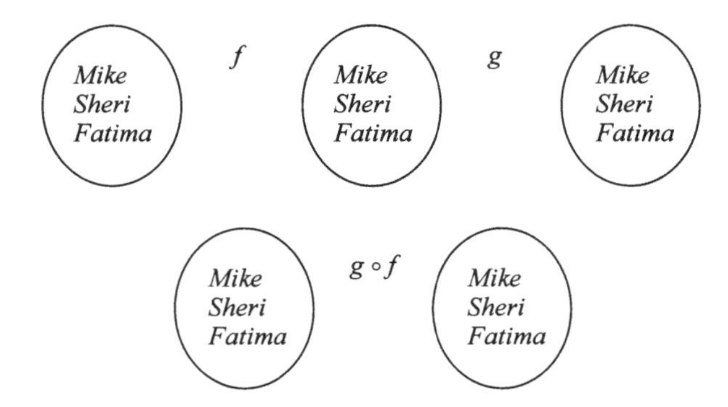

The two results are different. In composition of maps the order matters.

When I was little I had a large family, and in large families there are always many small chores to be done. So my mother would say to one of us: 'Wouldn't you like to wash the dishes?' But as we grew, two or more tasks were merged into one, so that my mother would say: 'Wouldn't you like to **wash and then rinse** the dishes?' or: '**scrape and wash** and then **rinse and dry** the dishes?' And you can't change the order. You'll make a mess if you try to dry before scraping. The 'associative law for tasks' says that the two tasks:

(**scrape** then **wash**) then (**rinse** then **dry**)

and

scrape then [(**wash** then **rinse**) then **dry**]

accomplish the same thing. All that matters is the order, not when you take your coffee break. All the parentheses are unnecessary; the composite task is:

scrape then **wash** then **rinse** then **dry**

Think about this and see if it suggests an explanation for the associative law. Then look back at the pictures, to see how you can *directly* draw the picture for a composite of several maps without doing 'two at a time'.

Several students have asked why some arrows disappear when you compose two maps, i.e. when you pass from the diagrams

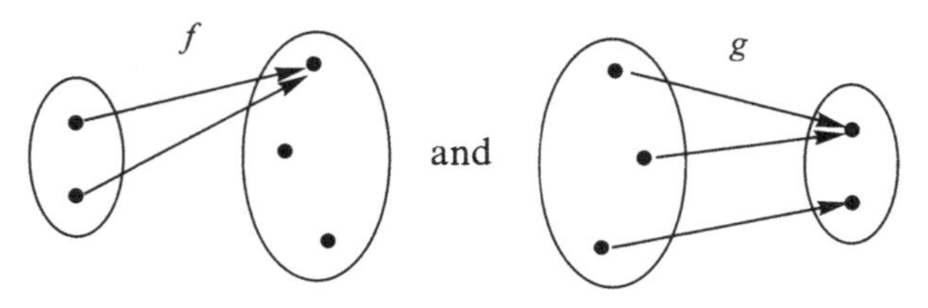

to the diagram for 'g following f'

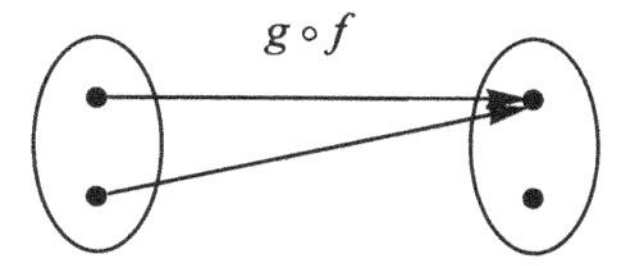

To understand this you should realize that the composite of two maps is supposed to be another map, so that it just has a domain, a codomain and a rule. The pasting together of two diagrams is not the composite map, it is just a rule to find the composite map, which can be done easily by 'following the arrows' to draw the diagram of the resulting (composite) map. The point of erasing all the irrelevant detail (like the extra arrows) is that the simplified picture really gives a different rule which defines the same map, but a simpler rule.

Suppose you carry a sleeping baby on a brief walk around town, first walking in the hot sun, then through the cool shade in the park, then out in the sun again.

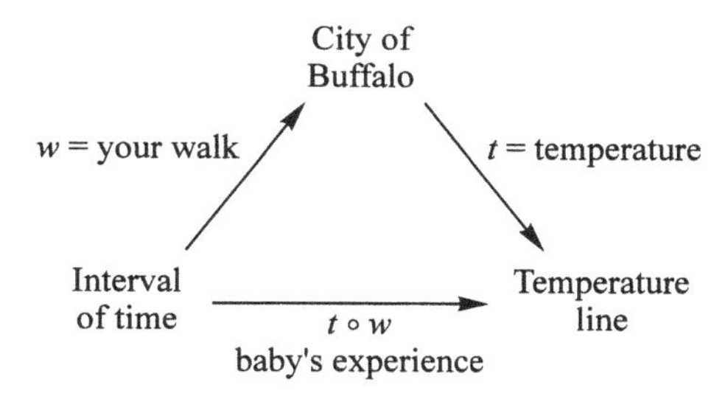

The map w assigns to each instant your location at that time, and the map t assigns to each spot in Buffalo the temperature there. ('Temperature line' has as its points physical temperatures, rather than numbers which measure temperature on some scale; a baby is affected by temperature before learning of either Fahrenheit or Celsius.) The baby was hot, then cool, then hot again, but doesn't know the two maps that were composed to get this one map.

2. An example of different rules for a map

The measurement of temperature provides a nice example of different rules for a 'numerical' map. If one looks at a thermometer which has both scales, Celsius and Fahrenheit, it becomes obvious that there is a map,

$$\text{Numbers} \xrightarrow{\text{change from Fahrenheit to Celsius}} \text{Numbers}$$

which sends the measure in degrees Fahrenheit of a temperature to the measure in degrees Celsius of the same temperature. In other words, it is the map that fits in the diagram

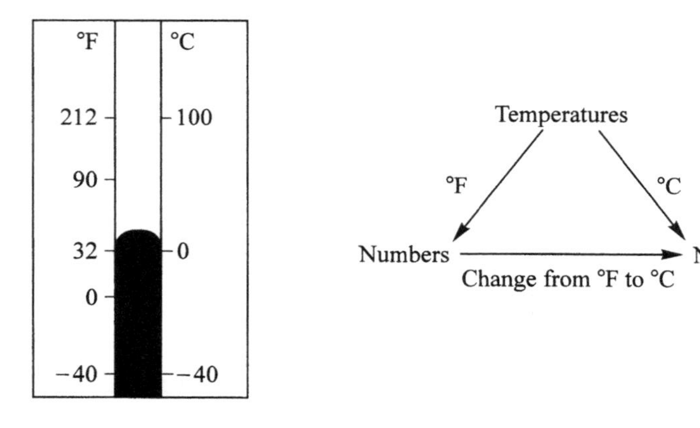

How is this map calculated? Well, there are several possible *rules*. One of them is: 'subtract 32, then multiply by 5/9.' Another is: 'add 40, multiply by 5/9, then subtract 40.' Notice that each of these rules is itself a composite of maps, so that we can draw the following diagram:

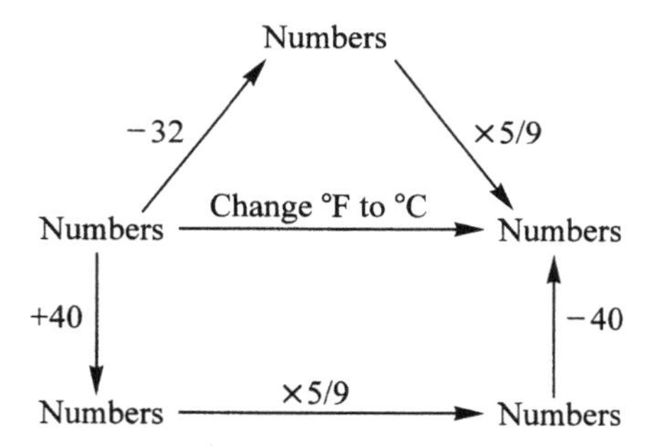

The above example illustrates that a single map may arise as a composite in several ways.

3. External diagrams

The pasting of the diagrams to calculate composition of maps is nice because from it you can read what f does, what g does, and also what the composite $g \circ f$ does. This is much more information than is contained in $g \circ f$ alone. In fact internal diagrams aren't always drawn. We use schematic diagrams like those in our 'temperature' example, or this:

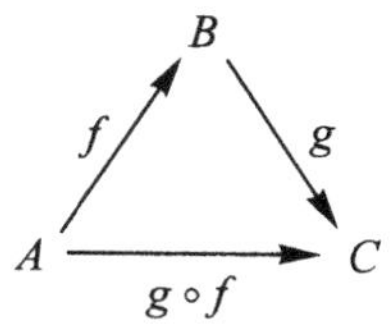

These are called **external diagrams** because they don't show what's going on inside. In Session 1 we met an external diagram when discussing Galileo's ideas:

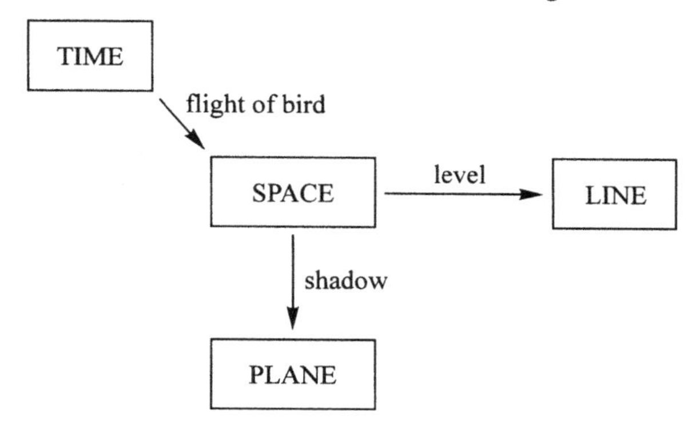

4. Problems on the number of maps from one set to another

Let's work out a few problems that are not in Article I. How many maps are there from the set A to the set B in the following examples?

(1) $A =$ (*Sheri*, *Omer*, *Alysia*, *Mike*) $B =$ (*Emilio*)

Answer: Exactly one map; all elements of A go to *Emilio*.

(2) $A =$ (*Emilio*) $B =$ (*Sheri*, *Omer*, *Alysia*, *Mike*)

Answer: There are four maps because all a map does is to tell where Emilio goes, and there are four choices for that.

(3) Now the set A is ... What shall I say? Ah! The set of all purple people-eaters in this room, and B is as before:

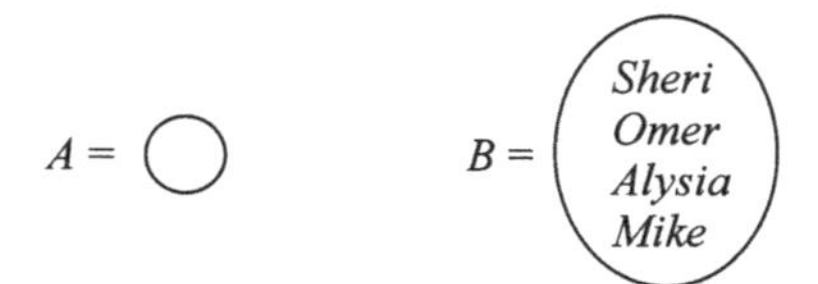

Answer: There is precisely one map, and its internal diagram is

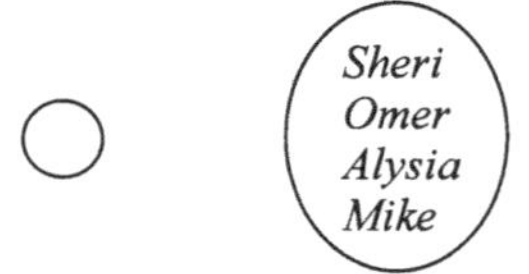

This diagram doesn't have any arrows, but it doesn't need any. An internal diagram needs one arrow for each element of the domain, and in this case the domain has no element. Try to convince yourself that this is right, but without giving yourself a headache!

(4) Now we reverse the previous example, that is:

Answer: Zero. We have four tasks, and each of them is impossible.

(5) Both A and B are empty, i.e.:

Answer: There is one map, and its internal diagram is

○ ○

which is a valid diagram for the same reason that the one in (3) is valid. Why does the reasoning in (4) not apply here?

Don't worry *too* much about these extreme cases. The reason I mention them is that as you learn the general setting you will see that they fit in quite nicely.

SESSION 3

Composing maps and counting maps

Let's look at some of the exercises from Article I, starting with Exercises 2 and 3. Can you explain why the results $h \circ (g \circ f)$ and $(h \circ g) \circ f$ always come out the same? What can you say about four maps in a row, like these?

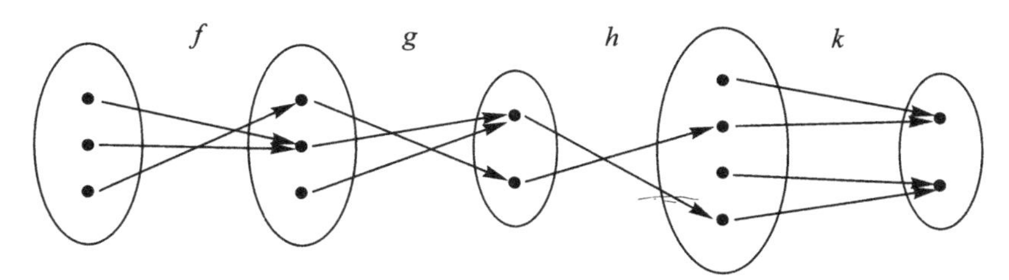

Clarification of these questions is what I was aiming at with the story of my mother and the tasks of scraping, washing, rinsing, and drying the dishes. The tasks were meant as an analog of maps, so that the four-step task corresponds to the composite map. When we first explained composition of maps, we said that the basic thing is to compose two maps, for example those in the diagram

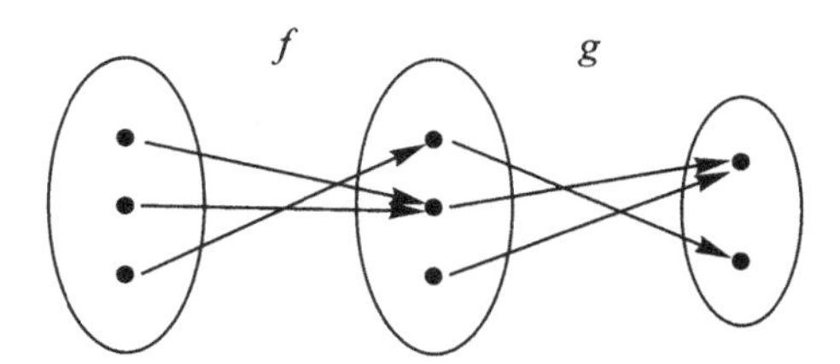

This diagram, as we said in last session, can itself be regarded as a rule to calculate the composite map $g \circ f$, namely the rule: '**Look at this diagram and follow the arrows.**' The internal diagram of $g \circ f$,

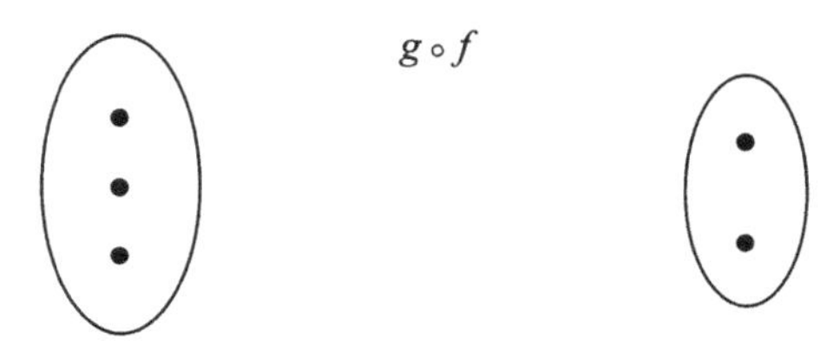

is just a simplified rule to calculate the same map. If we do the same thing with h and k, we can pass by steps from

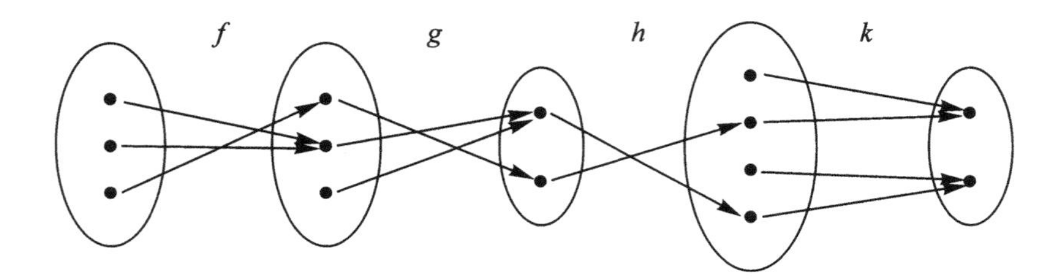

to

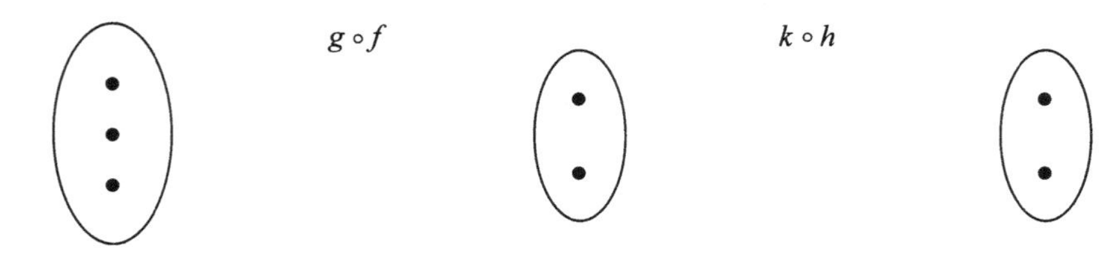

(Fill in any missing arrows yourself.) Then, repeating the process, we get

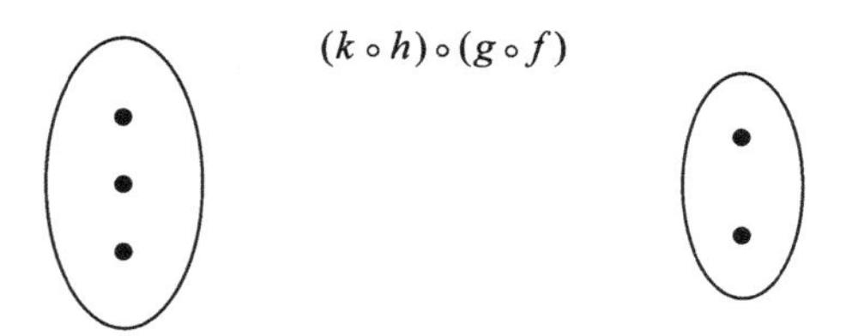

But this piecemeal work is unnecessary. The analogy of *scrape*, then *wash*, then *rinse*, then *dry* is meant to suggest that we can go from the beginning to the end in one step, if we stick to the idea that the diagram

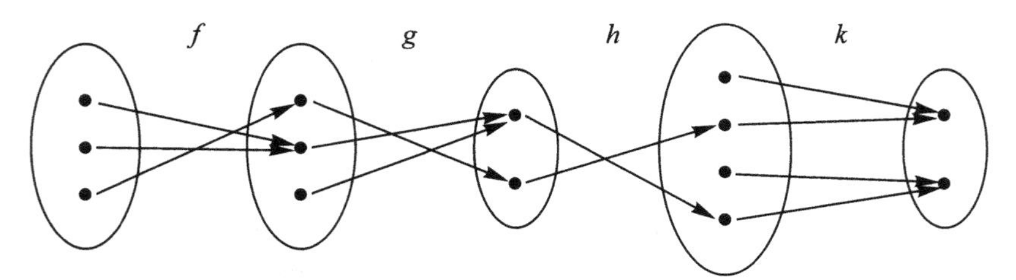

itself gives a good rule for calculating the composite $k \circ h \circ g \circ f$. Just 'look at the whole diagram and follow the arrows'; for example:

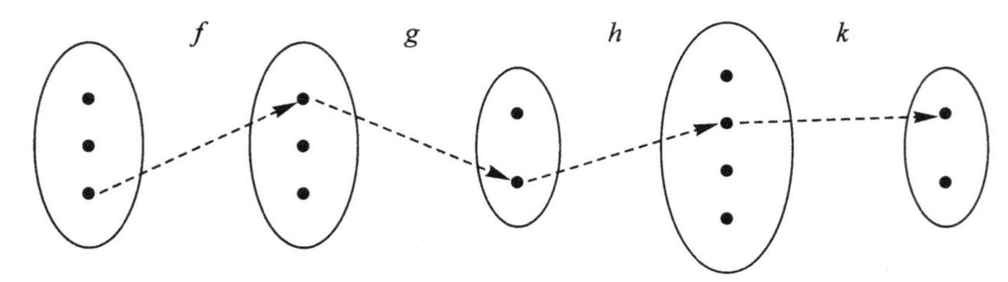

Now let's see if we can find a way to tell the number of maps between any two finite sets. For that we should start by working out simple cases. For example, Exercise 4 is to find the number of maps from a three-element set to a two-element set. How can we do this? The most immediate way I can think of is to draw them (taking care not to repeat any and not to omit any), and then count them. Say we begin with

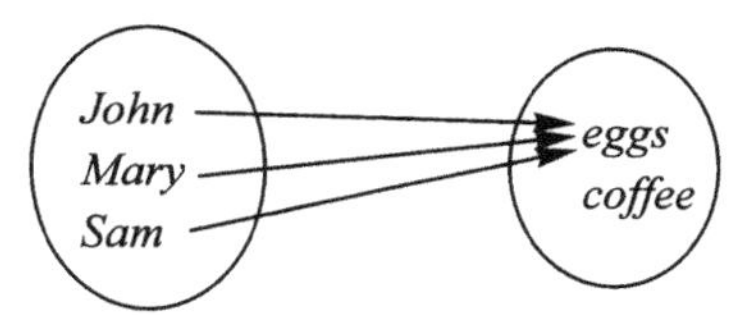

Then we can do something else,

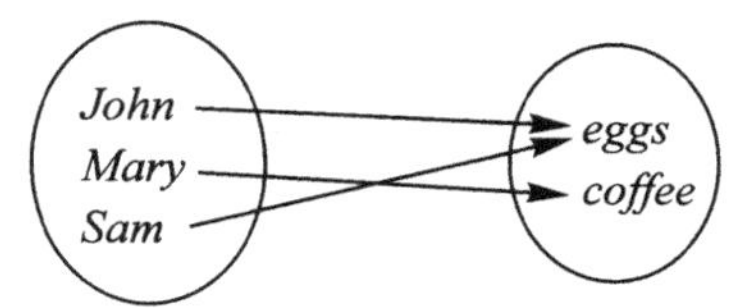

and then perhaps

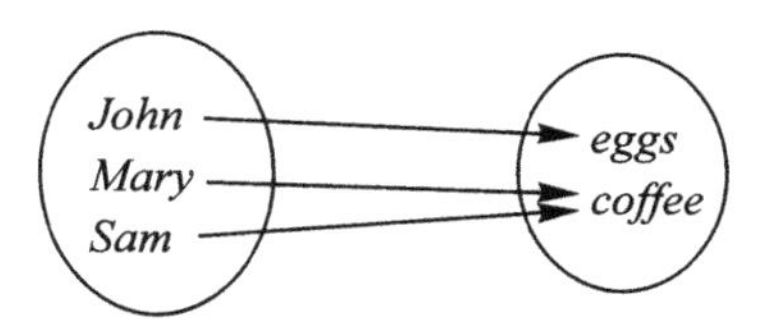

and let's see .... Do we have all the maps that send John to eggs? Right, we need one more, sending Mary to eggs and Sam to coffee. So there are four maps that send 'John' to 'eggs', and I hope it is clear that there are also four maps that send 'John' to 'coffee', and that their diagrams are the same as the four above, but changing the arrow from 'John'. Thus the answer to this exercise is 8 maps. The same method of drawing all possibilities should give you the answers to Exercises 5, 6, and 7, so that you can start to fill in a table like this:

Number of DOMAIN	3	3	2	2	
Number of CODOMAIN	2	3	3	2	
Number of MAPS	8	27	9	4	

hoping to find a pattern that may allow you to answer other cases as well.

ALYSIA: It seems that the number of maps is equal to the number of elements of the codomain raised to a power (the number of elements of the domain.)

That's a very good idea. One has to discover the reason behind it. Let's see if it also works with the extreme cases that we found at the end of last session.

Adding those results to our table we get:

Number of DOMAIN	3	3	2	2	4	1	0	4	0
Number of CODOMAIN	2	3	3	2	1	4	4	0	0
Number of MAPS	8	27	9	4	1	4	1	0	1
	2^3	3^3	3^2	2^2	1^4	4^1	4^0	0^4	0^0

and

n	1	0	$n \neq 0$
1	n	n	0
1	n	1	0
1^n	n^1	n^0	0^n

where n is any natural number, with the only exception that in the last column it must be different from zero. Now you should think of some reason that justifies this pattern.

CHAD: For every element of the domain there are as many possibilities as there are elements in the codomain, and since the choices for the different elements of the domain are independent, we must multiply all these values, so the number of maps is the number of elements of the codomain multiplied by itself as many times as there are elements in the domain.

Chad's answer seems to me very nice. Still we might want a little more explanation. Why multiply? What does 'independent' mean? If John has some apples and Mary has some apples, aren't Mary's apples independent of John's? So, if you put them all in a bag do you add them or multiply them? Why?

Going back to Alysia's formula for the number of maps from a set A to a set B, it suggests a reasonable notation, which we will adopt. It consists in denoting the set of maps from A to B by the symbol B^A, so that our formula can be written in this nice way

$$\#(B^A) = (\#B)^{(\#A)} \quad \text{or} \quad |B^A| = |B|^{|A|}$$

where the notations $\#A$ and $|A|$ are used to indicate the number of elements of the set A. The notation $\#A$ is self-explanatory since the symbol # is often used to denote 'number', while $|A|$ is similar to the notation used for the absolute value of a number. The bars indicate that you forget everything except the 'size'; for numbers you forget the sign, while for sets you forget what the elements are, and remember only how many of them there are. So, for example, if

$P =$ {*Ian*, *Chad*} $\qquad R =$ {*living room*, *dining room*}

then we wouldn't say $P = R$, but rather $|P| = |R|$. To remember which set goes in the base and which one in the exponent you can imagine that the maps are lazy, so that they go down from the exponent to the base. Another way to remember this is to think of an especially simple case, for instance the case in which the codomain has only one element, and therefore the set of maps has also only one element (and, of course, remember that $1^n = 1$).

In Exercise 9, we don't ask for the total number of maps from one set to another, but only the number of maps g

such that $g \circ g = g$. Can you think of one? Right,

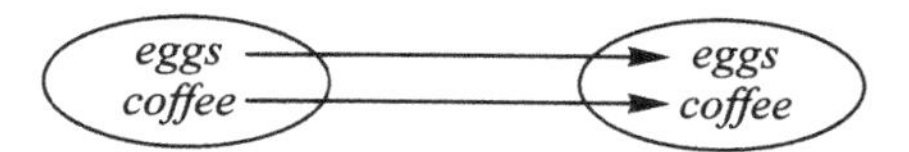

This is the first example anybody would think of. Remember from Article I that this map is called an **identity map**. Any set B has an identity map, which is denoted

$$B \xrightarrow{1_B} B$$

and sends each element of the domain to itself. This map certainly satisfies $1_B \circ 1_B = 1_B$. In fact it satisfies much more; namely, for any map $A \xrightarrow{f} B$, and any map $B \xrightarrow{g} C$,

$$1_B \circ f = f \quad \text{and} \quad g \circ 1_B = g$$

(These two equations give two different proofs of the property $1_B \circ 1_B = 1_B$: one by taking $f = 1_B$ and one by taking $g = 1_B$.) These properties of the identity maps are like the property of the number 1, that multiplied by any number gives the same number. So, identity maps behave for composition as the number 1 does for multiplication. That is the reason a '1' is used to denote identity maps. What's another map g

which satisfies $g \circ g = g$? What about the map

This map also has the property, since the composite

is

Now try to do the exercises again if you had difficulty before. One suggestion is to look back and use the special diagrams available only for endomaps explained in Article I.

Here are some exercises on the 'bookkeeping rules' about domains and codomains of composites.

Exercise 1:
A, B, and C are three different sets (or even three different objects in any category); f, g, h, and k are maps with domains and codomains as follows:

$$A \xrightarrow{f} B, \quad B \xrightarrow{g} A, \quad A \xrightarrow{h} C, \quad C \xrightarrow{k} B$$

Two of the expressions below make sense. Find each of the two, and say what its domain and codomain are:

(a) $k \circ h \circ g \circ f$ (b) $k \circ f \circ g$ (c) $g \circ f \circ g \circ k \circ h$

Exercise 2:
Do Exercise 1 again, first drawing this diagram:

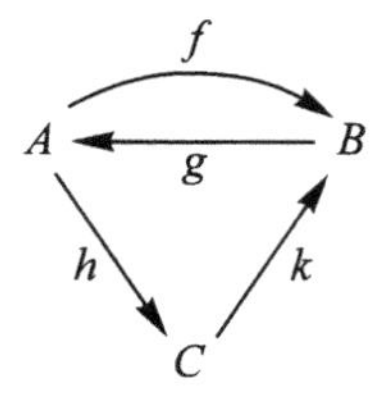

Now just read each expression from right to left; so (a) is 'f then g then h then k.' As you read, follow the arrows in the diagram with your finger, like this:

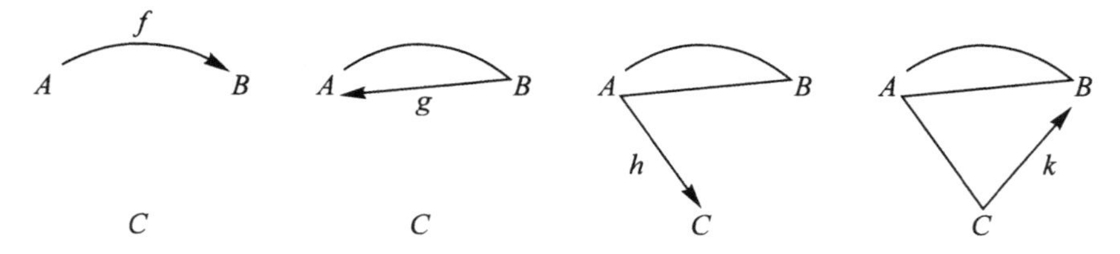

The composite makes sense, and goes from A to B. See how much easier this external diagram makes keeping track of domains, etc.

PART II

The algebra of composition

We investigate the analogy: If composition of maps is like multiplication of numbers, what is like division of numbers? The answers shed light on a great variety of problems, including (in Session 10) 'continuous' problems.

ARTICLE II

Isomorphisms

Retractions, sections, idempotents, automorphisms

1. Isomorphisms

It seems probable that before man learned to count, it was first necessary to notice that sometimes one collection of things has a certain kind of resemblance to another collection. For example, these two collections

are similar. In what way? (Remember that numbers had not yet been invented, so it is not fair to say 'the resemblance is that each has three elements.')

After some thought, you may arrive at the conclusion that the resemblance is actually given by choosing a *map*, for instance this one:

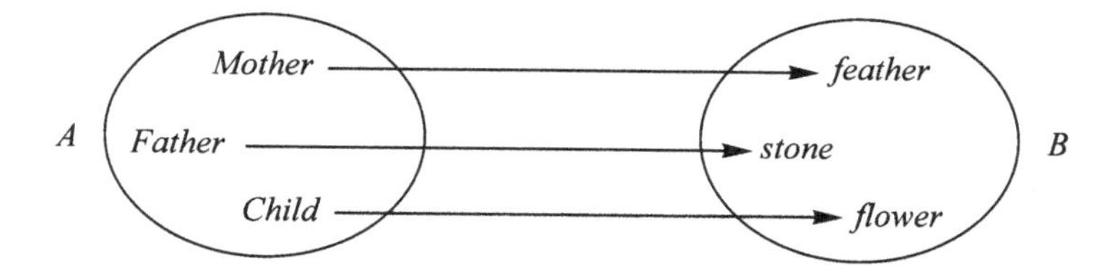

What special properties does this map f have? We would like them to be expressed entirely in terms of composition of maps so that we can later use the same idea in other categories, as well as in the category of finite sets. The properties should *exclude* maps like these:

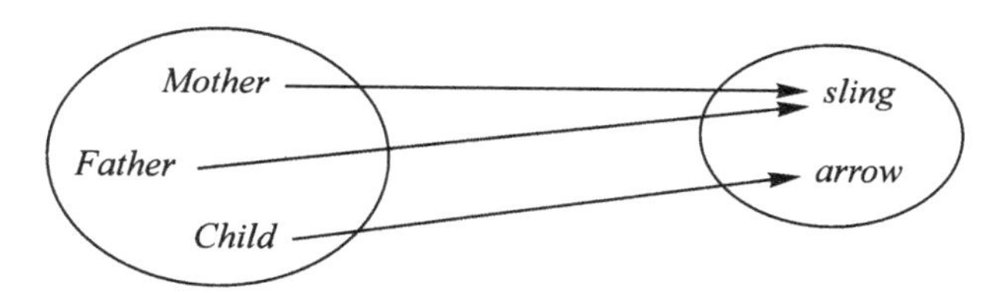

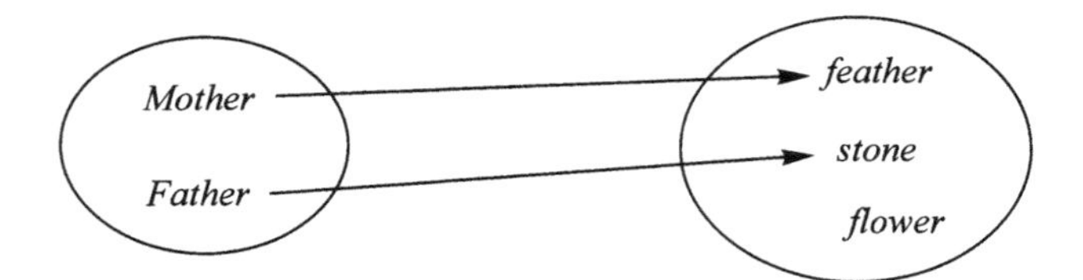

The crucial property that f has, and the other two maps do not have, is that there is an ***inverse map*** g for the map f. Here is a picture of g:

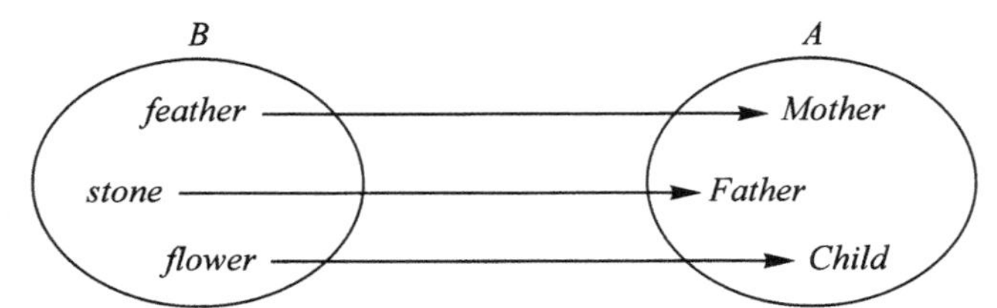

The important thing to notice is that g and f are related by *two equations*

$$g \circ f = 1_A \quad f \circ g = 1_B$$

As we will see, neither of these equations by itself will guarantee that A and B have the same size; we need both. This gives rise to the following concepts:

Definitions: *A map* $A \xrightarrow{f} B$ *is called an* **isomorphism**†, *or* **invertible map**, *if there is a map* $B \xrightarrow{g} A$ *for which* $g \circ f = 1_A$ *and* $f \circ g = 1_B$.
A map g related to f by satisfying these equations is called an **inverse for** *f.*
Two objects A and B are said to be **isomorphic** *if there is at least one isomorphism* $A \xrightarrow{f} B$

Notice that there are other isomorphisms from {*Mother*, *Father*, *Child*} to {*feather*, *stone*, *flower*}, for instance

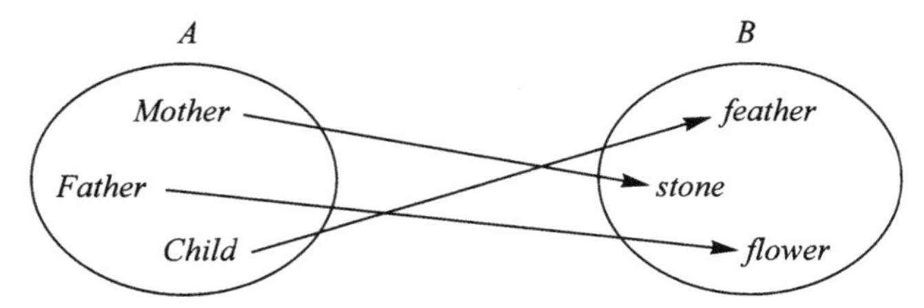

but to show that these two sets are isomorphic, we only need to find *one* of the many – how many? – isomorphisms from A to B.

Once mankind had noticed this way of finding 'resemblance' between collections, it was probably not too long before some names for the 'sizes' of small collections – words like *pair*, or *triple* – came about. But first a crucial step had to be made: one

†The word *isomorphism* comes from Greek: *iso* = same; *morph* = shape, form; though in our category of finite sets same *size* might seem more appropriate.

had to see that the notion of *isomorphic* or 'equinumerous' or 'same-size', or whatever it was called (if indeed it had any name at all yet), has certain properties:

Reflexive: A is isomorphic to A.
Symmetric: If A is isomorphic to B, then B is isomorphic to A.
Transitive: If A is isomorphic to B, and B is isomorphic to C, then A is isomorphic to C.

Surprisingly, all these properties come directly from the associative and identity laws for composition of maps.

Exercise 1:
(R) Show that $A \xrightarrow{1_A} A$ is an isomorphism.
(Hint: find an inverse for 1_A.)
(S) Show that if $A \xrightarrow{f} B$ is an isomorphism, and $B \xrightarrow{g} A$ is an inverse for f, then g is also an isomorphism.
(Hint: find an inverse for g.)
(T) Show that if $A \xrightarrow{f} B$ and $B \xrightarrow{k} C$ are isomorphisms, $A \xrightarrow{k \circ f} C$ is also an isomorphism.

These exercises show that the three properties listed before them are correct, but the exercises are more explicit: solving them tells you not just that certain maps *have* inverses, but how actually to *find* the inverses.

All this may seem to be a lot of fuss about what it is that all three-element sets have in common! Perhaps you will be partially persuaded that the effort is worthwhile if we look at an example from geometry, due to Descartes. P is the plane, the plane from geometry that extends indefinitely in all directions. $\mathbb{R}^2$ is the set of all lists of two real numbers (positive or negative infinite decimals like $\sqrt{3}$ or $-\pi$ or 2.1397). Descartes' analytic approach to geometry begins with an isomorphism

$$P \xrightarrow{f} \mathbb{R}^2$$

assigning to each point its coordinate-pair, *after* choosing two perpendicular lines in the plane and a unit of distance:

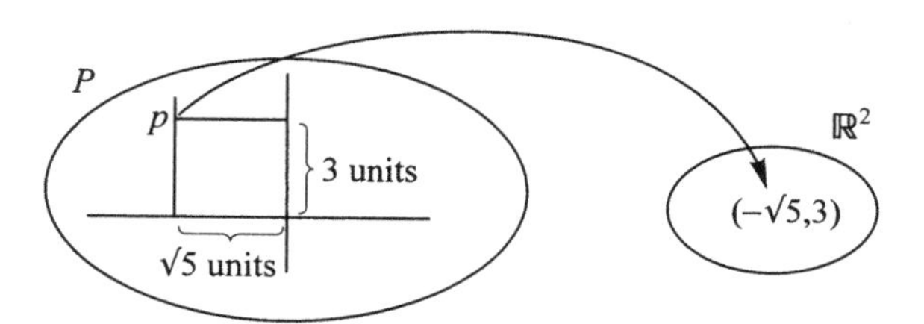

The map f assigns to each point p in the plane a pair of numbers, called the 'coordinates of p in the chosen coordinate system'. (What does the *inverse* map g do? It must assign to each pair of numbers, like $(\pi, 7)$, a point. Which point?)

By systematically using this kind of isomorphism, Descartes was able to *translate* difficult problems in geometry, involving lines, circles, parabolas, etc., into easier problems in algebra, involving equations satisfied by the coordinate-pairs of the points on the curves. We still use this procedure today, and honor Descartes by calling these coordinate systems 'cartesian coordinates'. Our notion of 'isomorphism' is what makes this technique work perfectly: we can 'translate' any problem about a plane – i.e. apply the map f to it – to a problem about pairs of numbers. This problem about pairs of numbers may be easier to solve, because we have many algebraic techniques for dealing with it. Afterwards, we can 'translate back' – i.e. apply the inverse map for f – to return to the plane. (It should be mentioned that Descartes' method has also proved useful in the opposite way – sometimes algebraic problems are most easily solved by translating them into geometry!)

You will notice that we have sneaked in something as we went along. Before, we talked of *an* inverse for f, and now we have switched to *the* inverse for f. This is justified by the following exercise, which shows that, while a map f may not have any inverse, it *cannot* have two different inverses!

Exercise 2:
Suppose $B \xrightarrow{g} A$ and $B \xrightarrow{k} A$ are *both* inverses for $A \xrightarrow{f} B$. Show that $g = k$.

Since the algebra of composition of maps resembles the algebra of multiplication of numbers, we might expect that our experience with numbers would be a good guide to understanding composition of maps. For instance, the associative laws are parallel:

$$f \circ (g \circ h) = (f \circ g) \circ h$$
$$3 \times (5 \times 7) = (3 \times 5) \times 7$$

But we need to take some care, since

$$f \circ g \neq g \circ f$$

in general. The kind of care we need to take is exemplified in our discussion of inverses. For numbers, the 'inverse of 5', or $\frac{1}{5}$, is characterized by: it is *the* number x such that $5 \times x = 1$; but for the inverse of a map, we needed *two* equations, not just one.

More care of this sort is needed when we come to the analog of division. For numbers, $\frac{3}{5}$ (or $3 \div 5$) is characterized as *the* number x for which

$$5 \times x = 3;$$

but it can also be obtained as

$$x = \tfrac{1}{5} \times 3$$

Thus for numbers we really don't need division in general; once we understand inverses (like $\frac{1}{5}$) and multiplication, we can get the answers to more general division problems by inverses and multiplication. We will see that a similar idea can be used for maps, but that not all 'division problems' reduce to finding inverses; and also that there are interesting cases of 'one-sided inverses', where $f \circ g$ is an identity map but $g \circ f$ is not.

Before we go into general 'division problems' for maps, it is important to master isomorphisms and some of their uses. Because of our earlier exercise, showing that a map $A \xrightarrow{f} B$ can have at most *one* inverse, it is reasonable to give a special name, or symbol, to that inverse (when there is an inverse).

Notation: If $A \xrightarrow{f} B$ has an inverse, then the (one and only) inverse for f is denoted by the symbol f^{-1} (read 'f-inverse', or 'the inverse of f'.)

Two things are important to notice:

1. To show that a map $B \xrightarrow{g} A$ satisfies $g = f^{-1}$, you must show that

$$g \circ f = 1_A \quad \textit{and} \quad f \circ g = 1_B$$

2. If f does *not* have an inverse, then the symbol 'f^{-1}' *does not stand for anything*; it's a nonsense expression like 'grlbding' or '$\frac{3}{0}$'.

Exercise 3:
If f has an inverse, then f satisfies the two cancellation laws:

(a) If $f \circ h = f \circ k$, then $h = k$.
(b) If $h \circ f = k \circ f$, then $h = k$.

Warning: The following 'cancellation law' is *not* correct, even if f has an inverse.

(c) (wrong): If $h \circ f = f \circ k$, then $h = k$.

When an exercise is simply a statement, the task is to prove the statement. Let's do part (a). We assume that f has an inverse and that $f \circ h = f \circ k$, and we try to show that $h = k$. Well, since $f \circ h$ and $f \circ k$ are the same map, the maps $f^{-1} \circ (f \circ h)$ and $f^{-1} \circ (f \circ k)$ are also the same:

$$f^{-1} \circ (f \circ h) = f^{-1} \circ (f \circ k)$$

But now we can use the associative law (twice – once on each side of our equation), so our equation becomes:

$$(f^{-1} \circ f) \circ h = (f^{-1} \circ f) \circ k$$

which simplifies to

$$1_A \circ h = 1_A \circ k \quad \text{(why?)}$$

which then simplifies to

$$h = k \quad \text{(why?)}$$

So we have finished: $h = k$ is what we wanted to show.

You will notice that this kind of calculation is very similar to algebra with numerical quantities. Our symbols $f, h, \ldots$ stand for *maps*, not for numbers; but since composition of maps satisfies *some* of the rules that multiplication of numbers does, we can often do these calculations almost by habit; we must only be careful that we never use rules, like the commutative law, that are not valid for maps.

Part (b) you should now be able to do yourself. Part (c), though, is a different story. How do you show that a general rule is *wrong*? To say it is wrong just means that there are cases (or really, at least one case) in which it is wrong. So to do part (c) select *one example* of a map f which has an inverse, and two maps h and k for which $h \circ f = f \circ k$; but not just any example, rather one in which h and k are *different* maps. The most interesting examples involve only one set, and three endomaps of that set. You should be able to find endomaps f, h, and k of a two-element set A, with f invertible and $h \circ f = f \circ k$ but $h \neq k$.

Here are some exercises with sets of *numbers*.'$\mathbb{R}$' stands for the set of all (real) numbers; '$\mathbb{R}_{\geq 0}$' for all the (real) numbers that are ≥ 0. To describe a map with an infinite set, like $\mathbb{R}$, as domain, it is not possible to list the output of f for each input in the domain, so we typically use *formulas*. For instance:

1. $\mathbb{R} \xrightarrow{f} \mathbb{R}$ $\qquad f(x) = 3x + 7$
2. $\mathbb{R}_{\geq 0} \xrightarrow{g} \mathbb{R}_{\geq 0}$ $\qquad g(x) = x^2$
3. $\mathbb{R} \xrightarrow{h} \mathbb{R}$ $\qquad h(x) = x^2$
4. $\mathbb{R} \xrightarrow{k} \mathbb{R}_{\geq 0}$ $\qquad k(x) = x^2$
5. $\mathbb{R}_{\geq 0} \xrightarrow{l} \mathbb{R}_{\geq 0}$ $\qquad l(x) = \dfrac{1}{x+1}$

Exercise 4:
For each of the five maps above: decide whether it is invertible; and if it is invertible, find a 'formula' for the inverse map.

2. General division problems: Determination and choice

In analogy with division problems for numbers (like $3 \times x = 21$, with exactly one solution: $x = 7$; or like $0 \times x = 5$, with no solutions; or like $0 \times x = 0$, with infinitely many solutions) we find *two* sorts of division problems for maps:

1. The '*determination*' (or 'extension') problem
   Given f and h as shown, what are all g, if any, for which $h = g \circ f$?

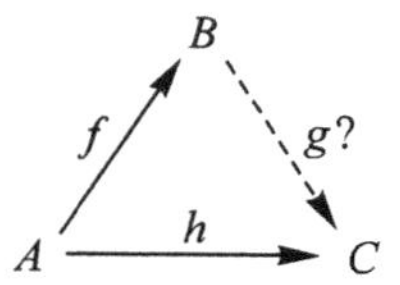

2. The '*choice*' (or 'lifting') problem
   Given g and h as shown, what are all f, if any, for which $h = g \circ f$?

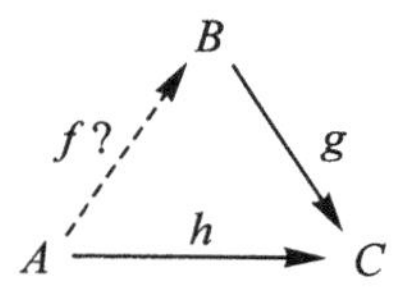

Let us study the determination problem first. If it has any solution g, we say that h is 'determined by' f, or h 'depends only on' f. (A particular solution g can be called/a 'determination' of h by f.) The same idea is often expressed by saying that h 'is a function of' f. After we have studied several examples, it will become clearer why this division problem is called the 'determination problem'.

Example 1, a 'determination' problem

When B is a one-element set, then the possibility of factoring a given $A \xrightarrow{h} C$ across B is a very drastic restriction on h. This is true because there is only one $A \xrightarrow{f} B$, whereas to choose a map $B \xrightarrow{g} C$ is the same as choosing a single element of C.

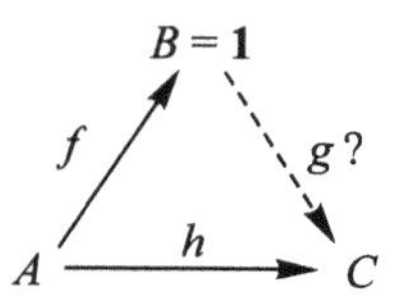

Therefore, denoting the element of B by b,

$$h(x) = (g \circ f)(x) = g(f(x)) = g(b)$$

for all x in A. Such a map h is called *constant* because it has constantly the same value even though x varies.

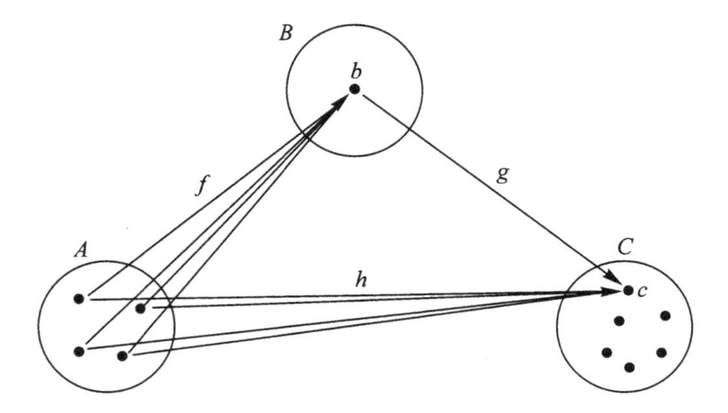

Example 2, a 'choice' problem

Now consider the following example in which B has three elements and $h = 1_A$ where $A = C$ has two elements, while $B \xrightarrow{g} C$ is a given map with the property that every element of C is a value of g, such as

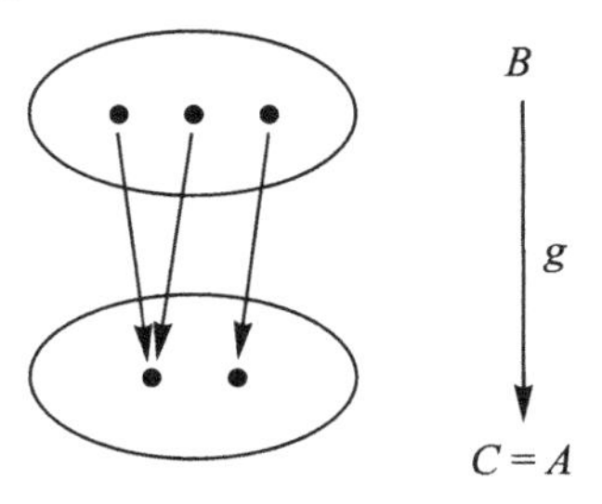

How many maps f can we find with $g \circ f = 1_A$? Such an f must be a map from $A = C$ to B and satisfy $g(f(x)) = x$ for both elements x. That is, f must 'choose' for each x an element z of B for which $g(z) = x$. From the picture we see that this determines the value of f at one x but leaves two acceptable choices for the value of f at the other x. Therefore there are exactly two solutions f to the question as follows:

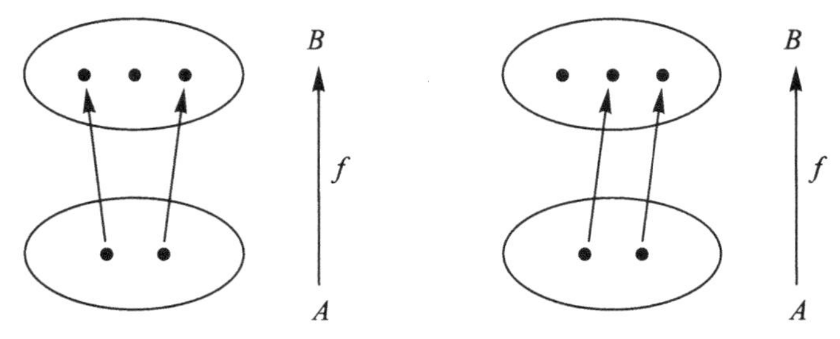

On the other hand, suppose the first of these f is considered given, and we ask for all maps g for which $g \circ f = 1_A$, a 'determination' problem. The equation $g(f(x)) = x$ can now be interpreted to mean that for each element of B which is of the form $f(x)$, g is forced to be defined so as to take it to x itself; there is one element of B to which that does not apply, so g can be defined to take *that* element to any of the two elements of A. Hence there are two such g, one of which is the g given at the beginning of the discussion of this example.

The fact that we got the same answer, namely 2, to both parts of the above example is due to the particular sizes of the sets involved, as seen by considering the two parts for the smaller example

two-element set

$$f\uparrow \quad \downarrow g \qquad g \circ f = 1_{\text{one}}$$

one-element set

and also for the larger pair of sets in the following exercise.

Exercise 5:
Given

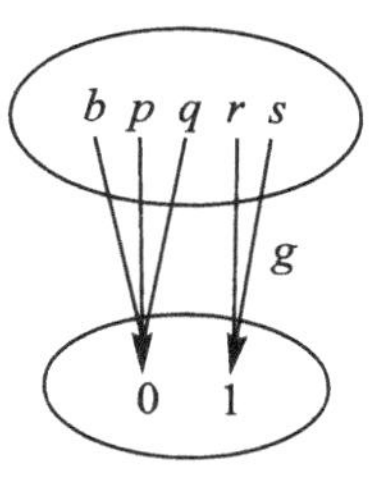

how many maps f are there with $g \circ f = 1_{\{0,1\}}$?
Choosing a particular such f, how many maps g (including the given one) satisfy the same equation?

Here are two more 'determination' examples.

Example 3

It surprised many people when Galileo discovered that the distance a dropped object falls in a certain time is determined by the time (in the absence of air resistance.) They had thought that the distance would depend also on the weight and/or density of the object.

Example 4, Pick's Formula

Imagine a grid of uniformly spaced points in the plane, and a polygonal figure with vertices among these points:

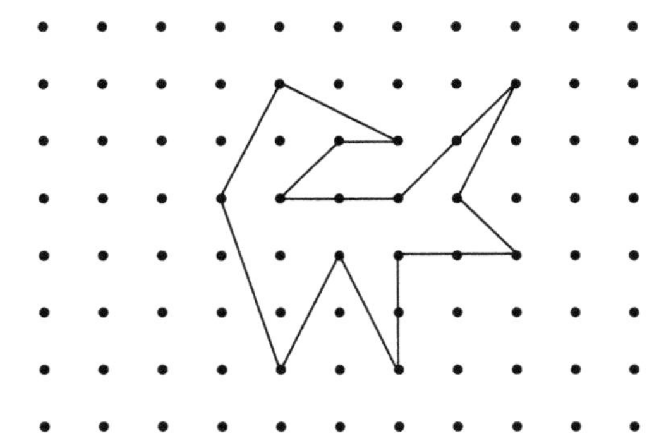

It turns out that the area (in square units) of such a polygon can be calculated from very little information: just knowing the number of interior dots and the number of boundary dots (in our example, 3 and 17) is enough. All the complicated details of the shape of the polygon are irrelevant to computing its area! Schematically

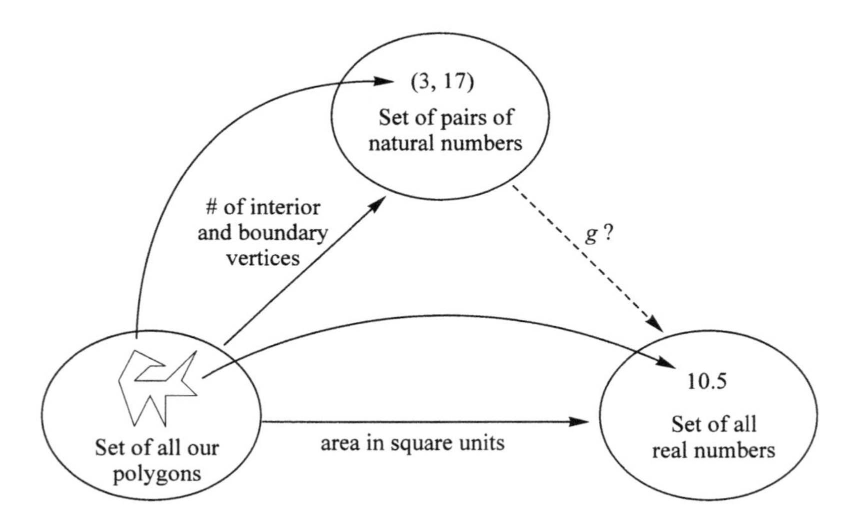

Once you guess there is such a map g, it is not too difficult to figure out a *formula* for g. (Try simple examples of polygons first, instead of starting with a complicated one like ours.)

The history of Galileo's problem was similar: once Galileo realized that the time of the fall determined the distance fallen, it did not take too many experiments before he found a *formula* for the distance in terms of the time; i.e. for g in

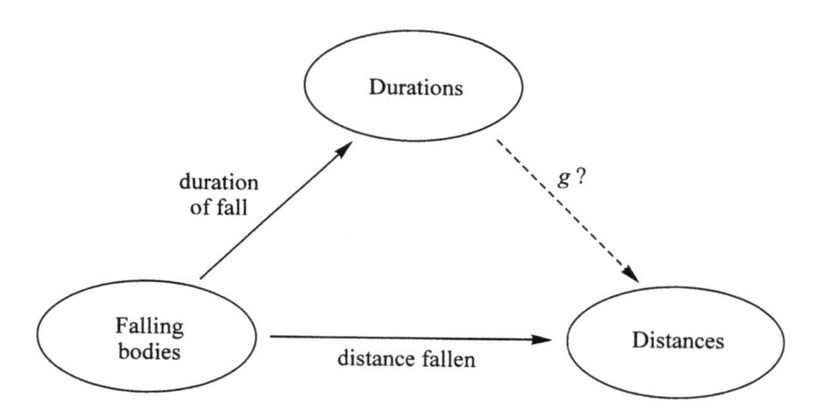

Further examples will be discussed in the sessions that follow.

3. Retractions, sections, and idempotents

The special cases of the determination and choice problems in which h is an identity map are called the 'retraction' and 'section' problems.

Definitions: *If* $A \xrightarrow{f} B$:
a **retraction for** f *is a map* $B \xrightarrow{r} A$ *for which* $r \circ f = 1_A$;
a **section for** f *is a map* $B \xrightarrow{s} A$ *for which* $f \circ s = 1_B$.

The retraction problem looks this way if we draw it as a 'determination' problem:

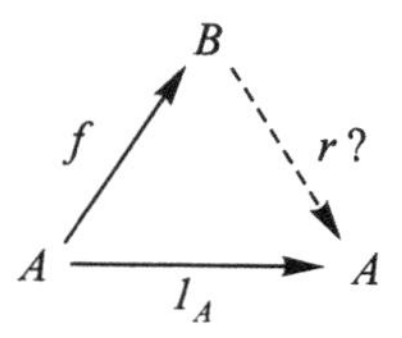

But since one of the maps is the identity map, it's simpler just to draw this

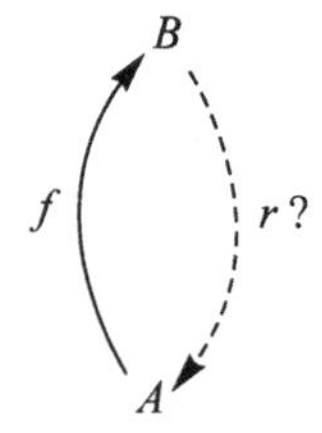

where we want r to satisfy $r \circ f = 1_A$.

Similarly, the section problem is a 'choice' problem:

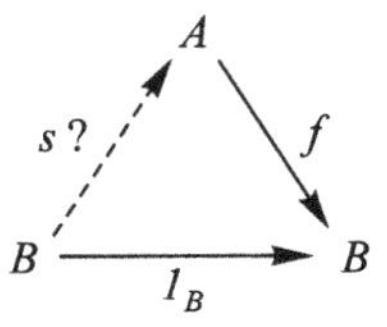

but it's simpler just to draw

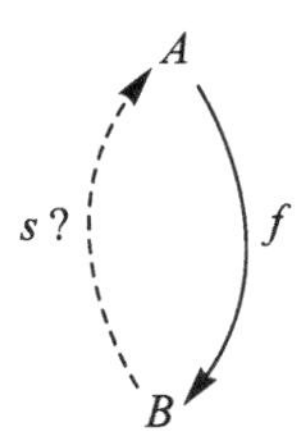

where we ask that s satisfy $f \circ s = 1_B$.

There is a slight advantage to drawing the triangular picture. It reminds us of the equation we want to satisfy, which just says the triangle 'commutes': the two ways of getting from the left corner to the right corner are equal.

From the examples just discussed we know that if a map has sections, it may have several, and another map may have several retractions. Moreover, some maps have retractions but no sections (or vice versa), and many have neither. There are some important conditions, which we can often check by looking at the map itself, that are necessary in order that a given map f could have sections or retractions. These conditions are stated in the following propositions.

The first proposition may be regarded as an analog for maps to the observation that once we have multiplication and 'reciprocals' (numbers like $x = \frac{1}{3}$ to solve equations like $3 \times x = 1$) we can then express the answers to more general division problems like $3 \times x = 5$ by $x = \frac{1}{3} \times 5$. The proposition says that if the single choice problem

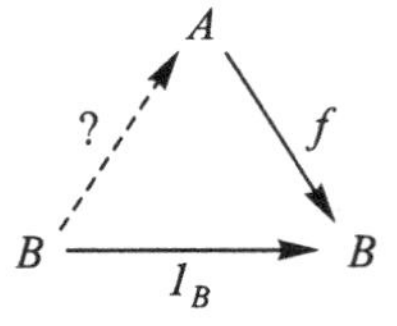

has a solution (a section for f), then *every* choice problem

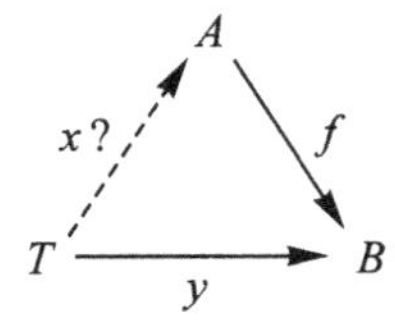

involving this same f has a solution.

Proposition 1: *If a map* $A \xrightarrow{f} B$ *has a section, then for any* T *and for any map* $T \xrightarrow{y} B$ *there exists a map* $T \xrightarrow{x} A$ *for which* $f \circ x = y$.

Proof: The assumption means that we have a map s for which $f \circ s = 1_B$. Thus for any given map y as below

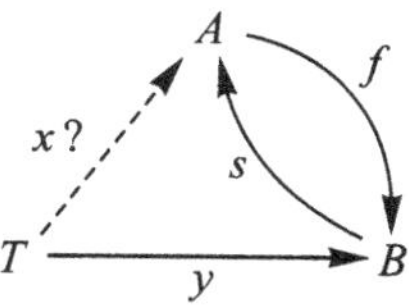

we see that we could define a map x with at least the correct domain and codomain by taking the composite s following y

$$x = s \circ y$$

Does this map x actually satisfy the required equation? Calculating

$$f \circ x = f \circ (s \circ y) = (f \circ s) \circ y = 1_B \circ y = y$$

we see that it does.

If a map f satisfies the *conclusion* of the above if ... then ... proposition (for any y there exists an x such that $fx = y$), it is often said to be 'surjective for maps from T.' Since among the T are the one-element sets, and since a map $T \xrightarrow{y} B$ from a one-element set is just an element, we conclude that if the codomain B of f has some element which is not the value $f(x)$ at any x in A, then f could not have any section s.

A section s for a map f is often thought of as a 'choice of representatives.' For example if A is the set of all US citizens and B is the set of all congressional districts, then a map f such as

$$A \xrightarrow{f = \text{residence}} B$$

divides the people up into clusters, all of those residing in a given district y constituting one cluster. If s means the congressional representative choice, then the condition $f \circ s = 1_B$ means that the representative of district y must reside in y. Clearly, there are theoretically a very large number of such choice maps s *unless* there happens to be some district which is uninhabited, in which case there will be no such maps s, as follows from Proposition 1.

There is a 'dual' to Proposition 1, which we'll call Proposition 1*. It says, as you might expect, that if the single determination problem

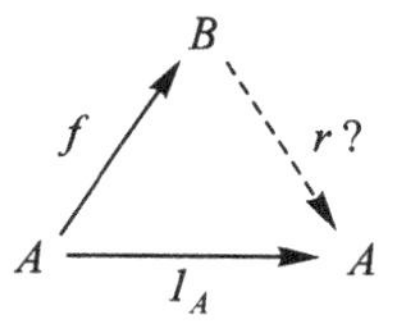

has a solution (a retraction for f), then *every* determination problem with the same f

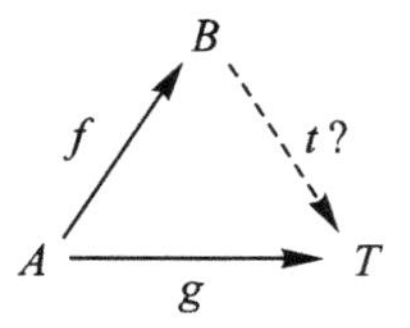

has a solution. Because the proof is so close to that of Proposition 1, we leave it as an exercise.

Exercise 6:
If the map $A \xrightarrow{f} B$ has a retraction, then for any map $A \xrightarrow{g} T$, there is a map $B \xrightarrow{t} T$ for which $t \circ f = g$. (This is Proposition 1*.)

Here is another useful property of those maps that have retractions.

Proposition 2: *Suppose a map* $A \xrightarrow{f} B$ *has a retraction. Then for any set* T *and for any pair of maps* $T \xrightarrow{x_1} A$, $T \xrightarrow{x_2} A$ *from any set* T *to* A

$$\text{if } f \circ x_1 = f \circ x_2 \text{ then } x_1 = x_2.$$

Proof: Looking back at the definition, we see that the assumption means that we have a map r for which $r \circ f = 1_A$. Using the assumption that x_1 and x_2 are such that f composes with them to get the same $T \longrightarrow B$, we can compose further with r as follows:

$$T \underset{x_2}{\overset{x_1}{\rightrightarrows}} A \xrightarrow{f} B \xrightarrow{r} A \qquad (r \circ f = 1_A)$$

$$x_1 = 1_A \circ x_1 = (r \circ f) \circ x_1 = r \circ (f \circ x_1) = r \circ (f \circ x_2) = (r \circ f) \circ x_2$$
$$= 1_A \circ x_2 = x_2$$

Definitions: *A map* f *satisfying the conclusion of Proposition 2 (for any pair of maps* $T \xrightarrow{x_1} A$ *and* $T \xrightarrow{x_2} A$, *if* $f \circ x_1 = f \circ x_2$ *then* $x_1 = x_2$*) is said to be* **injective for maps from** T.

If f *is injective for maps from* T *for every* T, *one says that* f *is* **injective**, *or is a* **monomorphism**.

Since T could have just one element, we conclude that if there were two elements x_1 and x_2 of A for which $x_1 \neq x_2$ yet $f(x_1) = f(x_2)$, then there could not be any retraction for f.

Notice that Proposition 2 says that if f has a retraction, then f satisfies the 'cancellation law' (a) in Exercise 3. Proposition 2 also has a 'dual' saying that if f has a section, then f satisfies the cancellation law (b) in Exercise 3.

Exercise 7:
Suppose the map $A \xrightarrow{f} B$ has a section. Then for any set T and any pair $B \xrightarrow{t_1} T$, $B \xrightarrow{t_2} T$ of maps from B to T, if $t_1 \circ f = t_2 \circ f$ then $t_1 = t_2$. (This is Proposition 2*.)

Definition: *A map f with this cancellation property (if $t_1 \circ f = t_2 \circ f$ then $t_1 = t_2$) for every T is called an* **epimorphism**.

Thus both 'monomorphism' and 'epimorphism' are 'cancellation' properties.

When we are given both f and r, and $r \circ f = 1_A$ then, of course, we can say both that r is a retraction for f and that f *is a section for* r. For which sets A and B can such pairs of maps exist? As we will see more precisely later, it means roughly (for non-empty A) that A is smaller (or equal) in size than B. We can easily prove the following proposition which is compatible with that interpretation.

Proposition 3: *If $A \xrightarrow{f} B$ has a retraction and if $B \xrightarrow{g} C$ has a retraction, then $A \xrightarrow{g \circ f} C$ has a retraction.*

Proof: Let $r_1 \circ f = 1_A$ and $r_2 \circ g = 1_B$. Then a good guess for a retraction of the composite would be the composite of the retractions *in the opposite order* (which is anyway the only order in which they can be composed)

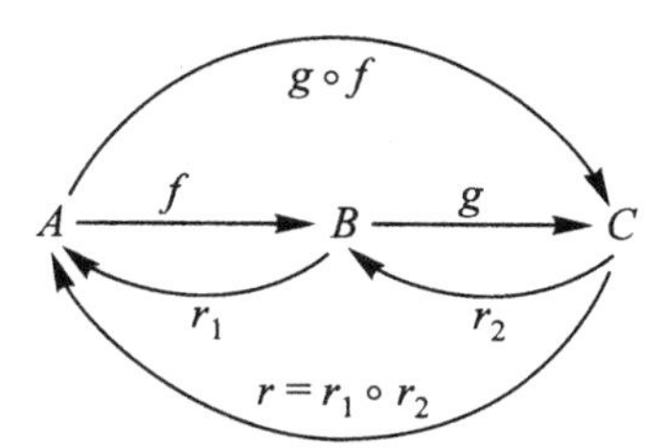

Does it in fact work?

$$r \circ (g \circ f) = (r_1 \circ r_2) \circ (g \circ f) = r_1 \circ (r_2 \circ g) \circ f = r_1 \circ 1_B \circ f$$
$$= r_1 \circ f = 1_A$$

proves that r is a retraction for $g \circ f$.

Exercise 8:
Prove that the composite of two maps, each having sections, has itself a section.

Definition: *An endomap e is called* **idempotent** *if* $e \circ e = e$.

Exercise 9:
Suppose r is a retraction of f (equivalently f is a section of r) and let $e = f \circ r$. Show that e is an idempotent. (As we'll see later, in most categories it is true conversely that all idempotents can 'split' in this way.) Show that if f is an isomorphism, then e is the identity.

A map can have many sections or many retractions, but if it has some of each they are all the same. That is, more exactly, we have:

Theorem (uniqueness of inverses): *If f has both a retraction r and a section s then* $r = s$.

Proof: From the definition we have, if $A \xrightarrow{f} B$, both of the equations

$$r \circ f = 1_A \quad \text{and} \quad f \circ s = 1_B$$

Then by the identity laws and the associative law

$$r = r \circ 1_B = r \circ (f \circ s) = (r \circ f) \circ s = 1_A \circ s = s$$

4. Isomorphisms and automorphisms

Using 'section' and 'retraction', we can rephrase the definition of 'isomorphism'.

Definitions: *A map f is called an* **isomorphism** *if there exists another map* f^{-1} *which is both a retraction and a section for f:*

$$A \underset{f^{-1}}{\overset{f}{\rightleftarrows}} B \qquad \begin{aligned} f \circ f^{-1} &= 1_B \\ f^{-1} \circ f &= 1_A \end{aligned}$$

Such a map f^{-1} *is called* **the inverse map for** *f; since both of the two equations are required, the theorem of uniqueness of inverses shows that there is only one inverse.*

Exercise 10:
If $A \xrightarrow{f} B \xrightarrow{g} C$ are both isomorphisms, then $g \circ f$ is an isomorphism too, and $(g \circ f)^{-1} = f^{-1} \circ g^{-1}$

The important (and necessary) reversal of order in the statement of the last exercise can be explained in terms of shoes and socks. The act f of putting on socks can be followed by the act g of putting on shoes, together a composite act $g \circ f$. The inverse of an act 'undoes' the act. To undo the composite act $g \circ f$, I must take off my shoes, which is the act g^{-1}, then follow that by taking off my socks (the act f^{-1}), altogether performing $f^{-1} \circ g^{-1}$.

What is the relation of A to B if there is an isomorphism between them? In the category of finite sets this just says that A and B have the same number of elements. But this enables us to give a usable *definition* of 'same number' without depending on counting – a definition which is very significant even for infinite sets. That is, we say that A and B have the same number of elements if they are isomorphic in the category of sets, where (in any category) A and B are *isomorphic* means that there exists an isomorphism from A to B in the category. Categories other than sets usually involve objects that are more richly structured, and, correspondingly, isomorphic objects will be alike in much more than just 'number of elements' – they will have the 'same shape', 'same structure', or whatever the category itself involves.

Check the correctness of the above idea of equal number for finite sets:

Exercise 11:
If $A = \{Fatima, Omer, Alysia\}$ and $B = \{coffee, tea, cocoa\}$, find an example of an isomorphism $A \xrightarrow{f} B$. If $C = \{true, false\}$, can you find any isomorphism $A \longrightarrow C$?

Now, how many isomorphisms are there from A to B? This question relates immediately to another question: How many isomorphisms $A \xrightarrow{f} A$ are there? Such a map, which is both an endomap and at the same time an isomorphism, is usually called by the one word **automorphism**.

Exercise 12:
How many isomorphisms are there from $A = \{Fatima,\ Omer,\ Alysia\}$ to $B = \{coffee,\ tea,\ cocoa\}$? How many automorphisms of A are there? The answers should be less than 27 – why?

In general, *if* there are any isomorphisms $A \longrightarrow B$, then there are the same number of them as there are automorphisms of A. This fact we can prove without counting by remembering the definition of 'same number' given above. If we let $Aut(A)$ stand for the set of all automorphisms of A and $Isom(A, B)$ stand for the set of all isomorphisms from A to B, the definition says that we need only construct an isomorphism between those two sets. Now $Aut(A)$ is always non-empty since at least 1_A is an example of an isomorphism $A \longrightarrow A$. If there is an isomorphism $A \xrightarrow{f} B$, choose such an f and use it to construct

$$Aut(A) \xrightarrow{F} Isom(A, B)$$

by defining $F(\alpha) = f \circ \alpha$ for any automorphism α of A.

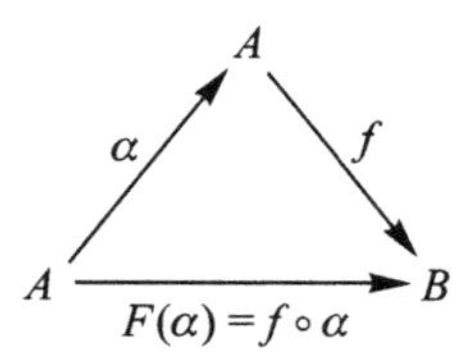

$F(\alpha)$ is indeed a member of $Isom(A, B)$ because of our previous proposition that any composite $f \circ \alpha$ of isomorphisms is an isomorphism. To show that F is itself an isomorphism, we have to construct an inverse

$$Isom(A, B) \xrightarrow{S} Aut(A)$$

for it, and this we can do using the same chosen f as follows:

$$S(g) = f^{-1} \circ g$$

for all isomorphisms g in $Isom(A, B)$

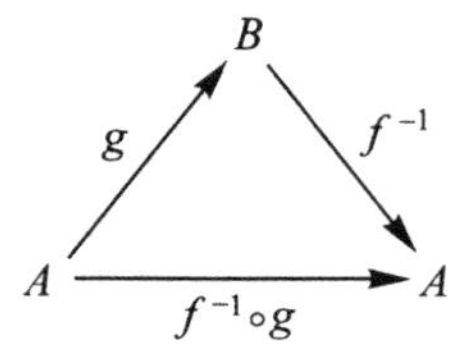

This $f^{-1} \circ g$ is an automorphism of A. Finally we have to show that S really is inverse to F, which involves showing two things:

$$(F \circ S)(g) = F(S(g)) = F(f^{-1} \circ g) = f \circ (f^{-1} \circ g) = (f \circ f^{-1}) \circ g$$
$$= 1 \circ g = g$$

for all g, so that

$$F \circ S = 1_{Isom(A,B)}$$

and also

$$(S \circ F)(\alpha) = S(F(\alpha)) = S(f \circ \alpha) = f^{-1} \circ (f \circ \alpha) = (f^{-1} \circ f) \circ \alpha$$
$$= 1 \circ \alpha = \alpha$$

for all α, showing that

$$S \circ F = 1_{Aut(A)}$$

An automorphism in the category of sets is also traditionally called a *permutation*, suggesting that it shifts the elements of its set around in a specified way. Such a specified way of shifting is one of the simple, but interesting kinds of *structure*, so we can use this idea to describe our second example of a category, the *category of permutations*. An *object* of this category is a set A *together with* a given automorphism α of A. A *map*

$$\text{from} \quad A^{\circlearrowright \alpha} \quad \text{to} \quad B^{\circlearrowright \beta}$$

is a map of sets $A \xrightarrow{f} B$, which 'respects' or 'preserves' the given automorphisms α and β in the sense that

$$f \circ \alpha = \beta \circ f$$

To compose maps f and g,

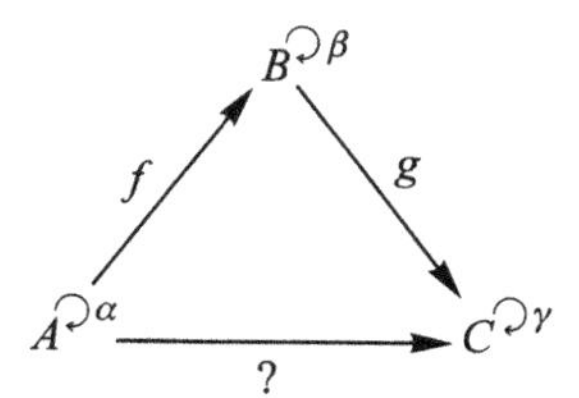

the natural thing would seem to be to compose them as maps of sets $A \xrightarrow{f} B \xrightarrow{g} C$, but we need to check that the composite as maps of sets is still a map in the category of permutations. That is, we suppose that f respects α and β and that g respects β and γ, and we must verify that $g \circ f$ respects α and γ. We are assuming

$$f \circ \alpha = \beta \circ f$$
$$g \circ \beta = \gamma \circ g$$

and so by associativity

$$(g \circ f) \circ \alpha = g \circ (f \circ \alpha) = g \circ (\beta \circ f) = (g \circ \beta) \circ f = (\gamma \circ g) \circ f$$
$$= \gamma \circ (g \circ f)$$

which completes the verification.

We will learn later that an object in the category of permutations has not only a total number of elements, but also a whole 'spectrum' of 'orbit lengths' and 'multiplicities' with which these occur. The only point which we want to preview here is that two objects between which there exists an isomorphism *in the sense of this category* will have their whole spectra the same.

5. Guide

We have discussed a number of important properties that a map may have, all related to division problems; these are summarized on the following page. Many examples will be presented in Sessions 4–9, followed by sample tests and review pages. Part II concludes, in Session 10, with an extended geometric example illustrating the use of composition of maps, and in particular the use of retractions.

Summary: Special properties a map 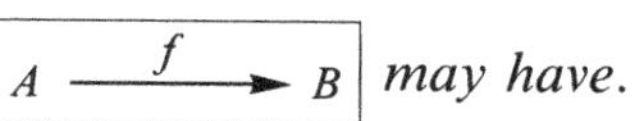 *may have.*

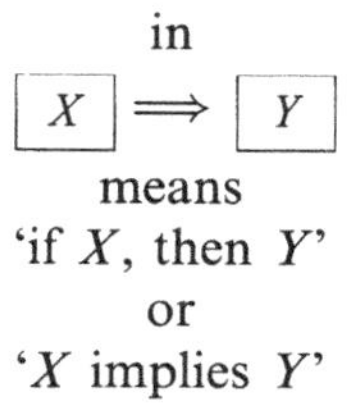

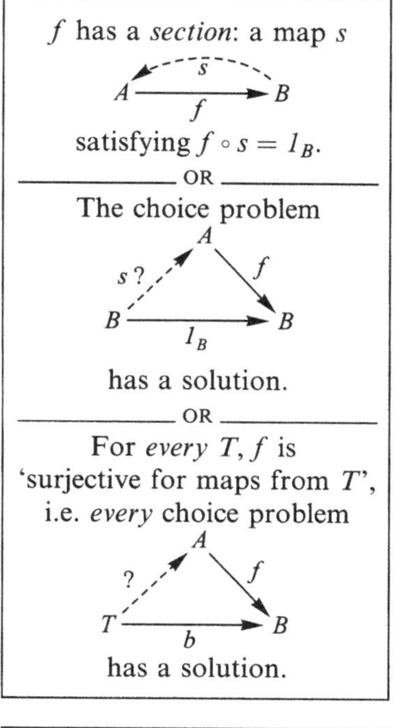

⟹

Cancellation

For *every* T, and *every*

$B \overset{t_1}{\underset{t_2}{\rightrightarrows}} T$

if $t_1 \circ f = t_2 \circ f$

then $t_1 = t_2$

(f is an 'epimorphism'.)

Inverse

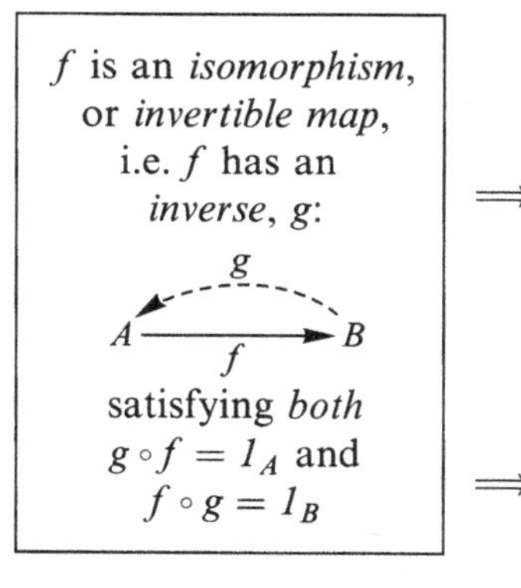

⟹

⟹

(Note: from either pair of 'diagonally opposite' properties,

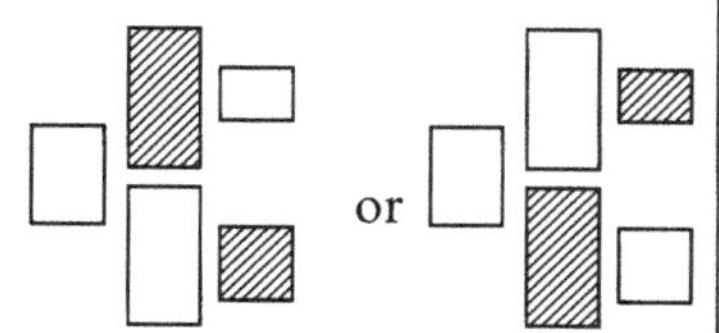

you can prove that f has an inverse!)

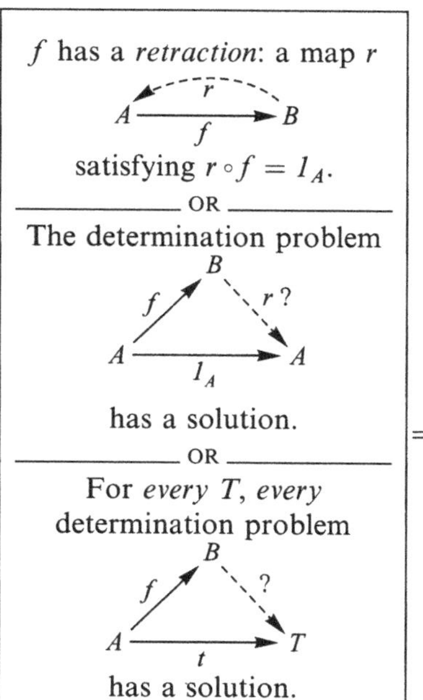

⟹

For *every* T, f is 'injective for maps from T', i.e. for *every*

$T \overset{a_1}{\underset{a_2}{\rightrightarrows}} A$,

if $f \circ a_1 = f \circ a_2$

then $a_1 = a_2$

(f is a 'monomorphism' or 'injective map'.)

(The three small boxes in each large box are just three ways to state the *same* property of f.)

SESSION 4

Division of maps: Isomorphisms

1. Division of maps versus division of numbers

Numbers	Maps
multiplication	composition
division	?

If composition of maps is analogous to multiplication of numbers, what is the analog of division of numbers? Let's first review the common features of composition and multiplication. Both operations are associative and have identities. (The identity for multiplication is the number 1.)

Multiplication of numbers	Composition of maps
For numbers x, y, z	For maps $A \xrightarrow{f} B \xrightarrow{g} C \xrightarrow{h} D$
$x \times 1 = x = 1 \times x$	$f \circ 1_A = f = 1_B \circ f$
$x \times (y \times z) = (x \times y) \times z$	$h \circ (g \circ f) = (h \circ g) \circ f$

Like most analogies this one is only partial because in multiplication of numbers the order doesn't matter, while in composition of maps it does. If we want both '$f \circ g$' and '$g \circ f$' to make sense and to have the same domain, we must have $A \xrightarrow{f} A$ and $A \xrightarrow{g} A$, and even then:

For all numbers x, y,	For most maps f, g,
$x \times y = y \times x$	$g \circ f \neq f \circ g$

Both multiplication of numbers and composition of maps are well-defined processes: you start with a pair (of numbers in one case, of maps in the other) and get a result. Usually, when you have a process like that there arises the question of reversing it, i.e. to find a new 'process' by which we can go from the output to the input, or from the result and one of the data, to the other datum. This reverse process may not give a unique answer.

For multiplication of numbers this reverse process is called the **division** problem, which is relatively simple because given one of the data and the result there is usually exactly one value for the other datum. For example, if multiplying a number by 3 we get 15,

$$3 \times ? = 15$$

we know that the number could only have been 5.

However, even in multiplying numbers we find problems for which there is no solution and problems for which there is more than one solution. This occurs when we multiply by zero. If we are told that multiplying a number by zero we get 7, we must reply that there is no such number, while if we are asked to find a number which multiplied by zero gives zero, we see that any number whatsoever is a solution.

$$0 \times ? = 7 \quad \text{no solution} \qquad 0 \times ? = 0 \quad \text{many solutions}$$

Such problems, which may be considered as exceptional in multiplication, are instead typical for composition of maps. For maps it usually happens that 'division' problems have several solutions or none. There is, however, one very useful case in which 'division of maps' produces exactly one solution, so we will treat this easier case first.

2. Inverses versus reciprocals

A 'reciprocal for the number 2' means a number satisfying $? \times 2 = 1$ (and therefore also $2 \times ? = 1$). As you know, 2 has precisely one reciprocal, 0.5 or 1/2. The corresponding notion for composition of maps is called 'inverse.'

Definitions: *If* $A \xrightarrow{f} B$, *an* **inverse for** f *is a map* $B \xrightarrow{g} A$ *satisfying both*

$$g \circ f = 1_A \quad \text{and} \quad f \circ g = 1_B$$

If f has an inverse, we say f is an **isomorphism**, *or* **invertible map**.

We really need both equations, as this example shows:

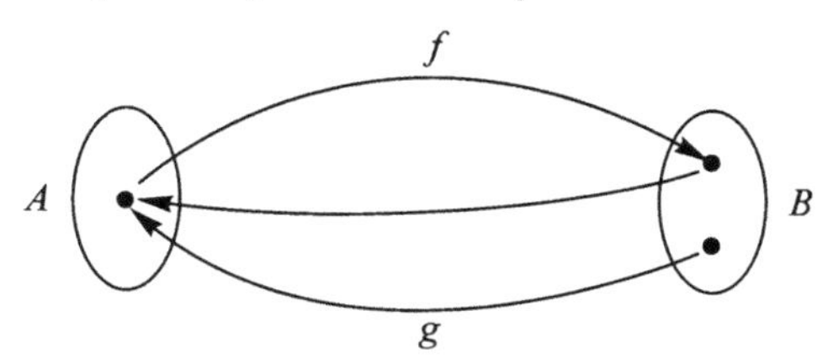

$$g \circ f = 1_A \quad \textit{but} \quad f \circ g \neq 1_B$$

You can make up more complicated examples of this phenomenon yourself. (What is the simplest example of maps f and g for which $f \circ g$ is an identity map, but $g \circ f$ is not?) The internal diagram of an isomorphism of sets looks pretty simple:

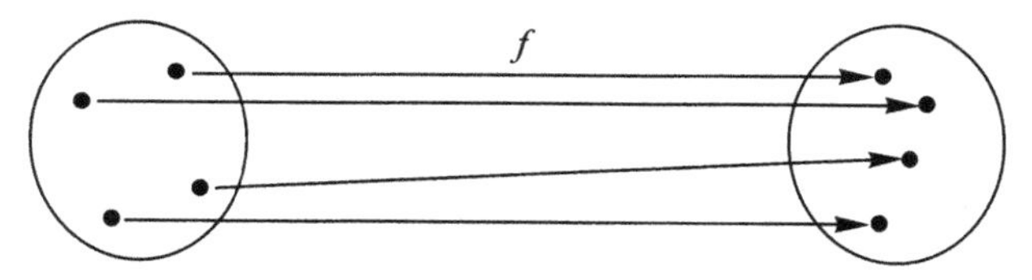

though it might be drawn in a less organized way:

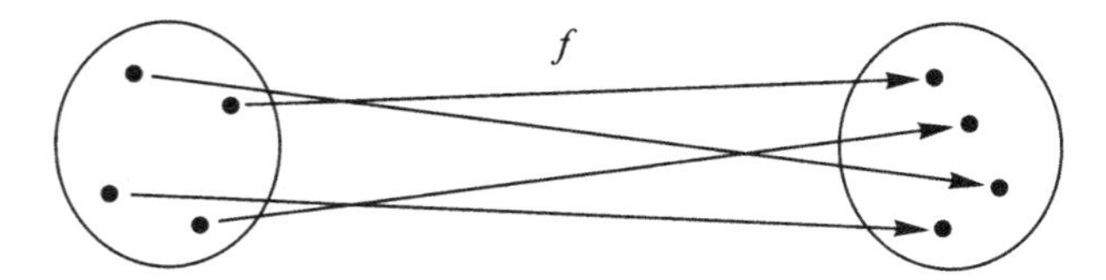

These pictures suggest that a map with an inverse has only *one* inverse: just 'reverse the arrows in the internal diagram.' This is true, and will be deduced from just the associative and identity laws for composition of maps:

Uniqueness of inverses: *Any map f has at most one inverse.*

Proof: Say $A \xrightarrow{f} B$, and suppose that both $B \xrightarrow{g} A$ and $B \xrightarrow{h} A$ are inverses for f; so

$$g \circ f = 1_A \quad \text{and} \quad f \circ g = 1_B$$
$$h \circ f = 1_A \quad \text{and} \quad f \circ h = 1_B$$

We only need two of these equations to prove that g and h are the same:

$$g = 1_A \circ g = (h \circ f) \circ g = h \circ (f \circ g) = h \circ 1_B = h$$

(Do you see the justification for each step? Which two of the four equations did we use? The easiest way to remember this proof is to start in the middle: the expression $h \circ f \circ g$, with f sandwiched between its two supposed inverses, simplifies two ways.)

There are two standard notations for the reciprocal of 2: 1/2 and 2^{-1}. For maps, only the second notation is used: if f has an inverse map, then its one and only inverse is denoted by f^{-1}. In both cases, numbers and maps, it makes no sense to use these symbols if there is no reciprocal or inverse: '0^{-1}', '1/0', and 'f^{-1}' if f is the map

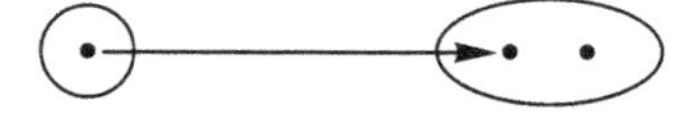

are nonsense-expressions that don't stand for anything. One further small caution: Whether a number has a reciprocal depends on what your 'universe of numbers' is. If by 'numbers' you mean only integers (whole numbers), i.e. $\ldots, -2, -1, 0, 1, 2, 3, \ldots$, then only -1 and 1 have reciprocals; 2 does not. But if by numbers you mean real numbers (often represented by infinite decimal expansions), then every number except 0 has a reciprocal. In exactly the same way, whether a map has an inverse depends on what 'universe of maps' (category) you are in. We'll take the category of abstract sets (and all maps) for now, but much of what we say will depend only on the associative and identity laws for composition of maps, and therefore will be valid in any category.

3. Isomorphisms as 'divisors'

If you have ever arrived a few minutes late to a movie, you have no doubt struggled to determine which isomorphism of sets

$$\text{Names of characters} \longrightarrow \text{Characters on screen}$$

is involved. When two characters discuss 'Titus,' you try to gather clues which may indicate whether that is the tall bald guy or the short dark-haired one. Later, if you particularly liked the film but the actors were unfamiliar to you, you learn the isomorphism

$$\text{Characters on screen} \longrightarrow \text{Actors in film}$$

or if the actors are familiar, but you cannot recall their names, you learn the isomorphism

$$\text{Actors in film} \longrightarrow \text{Professional names of cast}$$

(An unfortunate recent practice is to show you at the end of the film only the composite of these three isomorphisms, called 'cast of characters'.)

After you have grasped all of these isomorphisms, it is remarkable how easily you compose them. You perform a sort of mental identification of the name 'Spartacus' and the slave who led the revolt and the actor with the cleft chin and the name 'Kirk Douglas,' even while you are aware that each of these isomorphisms of sets resulted from many choices made in the past. Different actors could have been selected for these roles, the actors might have selected different professional names, etc.; each arrow in the internal diagram of any one of the isomorphisms may represent a story of its own. At the same time, these four sets are kept quite distinct. You do not imagine that the slave dined on Hollywood Boulevard, nor that the cleft-chinned actor contains nine letters. Each set is an island, communicating with other sets only by means of maps.

In spite of this seeming complexity, you use these isomorphisms of sets, and composites of these and their inverses, so freely in discussing the film that it seems almost miraculous. Apparently an isomorphism is easier to master than other maps, partly because of its 'two-way' character: with each isomorphism comes its inverse, and passing back and forth a few times along each arrow in the internal diagram cements it firmly in your mind. But the ease in composing them comes also from the simplicity of the *algebra* of composition of isomorphisms. *The process of following* (or preceding) *maps by a particular isomorphism is itself a 'reversible' process*, just as the process of multiplication by 3 is reversed by multiplication by 1/3. There is only one small difference. Because the order of composition matters, there are *two* types of division problems for maps. Each has exactly one solution if the 'divisor' is an isomorphism:

Problem:

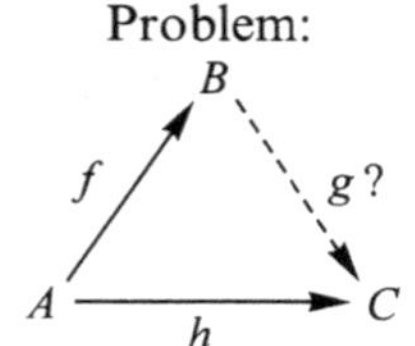

Given f and h, find all g for which $g \circ f = h$.
(Analogous to: $? \times 3 = 6$)
Solution, *if* the 'divisor' f is an isomorphism:
There is exactly one map g for which $g \circ f = h$;
it is $g = h \circ f^{-1}$.
(Analogous to: $? = 6 \times \frac{1}{3}$)

Problem:

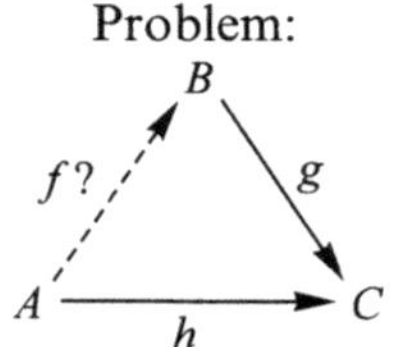

Given g and h, find all f for which $g \circ f = h$.
(Analogous to: $3 \times ? = 6$)
Solution, *if* the 'divisor' g is an isomorphism:
There is exactly one map f for which $g \circ f = h$;
it is $f = g^{-1} \circ h$.
(Analogous to: $? = \frac{1}{3} \times 6$)

Please don't bother memorizing these formulas. It's easier, and more illuminating, to learn the proof; then you can instantly get the formulas whenever you need them. Here it is for the left-hand column. If g were a solution to $g \circ f = h$, then (trying to get g by itself on the left-hand side) $(g \circ f) \circ f^{-1} = h \circ f^{-1}$, but now the left side simplifies (how?) to g, so $g = h \circ f^{-1}$. Caution: All that we have shown is that the only possible solution to our equation is the candidate we found, $h \circ f^{-1}$. We still must make sure that this candidate is really a solution. Is it true that $(h \circ f^{-1}) \circ f = h$? Simplify the left-hand side yourself to see that it's so. Notice that in the first simplification we used $f \circ f^{-1} = 1_B$, and in the second simplification we used $f^{-1} \circ f = 1_A$; we needed both. Now work out the proof for the right-hand column too, and you will have mastered this technique.

4. A small zoo of isomorphisms in other categories

To appreciate isomorphisms you need to look at examples, some familiar, some more exotic. This is a bit of a leap ahead, because it involves exploring categories other than the category of sets, but you can manage it. We'll start with the more familiar, but perhaps get you to take a fresh viewpoint.

In *algebra*, we often meet a set (usually of numbers) together with a rule (usually addition or multiplication) for combining any pair of elements to get another element. Let's denote the result of combining a and b by '$a * b$', so as not to prejudge whether we are considering addition or multiplication. An *object* in our algebraic category, then, is a set A together with a combining-rule *. Here are some examples:

$(\mathbb{R}, +)$ Real numbers (usually represented by infinite decimal expansions, like 3.14159..., or −1.414..., or 2.000...) with addition as the combining-rule.

$(\mathbb{R}, \times)$ Same, but with multiplication as the combining-rule.

$(\mathbb{R}_{>0}, \times)$ Only positive real numbers, but still with multiplication.

A *map* in this category from an object $(A, *)$ to an object $(A', *')$ is any map of sets $A \longrightarrow A'$ which 'respects the combining-rules,' i.e.

$$f(a * b) = (fa) *' (fb) \quad \text{for each } a \text{ and } b \text{ in } A$$

Here are some examples of maps in this category:

1. $(\mathbb{R}, +) \xrightarrow{d} (\mathbb{R}, +)$ by 'doubling': $dx = 2x$. We see that d is a map in our category, since

$$d(a + b) = (d\,a) + (d\,b)$$

i.e.

$$2(a + b) = (2a) + (2b)$$

2. $(\mathbb{R}, \times) \xrightarrow{c} (\mathbb{R}, \times)$ by 'cubing': $c\,x = x^3$. We see that c is a map in our category since

$$c(a \times b) = (c\,a) \times (c\,b)$$

i.e.

$$(a \times b)^3 = (a^3) \times (b^3)$$

3. $(\mathbb{R}, +) \xrightarrow{\exp} (\mathbb{R}_{>0}, \times)$ by 'exponentiation': $\exp x = e^x$, and exp is a map in our category since

$$\exp(a + b) = (\exp a) \times (\exp b)$$

i.e.

$$\mathrm{e}^{(a+b)} = (\mathrm{e}^a) \times (\mathrm{e}^b)$$

(If you don't know the number $\mathrm{e} = 2.718\ldots$, you can use 10 in its place.)

These examples of maps in our algebraic category were specially chosen: each of them is an *isomorphism*. This requires some proof, and I'll only do the easiest one, the doubling map d. You'll guess right away the inverse for the doubling map, the 'halving map:'

$$(\mathbb{R}, +) \xrightarrow{h} (\mathbb{R}, +) \quad \text{by 'halving': } h\,x = \tfrac{1}{2}x$$

Of course, we should check that h is a map in our category, from $(\mathbb{R}, +)$ to $(\mathbb{R}, +)$. Is it true, for all real numbers a and b, that

$$h(a + b) = (h\,a) + (h\,b)\ ?$$

Yes. Now we still must check the *two* equations which together say that h is the inverse for d: Are $h\,d$ and $d\,h$ identity maps?

Exercise 1:
Finish checking that d is an isomorphism in our category by showing that $h \circ d$ and $d \circ h$ are indeed identity maps.

We can find examples of objects in our algebraic category which aren't sets of numbers. You have probably noticed that adding an even whole number to an odd one always produces an odd result: $odd + even = odd$. Also, $odd + odd = even$, and so on. So the two-element set $\{odd, even\}$ with the 'combining-rule,' $+$, now has become an object in our algebraic category. Also, you know that multiplying positive numbers produces a positive result, while $positive \times negative = negative$, and so on. In this way, the set $\{positive, negative\}$ with the combining-rule $\times$ is also an object in our category. Our next exercise is an analog of the remarkable example (3) above, which showed that addition of real numbers and multiplication of positive numbers have the 'same abstract form.'

Exercise 2:
Find an isomorphism

$$(\{odd, even\}, +) \xrightarrow{f} (\{positive, negative\}, \times)$$

Hint: There are only two invertible maps of sets from $\{odd, even\}$ to $\{pos., neg.\}$. One of them 'respects the combining rules', but the other doesn't.

We should also get some experience in recognizing when something is *not* an isomorphism; the next exercise will challenge you to do that.

Exercise 3:
An unscrupulous importer has sold to the algebraic category section of our zoo some creatures which are *not* isomorphisms. Unmask the impostors.

(*a*) $(\mathbb{R}, +) \xrightarrow{p} (\mathbb{R}, +)$ by 'plus 1': $p\,x = x + 1$.
(*b*) $(\mathbb{R}, \times) \xrightarrow{sq} (\mathbb{R}, \times)$ by 'squaring': $sq\,x = x^2$.
(*c*) $(\mathbb{R}, \times) \xrightarrow{sq} (\mathbb{R}_{\geqslant 0}, \times)$ by 'squaring': $sq\,x = x^2$.
(*d*) $(\mathbb{R}, +) \xrightarrow{m} (\mathbb{R}, +)$ by 'minus': $m\,x = -x$.
(*e*) $(\mathbb{R}, \times) \xrightarrow{m} (\mathbb{R}, \times)$ by 'minus': $m\,x = -x$.
(*f*) $(\mathbb{R}, \times) \xrightarrow{c} (\mathbb{R}_{>0}, \times)$ by 'cubing': $c\,x = x^3$.

Hints: Exactly one is genuine. Some of the cruder impostors fail to be maps in our category, i.e. don't respect the combining-rules. The crudest is not even a map of *sets* with the indicated domain and codomain.

If you have always found the algebraic rules that came up in discussing these examples somewhat mysterious, you are in good company. One of our objectives is to demystify these rules by finding their roots. We will get to that, and after we nourish the roots you will be surprised how far the branches extend. For now, though, it seemed fair to use the algebraic rules as sources of examples. The rest of the isomorphisms in our zoo will be easier to picture, and won't require algebraic calculations. Since this is only a sightseeing trip, we will be pretty loose about the details.

In *geometry*, a significant role is played by 'Euclid's category.' An *object* is any polygonal figure which can be drawn in the plane, and a *map* from a figure F to a figure F' is any map f of sets which 'preserves distances': if p and q are points of F, the distance from fp to fq (in F') is the same as the distance from p to q. (Roughly, the effect of this restriction on the maps is to ensure that if F were made of some perfectly rigid material you could pick it up and put it down again precisely onto the space occupied by F'; but notice that any idea of actually *moving* F is not part of the definition.) Objects which are isomorphic in this category are called by Euclid 'congruent' figures. Here is an example.

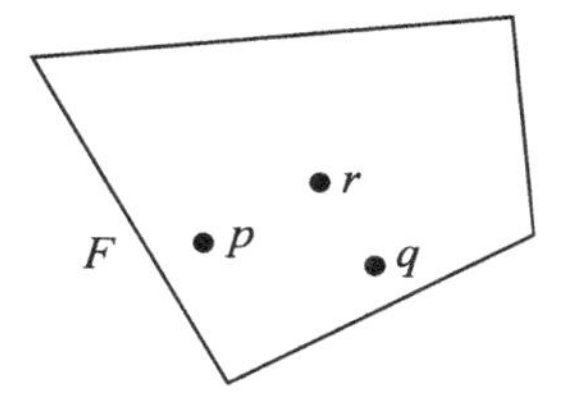

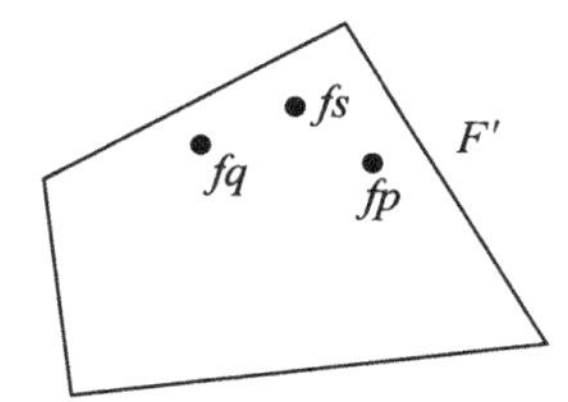

Isomorphic objects in Euclid's category

Do you see what the map f is, and what its inverse is? If so, you should be able to locate fr and s in the picture. We might enlarge Euclid's category to include solid figures, and to allow curved boundaries. Then if you are perfectly symmetric, your left hand is isomorphic to your right hand when you stand at attention, and your twin's right hand is isomorphic to both of these.

In *topology*, sometimes loosely referred to as 'rubber-sheet geometry,' maps are not required to preserve distances, but only to be 'continuous': very roughly, if p is close to q then fp is close to fq. Objects which are isomorphic in such a category are said to be 'homeomorphic.' The physique of a tall thin man is homeomorphic to that of a short stout one unless accident or surgery has befallen one of them.

A radiologist examining images of the human body from X-rays needs to make sharper distinctions, and so may use a more refined category. An *object* will have as additional structure a map associating to each point a density (measured by the darkness of its image); and a *map* in the radiologist's category, in addition to being continuous, must have the property that if p and q are nearby and the density at p is greater than that at q, then correspondingly the density at fp is greater than that at fq. Failure to find an isomorphism in this category from your body to an 'ideal' body is regarded as an indication of trouble. (This example is not to be taken *too* seriously; it is intended to give you an idea of how one tries to capture important aspects of any subject by devising appropriate categories.)

SESSION 5

Division of maps: Sections and retractions

1. Determination problems

Many scientific investigations begin with the observation that one quantity f *determines* another quantity h. Here is an example. Suppose we have a cylinder, with a weighted piston pushing down on a trapped sample of gas. If we heat the system, the volume of the trapped gas will increase, raising the piston. If we then cool the system to its original temperature, the gas returns to its original volume, and we begin to suspect that the temperature *determines* the volume. (In the diagram below, f assigns to each state of the system its temperature, and h assigns to each state its volume.)

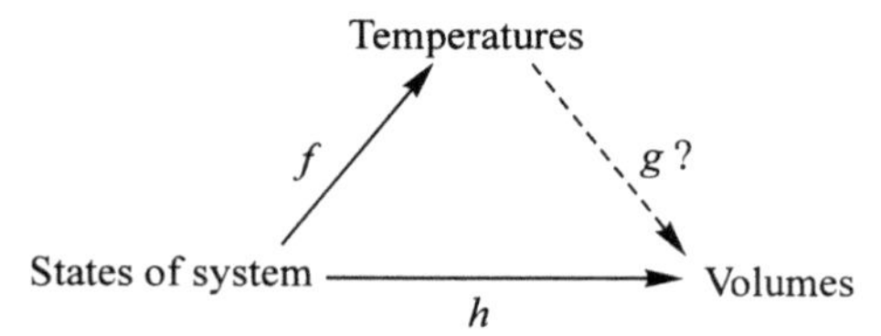

Our suspicion is that there is a map g which makes $h = g \circ f$; such a g is called a *determination* of h from f. The problem for the scientist is then to find one g (or all g, if there is more than one) which makes $h = g \circ f$ true. (In this example, it turns out that there is exactly one such g. If we choose the zero for temperatures appropriately, g even has a very simple form: multiplication by a constant.)

Let's put all this more generally. Suppose that we have a map of sets $A \xrightarrow{f} B$ and a set C. Then every map from B to C can be composed with f to get a map $A \longrightarrow C$. Thus f gives us a process that takes maps $B \longrightarrow C$ and gives maps $A \longrightarrow C$:

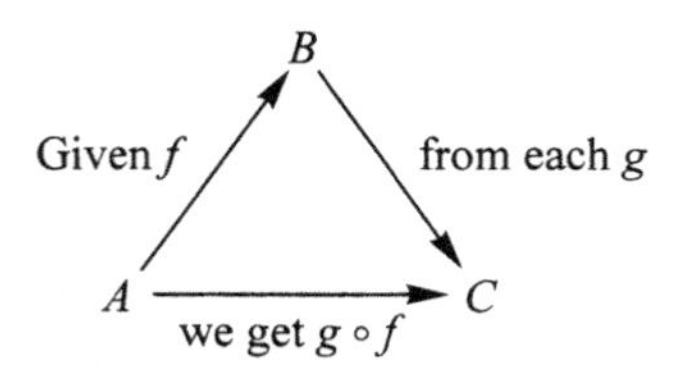

and we are interested in reversing this process. The *determination problem* is: *Given maps f from A to B and h from A to C, find all maps g from B to C such that $g \circ f = h$.* (See diagram below.)

This problem asks: '*Is h determined by f?*' and more precisely asks for all *ways* of determining h from f, as shown in the diagram

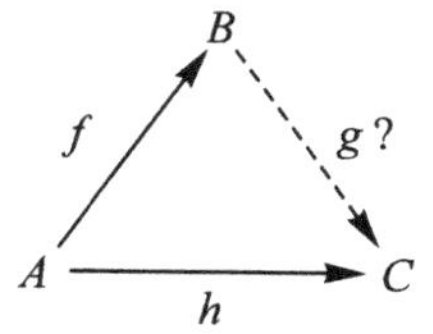

Here is an example with finite sets. Let A be the set of students in the classroom and B the set of genders 'female' and 'male'; and let $A \xrightarrow{f} B$ be the obvious map that gives the gender. If C is the set with elements *yes* and *no*, and h is the map which answers the question 'Did this student wear a hat today?', then depending on who wore a hat today there are many possibilities for the map h. But since there are so few maps

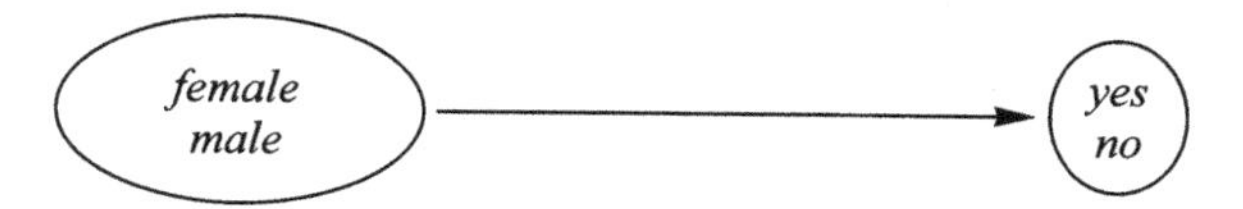

(how many?), it is very unlikely that a given h is equal to f followed by one of the maps from B to C.

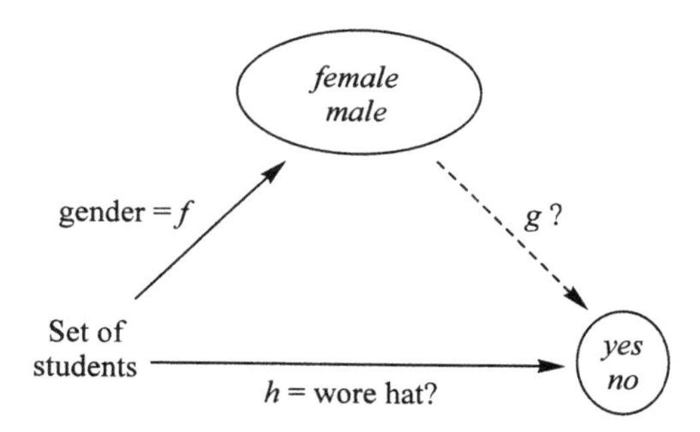

Let's try to figure out what special properties a map $A \xrightarrow{h} C$ has if it is equal to $g \circ f$ for some

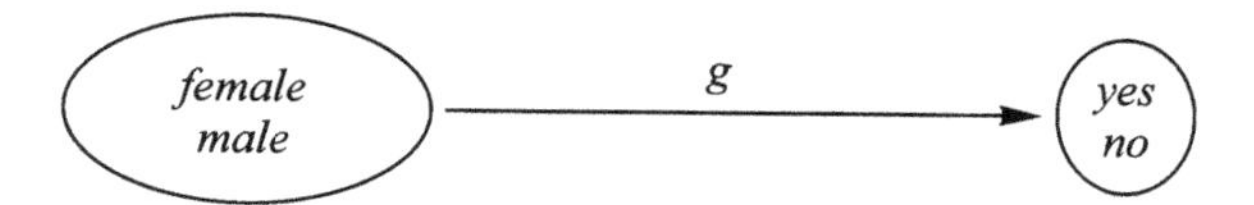

Obviously that means that by knowing whether a student is female or male you can tell whether the student wore a hat or not. In other words, either all females wore hats today or none did, and either all the males wore hats or none did. The existence of a map g such that $h = g \circ f$ would mean that h (whether a student wore a hat today) is determined by f (the gender of the student).

Incidentally, the survey of our class revealed that Ian wore a hat today and Katie did not. This much information alone would force g to be as shown below

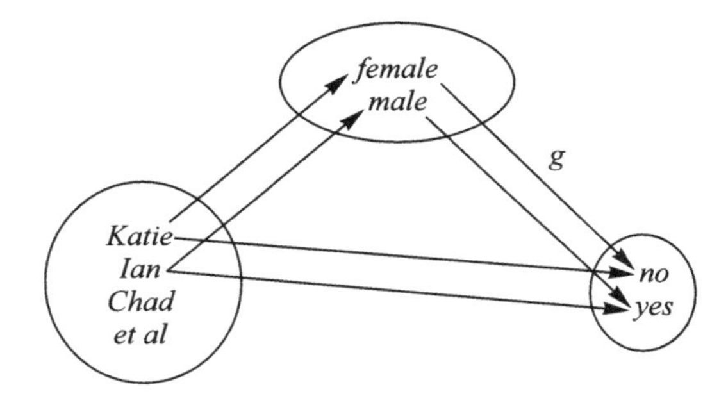

But even this g does not satisfy $g \circ f = h$, since Chad is male but he did not wear a hat:

$$(g \circ f)(Chad) = g(f(Chad)) = g(male) = yes \qquad h(Chad) = no$$

For a general idea of how a map f must be related to a map h in order that it be possible to find an explicit 'proof' g that h is determined by f, try the following exercise. (Recall that '**1**' is any singleton set.)

Exercise 1:
(a) Show that if there is a map g for which $h = g \circ f$, then for any pair a_1, a_2 of points $\mathbf{1} \longrightarrow A$ of the domain A of f (and of h) we have:

$$\text{if } fa_1 = fa_2 \text{ then } ha_1 = ha_2$$

(So, if for some pair of points one has $fa_1 = fa_2$ but $ha_1 \neq ha_2$, then h is not determined by f.)
(b) Does the converse hold? That is, if maps (of sets) f and h satisfy the conditions above ('for any pair ... then $ha_1 = ha_2$'), must there be a map $B \xrightarrow{g} C$ with $h = g \circ f$?

2. A special case: Constant maps

Let's suppose now that B is a one-element set, so f is already known: it takes all elements of A to the only element of B. For which maps h does our determination problem have a solution?

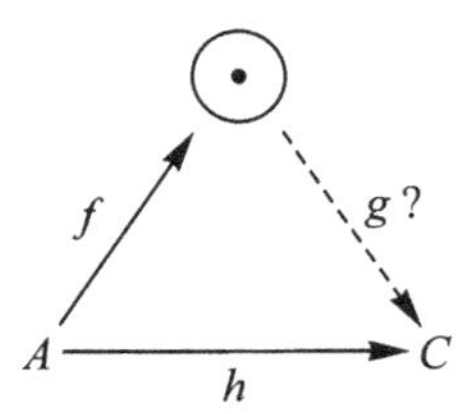

According to Exercise 1, such a map h must send all elements of A to the same element of C. This conclusion can also be reached directly: since B has only one element, a map g from B to C is the same as a choice of an element in C; and the composite $g \circ f$ will send all elements of A to that element of C. Such a map (which takes only one value) is called a *constant* map.

Definition: *A map that can be factored through* **1** *is called a* **constant map**.

3. Choice problems

Another division problem for maps consists in looking for the other factor, i.e. looking for f when g and h are given, like this:

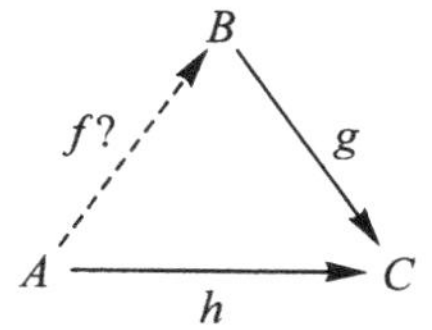

This is called the *choice problem* because in order to find a map f such that $g \circ f = h$, we must choose for each element a of A an element b of B such that $g(b) = h(a)$.

Here is a choice problem. Let C be a set of towns, A the set of people living in those towns, and let h be the map from A to C assigning to each person his or her town of residence. Let's take as the set B the set of all supermarkets and as map g the location of the supermarkets:

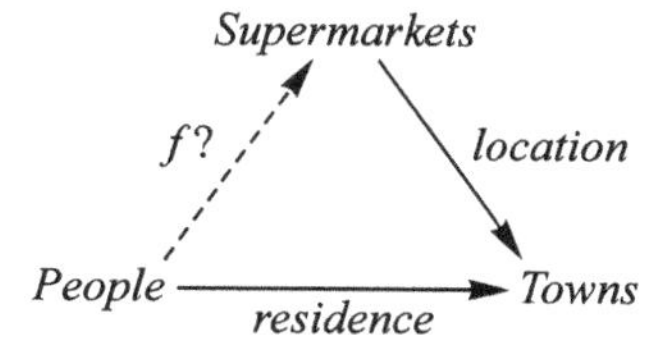

To get a solution to this problem, each person must choose a supermarket located in his or her town of residence. It should be clear that as long as there are no inhabited towns without supermarkets, the problem has a solution, and usually more than one.

As with the determination problem (Exercise 1), there is a criterion for the existence of 'choice' maps:

Exercise 2:
(a) Show that if there is an f with $g \circ f = h$, then h and g satisfy: For any a in A there is at least one b in B for which $h(a) = g(b)$.

(b) Does the converse hold? That is, if h and g satisfy the condition above, must there be a map f with $h = g \circ f$?

4. Two special cases of division: Sections and retractions

An important special case of the choice problem arises if the set A is the same as C, and the map h is its identity map.

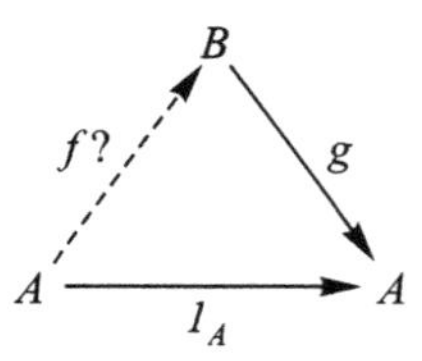

This asks for a map $A \xrightarrow{f} B$ which chooses for each element a of A an element b of B for which $g(b) = a$. This is less than being an inverse for g, since only one of the two conditions demanded of an inverse is required here. Still, this relationship of f to g is of such importance that we have given it a name:

Definition: $A \xrightarrow{f} B$ *is a* **section of** $B \xrightarrow{g} A$ *if* $g \circ f = 1_A$.

One of the important applications of a section is that it permits us to give a solution to the choice problem for any map $A \xrightarrow{h} C$ whatsoever. How? It's a variant of 'If you have 1/2 you don't need *division* by 2; *multiply* by 1/2 instead.' Suppose that we have a choice problem, such as the one of the supermarkets, and let's suppose that the given $B \xrightarrow{g} C$ has a section s. If we draw all the maps we have, in a single external diagram,

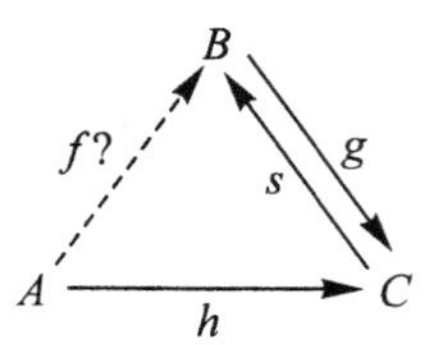

we see that there is a way to go from A to B: the composite $s \circ h$. Let's check whether putting $f = s \circ h$ gives a solution; that is, whether $g \circ f = h$. This is easily checked with the following calculation

$$\begin{aligned} g \circ f &= g \circ (s \circ h) \quad (\text{since } f = s \circ h) \\ &= (g \circ s) \circ h) \quad (\text{associative law}) \\ &= 1_C \circ h \quad (\text{since } s \text{ is a section of } g) \\ &= h \quad (\text{identity law}) \end{aligned}$$

This calculation is another example of the algebra of composition of maps, but it should look familiar. It was half of the calculation by which we showed that a choice problem with an *invertible* divisor g has *exactly one* solution. So, each section of g gives a solution to any choice problem with g as divisor. However, usually there are other solutions to the choice problem besides those given by the sections of g (and different sections may give the same solution), so that the number of sections often differs from the number of solutions of the choice problem.

FATIMA: How would that apply to the example of the supermarkets?

Well, a section for the map $g =$ *location of the supermarkets* assigns to every town a supermarket *in that town*. For example, imagine that there is a chain that has one supermarket in each town. Then one solution to the choice problem (the solution which comes from the chain's *section* of g) is that everybody chooses to shop in the supermarket of that chain located in his or her own town. You'll notice that those solutions to choice problems which come from sections are pretty boring: in each town, everybody shops in the same supermarket. The same thing will happen for determination problems and retractions. Retractions give solutions of determination problems, as Exercise 6 of Article II shows, but the interesting cases of determination are usually those which do not come from retractions.

OMER: For the identity map it seems that the order should not matter, or should it?

I'm glad you asked, because it is easy to make this mistake, and we should clear it up so that we will have it all neatly organized. Let's compare a choice problem for the identity map, which we just looked at, with a determination problem for the identity map. It's clearer if we don't give the sets and maps any names (since every time you use these ideas the maps involved may have different names) and just draw the schematic external diagrams:

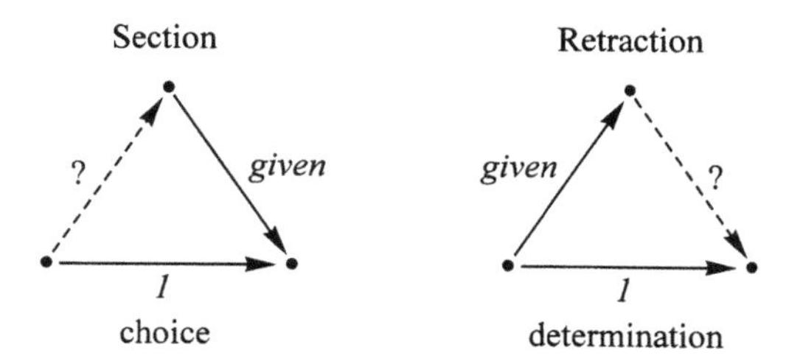

You can see why confusion might arise; the only difference is which map is regarded as given. Let's review.

Say $A \xrightarrow{f} B$ is a map.

(*a*) A *section of* f is any map s such that $f \circ s = 1_B$.
(*b*) A *retraction of* f is any map r such that $r \circ f = 1_A$.

Comparing the definitions, we see that a section of f is not the same as a retraction of f. The symmetry comes in noticing that a single relationship between two maps can be described in two ways: if $g \circ f$ is an identity map we can either say that g is a retraction of f or that f is a section of g. The relationship among maps of 'section' to 'retraction' is nearly the same as the relationship among women of 'aunt' to 'niece'. Just be careful not to use these words in isolation. You cannot ask whether a map is an inverse, or section, or retraction. It only makes sense to ask that it be an inverse (or section or retraction) *of a specified map.*

Try Exercises 6 and 7 of Article II to see how a retraction gives solutions to determination problems, and how a section gives solutions to choice problems.

5. Stacking or sorting

To find all the sections for a given map $A \xrightarrow{f} B$, it is useful to view the map f as 'stacking' or 'sorting' the elements of A. Here is an example. Let's suppose that A is the set of all the books in the classroom and B is the set of people in the same classroom. We have a map $A \xrightarrow{\text{belongs to}} B$ which assigns to each book the person who brought it into the classroom. One way to picture this map is the internal diagram we have been using,

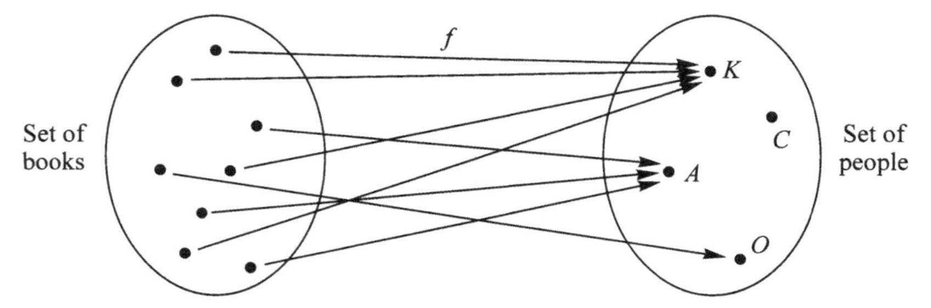

But another picture can be drawn in which we arrange all the people in a row and stack on top of each one in a column all the books that belong to him or her:

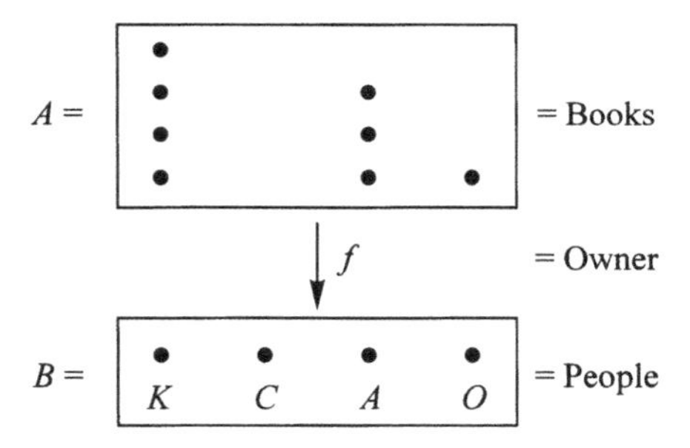

In this picture we can read off easily what f does, and at the same time we clearly see the stack of books that belongs to each student. It might involve a lot of work to arrange the domain and codomain so as to get the 'stacks picture' of a particular map, but once it is done, it is a very useful picture and, in principle, every map can be viewed this way.

Coming back to the sections, let's see how the stacks picture of a map can help us to find all the sections of that map. What would be a section of the map f which assigns to every book the person who brought it into the classroom? It would be a map assigning to every person one of his or her books, such as:

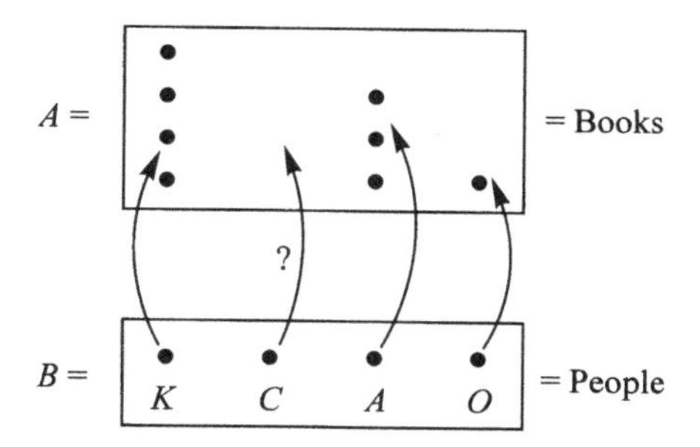

Woops! Chad didn't bring a book. There is no way of assigning one of Chad's books to him, so there is no section for f. Thus, this stacks arrangement permits us to see right away that this particular map has no sections. In general, in order that a map $f: A \longrightarrow B$ can have a section it is necessary that for every element of B its corresponding stack is not empty. In other words, for every element b of B there should be an element a of A such that $f(a) = b$.

The stacks picture of a map allows one to find a formula for the number of sections of a map. Suppose that f is the following:

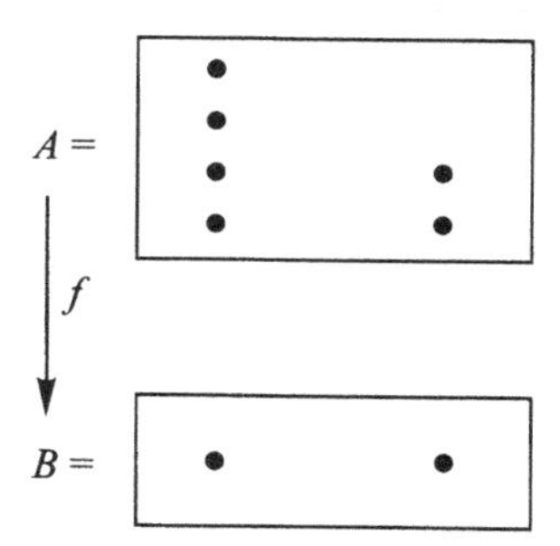

Exercise 3:
Draw the internal diagrams of all the sections of f.

You should get eight sections, many fewer than the total number of maps from B to A, which according to Alysia'a formula is what? Right, $6^2 = 36$. Any guess as to the number of sections of an arbitrary map?

CHAD: You multiply the number of elements in one stack by the number of elements in the next and so on.

That's right. You multiply them because the choice you make in any stack is independent of the choices you made in the other stacks.

SHERI: So, if one point has its stack empty what do you do?

You count it too. If that stack has zero elements, one of the numbers being multiplied is zero. And, of course, the result is zero: there are no sections.

Now, as we saw, the same equation that says that s is a section for f means that f is a retraction for s, so that whenever we have a 'commutative diagram' (i.e. the two ways of getting from B to B give the same result)

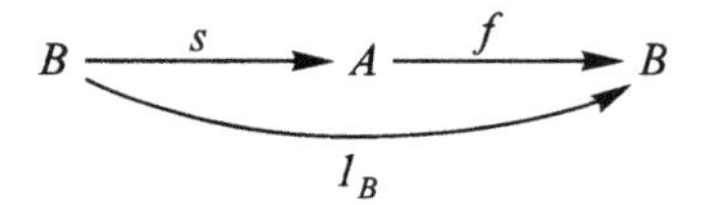

we are talking about a pair section-retraction.

DANILO: If you want to expand that diagram to include the retraction, would you have to put the identity of A?

No. The diagram as it stands means both things: that s is a section for f and that f is a retraction for s. The identity of A would be involved only if we had a retraction for f. We saw that when both diagrams commute, i.e. if we *also* have $s \circ f = 1_A$

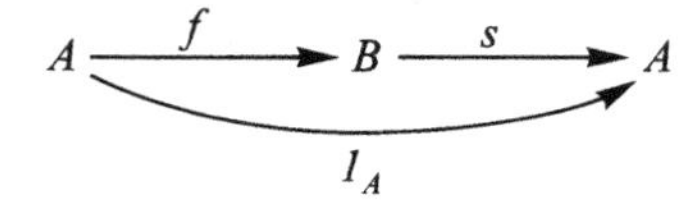

then s is the only section of f, and it is called the *inverse* of f.

6. Stacking in a Chinese restaurant

Let me explain an interesting example of stacking based on the practice of a Chinese restaurant in New York City that we used to visit after the mathematics seminar. The example illustrates that the use of the category of sets can be more direct than translating everything into the more abstract numbers.

In this restaurant the stacking of plates according to shape is consciously used systematically in order to determine the total bill for each given table of customers without having to make any written bill at all.

In any restaurant there is the basic map

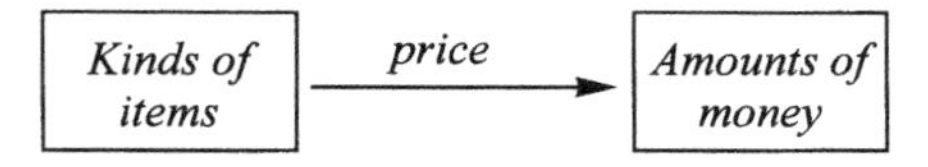

which may assign five dollars to 'moo shu pork', a dollar to 'steamed rice', etc.

Each particular group of customers at a particular table on a particular occasion gives rise (by ordering and consuming) to another map

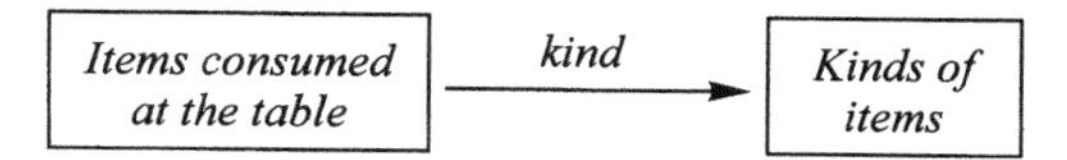

which is neither injective nor surjective because more than one item of the same kind may have been consumed and also some possible kinds were not actually ordered at all. The prices of the items consumed at the table on that occasion are given by the composed map $f = price \circ kind$:

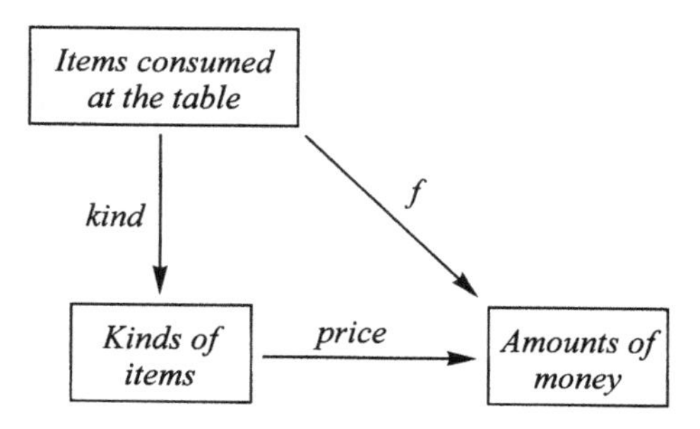

The total bill for the table is obtained as the sum ($\sum$) of products

$$\sum_k price(k) \cdot (\text{size of the stack of } kind \text{ over } k)$$

where k ranges over all kinds of items. But knowing f, the total bill for the table can also be obtained using f alone, as the sum of products

$$\sum_x x \cdot (\text{size of the stack of } f \text{ over } x)$$

where x ranges over amounts of money. In most restaurants the specification of f is recorded in writing on a slip of paper, and the arithmetic is done by the waiter and checked by the cashier.

In this particular Chinese restaurant, the problem of achieving rapid operation, even though cooks, customers, servers, and cashier may all speak different languages, is neatly solved without any writing of words and numbers (and without any slips of paper at all); the map f is instead recorded in a direct physical way by stacking plates.

In fact f is calculated via another map $\overline{f}$, constructed as the composite of two maps $\overline{price}$ and $\overline{kind}$. The key to the plan is to have several different shapes of plates: small round bowls, large round bowls, square plates, round plates, triangular plates, elliptical plates, etc. (so that it is hard to stack one plate on top of a plate of different shape), and the cooks in the kitchen always put a given kind of food onto plates of a definite shape. Thus a map

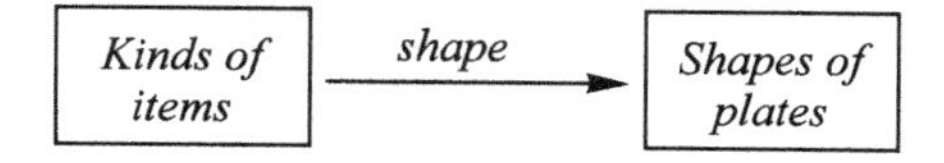

is set up, but not arbitrarily: it is done in such a way that the price of an item is *determined* by the shape of the plate on which it is served. That is, there is a map $\overline{price}$ for which $price = \overline{price} \circ shape$:

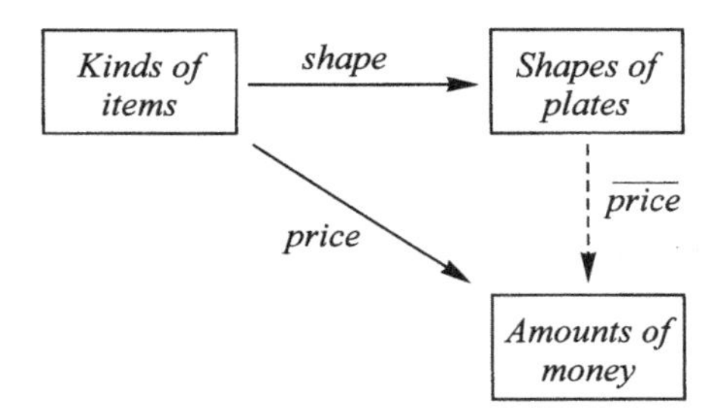

The cashier knows the map $\overline{price}$, but doesn't need to know the maps *shape* or *price*.

The servers take big trays of many different dishes from the cooks and circulate through the restaurant, the diners at the tables selecting all the dishes that appeal to them without anyone's writing down any record. Empty plates are stacked at the table according to shape after use.

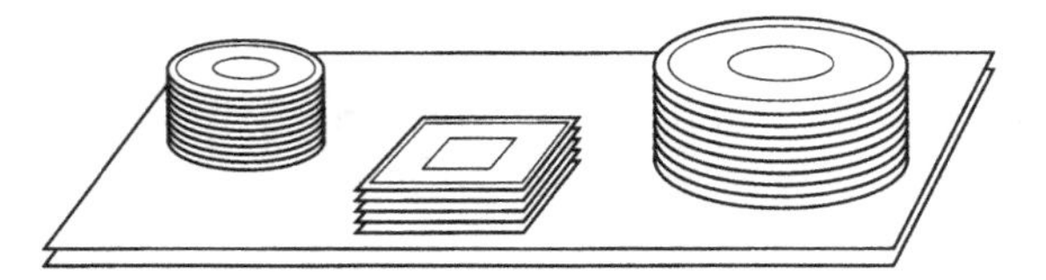

Thus when the diners at a table have finished with their dinner, there remain the empty plates stacked according to shape, as shown in the picture. This defines a map whose external diagram is

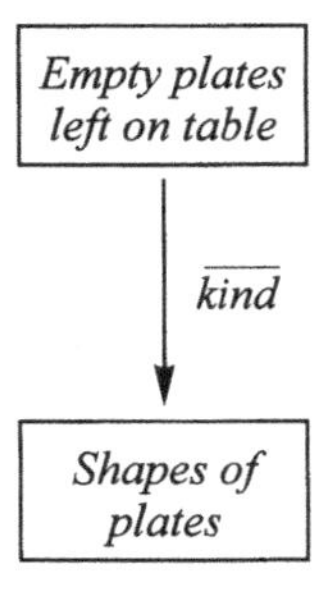

This map, resulting from the particular choices made by the customers at the particular table, can be composed with the map $\overline{price}$ resulting from the general organization of the restaurant, to yield a map $\overline{f}$ (with its own abstract 'stack' structure)

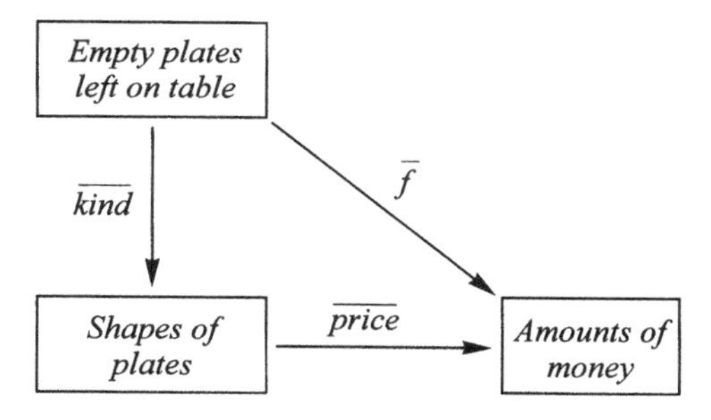

A glance at the table is sufficient for the cashier quickly to calculate the total bill as the sum of products

$$\sum_s \overline{price}(s) \cdot (\text{size of the stack of } \overline{kind} \text{ over } s)$$

(where s ranges over all shapes of plates). The total can also be calculated using only the map $\overline{f}$ since the total bill is also given by

$$\sum_s \overline{price}(s) \cdot (\text{size of the stack of } \overline{kind} \text{ over } s)$$
$$= \sum_x x \cdot (\text{size of the stack of } \overline{f} \text{ over } x)$$

where x ranges over possible amounts of money.

To prove that the last formula in terms of $\overline{f}$ gives the same result as the earlier, more commonly used formula in terms of f, we need only see that for each amount x, the 'stack' sizes of f and $\overline{f}$ are the same. But that follows from the more basic fact that f and $\overline{f}$ are themselves 'isomorphic' as we will explain from the following diagram showing all our maps.

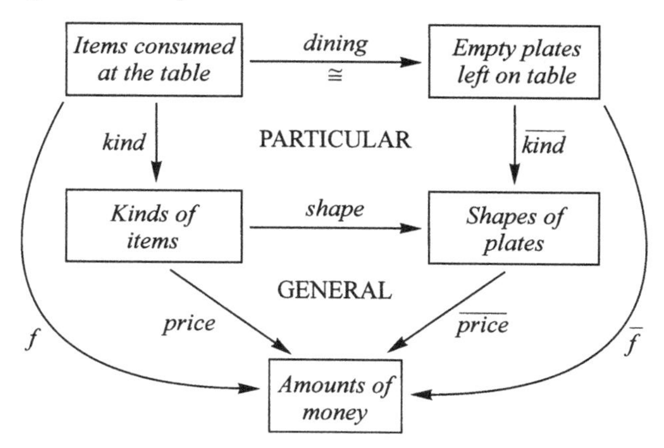

Here we have explicitly introduced the map *dining*, which transforms each item consumed at the table into an empty plate. Then clearly

$$shape \circ kind = \overline{kind} \circ dining$$

in the 'particular' square and $price = \overline{price} \circ shape$ in the 'general' triangle. The map *shape* which occurs in both these equations is the restaurant's key contact between the general and the particular. It is also pivotal in the proof, by associativity and by the definition of f and $\overline{f}$, that

$$f = \overline{f} \circ dining$$

But for every empty plate on the table there was exactly one item consumed, so the map *dining* has an inverse. We can say that the two maps f and $\overline{f}$ (with codomain *amounts of money*) are isomorphic, which implies that their stack-sizes over each x are the same.

While the detailed explanation of these relationships may take a little time to master, in practice the servers can work with a speed that is amazing to see, and the diners are well satisfied too. Moreover, the cashier can perform the $\overline{f}$-summation at least as fast as cashiers in other restaurants perform the f-summation, the French expression 'l'addition s'il vous plaît' taking on a surprising Chinese twist.

We can see that though abstract sets and maps have more information than the more abstract numbers, it is often more efficient to use them directly.

SESSION 6

Two general aspects or uses of maps

1. Sorting of the domain by a property

The abstract sets we are talking about are only little more than numbers, but this little difference is enough to allow them to carry rich structures that numbers cannot carry. In the example of the Chinese restaurant that we discussed in Session 5, I used the word 'stacking.' Now I would like to introduce some other words which are often used for the same idea.

For a general map $X \xrightarrow{g} B$ we can say that g gives rise to a **sorting** of X into B 'sorts', or that the map g is a sorting of X by B. (Note that we are speaking of 'B' as if it were a number.) Once g is given, every element b of B determines which elements of X are of the **sort** b, namely those elements mapped by g to b. For example, suppose that B has three elements. Then, without changing the map g, we can arrange the elements of X into the three different sorts so that the picture of g may look like:

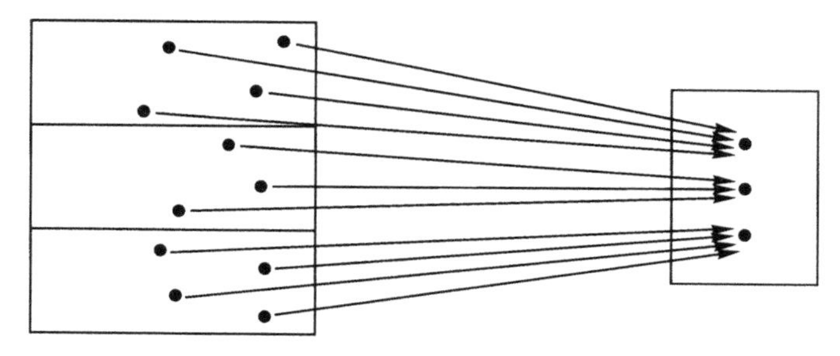

(For other maps g some of the bunches may be empty.) Here we have put in the same bunch all elements of X that go to the same element (sort) in B.

This way of viewing a map can also be described by saying that the map is a B-valued **property** on X. This means the same as saying that g is a **stacking** of the elements of X into B stacks. The number of stacks is always equal to the number of elements of B, while it is the elements of X that get stacked. An example is the obvious map from the set of presidents of the United States to the set of political parties that have existed in this country. This map assigns to each president the party to which he belonged. In this way the presidents get sorted by the parties in the sense that to each political party there corresponds a sort of presidents, namely the presidents that belonged to that party. Some of the sorts are empty since there are some parties which never had a president.

Another word that is used to describe this point of view about a map is **fibering**, by the agricultural analogy in which a bunch is imagined in the shape of a line or fiber. We say that X is divided into B fibers. If one fiber is empty, the map has no sections. Furthermore, for maps between finite sets the converse is also true: if no sort is empty, then the map has a section. For such maps one also uses the word **partitioning**.

So, the terms 'stacking,' 'sorting,' and 'fibering' are here regarded as synonymous, while 'partitioning' has a more restricted meaning. All of these terms emphasize that a given map $X \longrightarrow B$ produces a 'structure' in the *domain* X, and when we want to emphasize this effect we may refer to the map itself as a B-valued 'property.' An example is hair color. This is a map from the set of people to the set of colors, assigning to each person the hair color of that person. People are sorted by the property of hair color.

Example:

Sorts can themselves be sorted. Let X be the set of all creatures and B the set of species. Then $X \xrightarrow{s} B$ assigns to each creature the species to which it belongs. We can go further: species are sorted in genera by a map $B \xrightarrow{g} C$ which assigns to each species its genus; and by composing the two maps we obtain a coarser sorting $h = g \circ s$ of X.

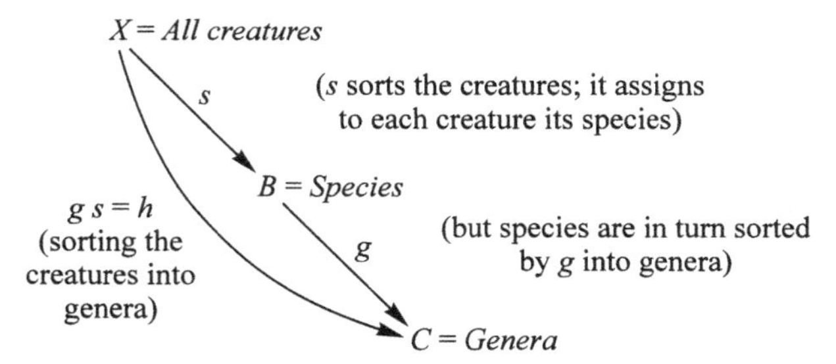

2. Naming or sampling of the codomain

All the words that we have discussed so far express one view of maps. But there is a second point of view that one can take about a map. Given a map $A \xrightarrow{f} X$, we can say that f is a family of A elements of X. For example, suppose that A has three elements. Then a map

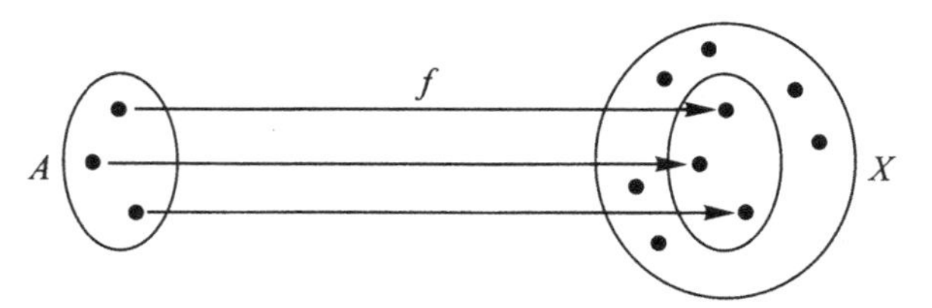

is a family of three elements of X. (Some of the three elements may coincide in other examples.) Again we are using A as if it were a number. Another word for this point of view (coming from geometry) is 'figure': a map from A to X is an A-shaped figure in X. We can also say 'A-element,' meaning the same as 'figure of shape A.' An ancient principle of mathematics holds that a figure is the locus of a varying element. An A-parameterized family $A \longrightarrow X$ is a varying element, in that (a) if we evaluate it at various $\mathbf{1} \longrightarrow A$, we will vary it through various points of X, but also in that (b) we can replace the special $\mathbf{1}$ by D, thus deriving from the given map $A \longrightarrow X$ a family of D-elements of X, one for each $D \longrightarrow A$. For example, we can take $D = A$, and the identity $D \longrightarrow A$, thus revealing that (c) the varying element, as a single thing, is a single figure or element itself.

We can also say that a map $A \longrightarrow X$ is a naming of elements of X by A, or a listing of elements of X by A. Let me give you an example of this. Suppose that we ask each student to point out a country on a globe. Then we get a map from the set of students to the set of countries, and in an accompanying discussion we might speak of 'Sheri's country,' 'Danilo's country,' etc. Not all the countries are necessarily named, and some country may be named more than once. The word 'listing' usually has the connotation of 'order'; this is *not* how it is meant in our discussion. Another couple of words for this point of view about maps are 'exemplifying' (in the sense of 'sampling') and 'parameterizing': we say that to give $f\colon A \longrightarrow X$ is to parameterize part of X by moving along A following f.

The above example of using students as 'names' for countries emphasizes that naming or listing is often done just for convenience and may have no permanent or inherent significance, in that we didn't ask 'why' each student chose the country he or she did. In other examples the naming may have more permanent meaning. For example, let A be the set of all fraction symbols, which are just pairs of whole numbers 3/5, 2/7, 13/4, 2/6, 1/3, ..., and let B be the set of all possible lengths. We can use the fraction symbols to name lengths with help of a chosen unit such as 'meter,' as follows. The map $A \xrightarrow{f} B$ assigns to the fraction 3/4 the length obtained by dividing the meter into 4 equal parts, then laying off 3 of these, whereas $f(3/5)$ is the length obtained by dividing the meter instead into 5 parts and laying off 3 of *those*, etc. Many names name the same length since $f(2/4) = f(3/6)$, but 2/4 and 3/6 are different names. Most lengths, such as $\sqrt{2}$ meters, are not named at all by f.

The terms 'naming,' 'listing,' 'sampling,' 'parameterizing' emphasize that a map $A \longrightarrow X$ produces a 'structure' in the *codomain* X, and when we want to emphasize this effect we may refer to the map itself as an A-shaped figure (or as an A-parameterized family) in the codomain.

The point of view about maps indicated by the terms 'naming,' 'listing,' 'exemplifying,' and 'parameterizing' is to be considered as 'opposite' to the point of view indicated by the words 'sorting,' 'stacking,' fibering,' and 'partitioning'. The sense in which this 'opposition' is meant can be explained philosophically in the following way.

3. Philosophical explanation of the two aspects

One explanation for these two aspects of a map comes from philosophy. **Reality** consists of fish, rivers, houses, factories, fields, clouds, stars, i.e. things in their motion and development. There is a special part of reality: for example, words, discussions, notebooks, language, brains, computers, books, TV, which are in their motion and interaction a part of reality, and yet have a special relationship with reality, namely, to **reflect** it.

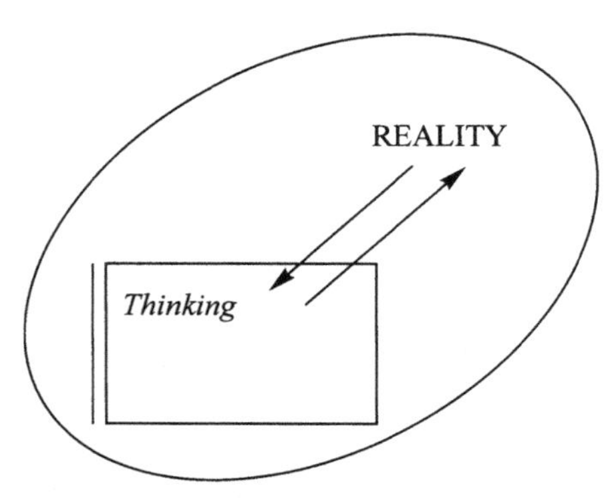

Thinking is going out and looking, manipulating, perceiving, considering, ...

The result of this reflective process is **knowledge**, and the totality of accumulated knowledge with its inner relationships is **science** (a purpose of which is to plan further manipulation of reality). Science is actually a complex of interrelated *sciences* focussing on different aspects of reality.

One of the particular sciences is **philosophy**, reflecting (as general knowledge) this particular relationship within reality, the relationship between thinking and reality. Thus within the complex of all scientific thinking there is the particular relation between **objective** and **subjective**:

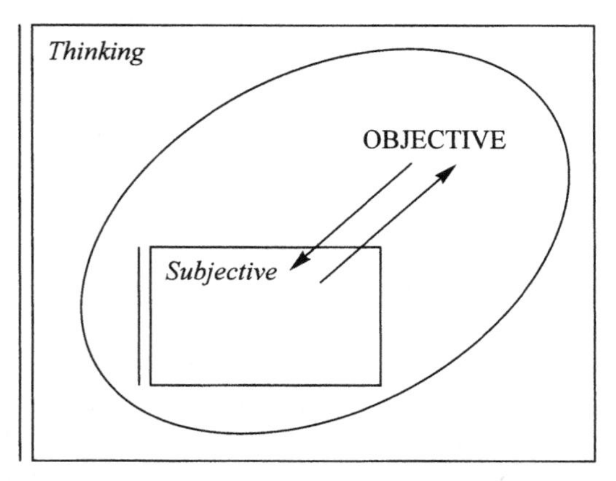

In the objective we strive to have as clear an image as possible of reality, as it is and moves in itself, independent of our particular thoughts; in the subjective we strive to know as clearly as possible the laws of thinking (as defined above) in itself, arriving at laws of grammar, of pure logic, of algebra, etc.

One further reflection within mathematical thinking of this relation between objective and subjective arises when within some given objective category (such as the category of sets) we choose some of the objects A, B (say, the sets with fewer than four elements) to use as subjective instruments for investigating the more general objects, such as the set of all creatures, all countries, etc. Then a chosen object A may be used as domain for listing elements of X, and also a chosen B can be used as codomain for properties of X. The composites of such listings and sortings become map-expressed structures in and among the chosen objects A, B, ... themselves, and these structures record as knowledge the results of investigating X.

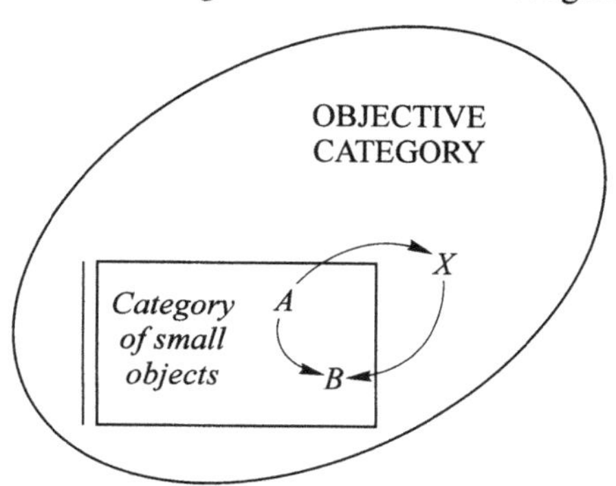

With this division of the category into 'small' objects among all objects, the two ways of considering a given map become no longer merely two 'attitudes', but a real difference: maps whose domain is small (listing) versus maps whose codomain is small (properties). Of course, if X itself happens to be small, we still have two aspects: a property of indices is the same as a list of values

$$I \xrightarrow{h} V$$

for example, the map

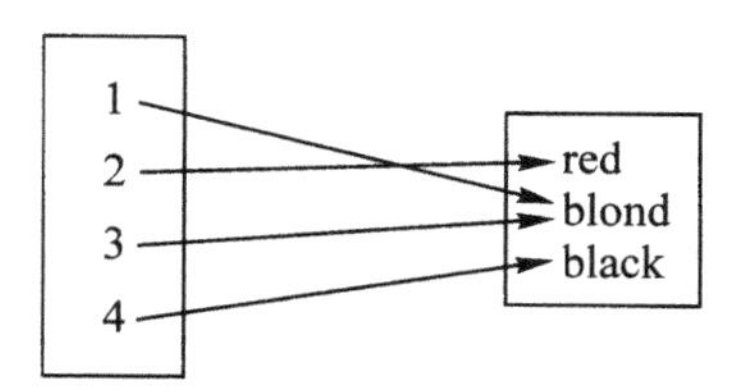

may be a record of composing two maps through some set X of actual people, whereby we sample I people among X, then observe their hair color; from this map alone (i.e. without further investigation, recorded by like maps) we can't tell – and it might be crucial in a criminal investigation – whether the first and third persons were the same or merely had the same hair color. The resulting 'listing h of values' has a repetition, or (equivalently) the 'property h of indices' has a sort with more than one element.

SESSION 7

Isomorphisms and coordinates

1. One use of isomorphisms: Coordinate systems

The idea of 'subjective contained in the objective,' or 'familiar contained in the general,' discussed in the last session, is especially simple if the 'naming' map is an isomorphism. That is, to have an isomorphism from a 'known' object A to an object X allows us to know X as well. To fit with the applications, let's give the isomorphism and its inverse these names:

$$A \underset{\text{coordinate}}{\overset{\text{plot}}{\rightleftarrows}} X$$

$$\textit{coordinate} \circ \textit{plot} = 1_A \quad \text{and} \quad \textit{plot} \circ \textit{coordinate} = 1_X$$

Here is an example. Imagine a geometrical line L, extending forever in both directions. It is often useful to *choose* an isomorphism from the set $\mathbb{R}$ of real numbers to the line L. The usual way to do this begins by choosing a point p on L, called an 'origin', and to decide that $\textit{plot}\,(0) = p$. Choose also a 'measuring stick', or unit of distance (foot, meter, light-year, etc.), and choose a direction on L to call the 'positive' direction. Having made these three choices, we get a map

$$\mathbb{R} \xrightarrow{\textit{plot}} L$$

in a way that is probably familiar to you. For instance, if our choices are as listed below, then *plot* (3.5) is the point q and *plot* (-4.3) is the point r.

——	chosen unit of distance
$\longrightarrow$	chosen positive direction
p	(below) chosen origin

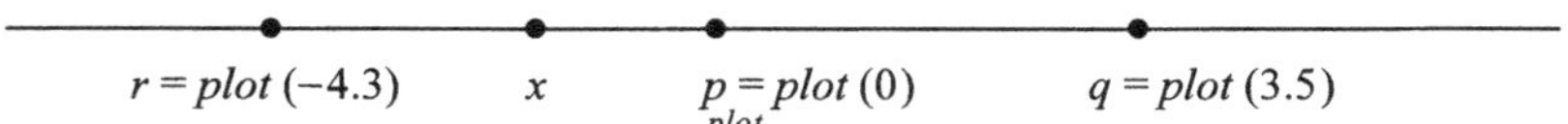

$r = \textit{plot}\,(-4.3)$ $\quad x \quad$ $p = \textit{plot}\,(0)$ $\quad$ $q = \textit{plot}\,(3.5)$

The remarkable utility of the map $\mathbb{R} \xrightarrow{\textit{plot}} L$ comes from its invertibility; there is an inverse (and hence exactly one inverse) for *plot*:

$$\mathbb{R} \xleftarrow[\textit{coordinate}]{} L$$

assigning to each point a number. (What, approximately, is $\textit{coordinate}(x)$ for the point x in the picture?)

We regard $\mathbb{R} \xrightarrow{plot} L$ as 'naming' the points on the line, and so the inverse map *coordinate* assigns to each point its numerical name. Of course the decision as to which objects have been incorporated into our 'subjective' realm is not eternally fixed. Euclid would have found it more natural to treat the geometric line L as known, and to use points as names for numbers.

There are other well-established isomorphisms with $\mathbb{R}$ as domain. When we say that Columbus sailed to America in 1492, we depend on having fixed an isomorphism from $\mathbb{R}$ to the 'time-line.' (What are the choice of origin, positive direction, and unit of 'distance' involved in specifying this isomorphism? Which of them would seem natural to an inhabitant of another planet?) If you have read popular accounts of relativity theory, you may doubt how well established even the time-line is, let alone an isomorphism from it to $\mathbb{R}$. Nevertheless, such an isomorphism has proved extremely useful; racing-car drivers, historians, and geologists are equally unwilling to part with it. Modern scientific theories of time still take our description as an excellent first approximation to a more refined theory.

Back to geometry. The cartesian (after René Descartes) idea of using an isomorphism from $\mathbb{R}^2$, the pairs (x, y) of real numbers, to a geometric plane P was sketched in Article II. (What choices need to be made in order to specify such an isomorphism?)

$$\mathbb{R}^2 \underset{\text{coordinate}}{\overset{\text{plot}}{\rightleftarrows}} P$$

If you type '*plot*(2, 1.5)' into a computer programmed for graphing, a dot will appear on the screen. The computer actually displays the output of the map *plot* at the input (2, 1.5). But *before* all this, you have to tell the computer which particular isomorphism *plot*, from pairs of numbers to the plane of the screen, you wish it to use. You must input your choices of origin, unit of distance, and even directions of axes, if you don't wish them to be horizontal and vertical. In this example, two additional maps, which can be called *first* and *second*, are relevant:

$$\mathbb{R}^2 \xrightarrow{first} \mathbb{R} \qquad \mathbb{R}^2 \xrightarrow{second} \mathbb{R}$$

$$first(x, y) = x \qquad second(x, y) = y$$

For instance, *first* (3.12, 4.7) = 3.12. Now if q is a point in the plane, we can compose these three maps

$$\mathbf{1} \xrightarrow{q} P \xrightarrow{coordinate} \mathbb{R}^2 \xrightarrow{first} \mathbb{R}$$

to get a number, $first \circ coordinate \circ q$, called, naturally enough, 'the first coordinate of q'.

Here is an example which doesn't involve $\mathbb{R}$. Tennis tournaments are usually arranged so that a loss of one match will eliminate the loser. For simplicity, let's take an eight-player tournament. 'Brackets' are set up as in the diagram below. The names of the eight players are to be listed in the left column. In the first round each 'bracketed pair' will play a match and the winner's name will be entered in the adjacent space in the second column, and then the whole process is repeated with the remaining four players, etc. Before the tournament can begin, though, there is the job of 'seeding' the players, i.e. choosing an isomorphism of finite sets (and thereby also its inverse):

$$\{1, 2, 3, \ldots, 8\} \underset{\text{seed}}{\overset{\text{rank}}{\rightleftarrows}} P = \textit{set of players}$$

For example, *rank* 1 may be Pete Sampras, so that the 'seed' of Sampras is the number 1. Then, no matter who the players are, they are bracketed according to the following scheme:

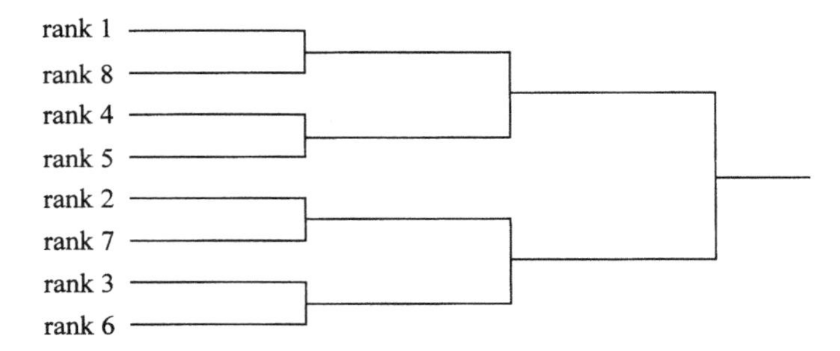

Every effort is made to seed fairly, so that the best player as judged by past performance is seeded number one, the next best number two, and so on. (You'll notice a 'particular versus general' aspect to this example. The assignment of numbers to positions in the chart above is general, applying to every eight-player tournament, while the isomorphism *seed* is particular to the past performances of the eight players who are involved in this one tournament.) Incidentally, can you figure out any rational explanation for the curious bracketing above? What would be a suitable bracketing by rank for four players, or for sixteen players?

The rest of our discussion applies to all examples. Once a *coordinate system*, a pair

$$A \underset{\text{coordinate}}{\overset{\text{plot}}{\rightleftarrows}} X$$

of maps inverse to each other, is established, we tend to pass freely back and forth between A and X as if they were the same object. In the plane example, we speak of 'the point (2, 3.7),' meaning 'the point *plot* (2, 3.7).' In the tennis tournament, we say, 'There has been an upset; number eight beat number one.' A practice so common, which seldom seems to cause confusion (but see 'Abuses' below), must have its explanation, and indeed it does. Once we have *fixed* an isomorphism $A \xrightarrow{f} X$, it is harmless to treat A and X as the same object, precisely because we have the maps f and f^{-1} to 'translate.' For example, if we want to specify a map $X \xrightarrow{g} Y$ we can instead specify a map $A \xrightarrow{G} Y$, and everyone who is aware of the chosen isomorphism will understand that we mean the composite map $X \xrightarrow{f^{-1}} A \xrightarrow{G} Y$. But why do we cause everyone the trouble of making this translation? We shouldn't, unless A is a 'better-known' object than X, i.e. an object incorporated into our 'subjective' category inside the large 'objective' category. Or, as in the tennis example, it may be that the object A is more familiar to our audience than X. Someone who understands tournaments in general, but hasn't followed tennis in recent years, might fail to be surprised if Becker beat Sampras, but still could understand that a defeat of number one by number eight is cause for comment. Notice, though, that this is only because the isomorphism *rank* from numbers to players was not arbitrary. In a friendly tournament at school, numbers might be assigned to players at random; then a defeat of number one by number eight would not be surprising. Our professional tournament seeding was not just an isomorphism of sets, but an isomorphism in the category of 'ordered sets', sets whose elements are arranged in an order which maps in the category are required to 'respect'. The study of various types of 'structure' and the categories to which they give rise will be a recurring theme in the rest of the book, and you will see how 'respecting structure' is made precise.

2. Two abuses of isomorphisms

Since a principal use of isomorphisms is to give coordinate systems, you would expect the main abuses of isomorphisms to stem from this use, and they do. There are two fundamental errors to avoid. Most often they occur when the 'familiar' object A is some set of numbers (or related to numbers, like $\mathbb{R}^2$ in our 'plane' example). Watch carefully for these abuses when you suspect that mathematics is being misapplied.

The first abuse is to assume that an isomorphism of sets $A \longrightarrow X$ means that some additional structure that A has, for instance by virtue of being a set of numbers, will be meaningful in X. An example was given above: it is neither an honor nor an advantage to be ranked number one in a tournament if the rankings were drawn from a hat. Similarly, identifying points on a line with numbers doesn't make adding two points to get a third point a reasonable operation.

The second abuse is subtler, involving one familiar object A and two objects X and Y coordinatized by A. I'll just give you one example, to which actual students have been subjected. (I hope not you!) The physicist Richard Feynman was pleased to see that his child's elementary-school textbook gave meaning to large numbers by listing the distances from the planets to the sun, the masses of the planets, and various other astronomical data. But then, to his dismay, followed exercises of this type: add the distance from Venus to the sun, the mass of Mars, and the .... Well, you see the point. It only appeared to make sense to add a distance to a mass because the objects 'distances' and 'masses' had each separately been identified with the object 'numbers,' by choosing a unit of measurement for each.

While these simple examples may appear ludicrous, errors of exactly these two types have often been made by people who should know better. Soon, when you have become familiar with some 'types of structure,' you should be in little danger of commiting these abuses. For now, the best advice I can give you is this.

To decide what calculations to do, *think in the large 'objective' category*. As we'll see, a surprising variety of calculations can actually be carried out in objective categories. But if it is necessary, *after* determining the calculations to be done, you can choose coordinate systems and calculate in the smaller 'subjective' category, and then translate the results back into the objective category. It will not occur to you to add two tennis players to get a third player; you could only make this mistake after identifying (objective) players with (subjective) numbers.

SESSION 8

Pictures of a map making its features evident

Let's start by doing Exercise 5 from Article II. Given the map g from the set A to the set B pictured below,

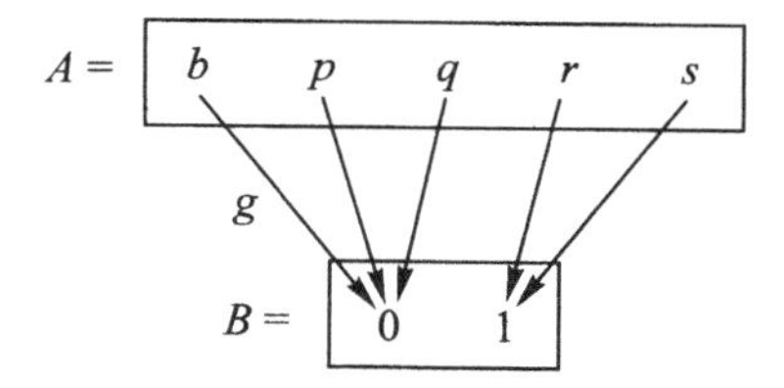

how many maps f are there with $g \circ f = 1_{\{0,1\}}$ (the identity map on $\{0,1\}$)?

Obviously such an f must go from B to A, so that schematically we may picture the maps f and g as

$$A \underset{f}{\overset{g}{\rightleftarrows}} B$$

A map f with this property is called a section of g, so that another way to phrase the problem is: *How many sections does the map g have*? Did anybody find any?

KATIE: Yes, I found two.

Tell me one of them.

KATIE: The one that sends 0 to q and 1 to r.

All right, so you have $f(0) = q$ and $f(1) = r$. To see whether this map is really a section for g, we have to check the equation $g \circ f = 1_{\{0,1\}}$. Now we have two maps $g \circ f$ and $1_{\{0,1\}}$ and we want to know whether they are equal. When are two maps equal?

FATIMA: They must have the same domain and the same codomain.

So, you are saying:

1. the domain of $g \circ f$ must be the same as the domain of $1_{\{0,1\}}$, and
2. the codomain of $g \circ f$ must be the same as the codomain of $1_{\{0,1\}}$.

Is that all? No. Let's review our test for equality of maps of sets. A map of sets f: $X \longrightarrow Y$ is specified by a rule which to each element of X (the *domain* of f) assigns exactly one element of Y (the *codomain* of f). The question is: If we have two rules, when do we say that they specify the same map? Let's call these two rules h and k. In order to verify that $h = k$ you have to check that *for each* particular input you get the same output with both rules. In summary: to say $h = k$ means *three* things:

1. the domain of h is the same as the domain of k,
2. the codomain of h equals the codomain of k, *and most importantly*,
3. for each x in the domain of h and k, we must have $h(x) = k(x)$.

In the third condition, the number of things one has to check is equal to the number of elements of the domain, because the condition has to be checked *for each* element of the domain.

So, let's see: f goes from B to A, and g goes from A to B, so $g \circ f$ goes from B to B, and $1_{\{0,1\}}$ also goes from $B = \{0, 1\}$ to B. So we do have

1. the domain of $g \circ f$ is $B = \{0, 1\}$, which equals the domain of $1_{\{0,1\}}$, and
2. the codomain of $g \circ f$ is $B = \{0, 1\}$, the same as the codomain of $1_{\{0,1\}}$.

All this writing is really not necessary, though. You can see directly from the diagram with the arrows f and g going back and forth between A and B that (1) and (2) are true. When you get used to this, conditions (1) and (2) aren't much of a fuss because you won't even ask if two maps are equal if they don't have the same domain and codomain. It is like asking whether two travellers followed the same *route*; you wouldn't *ask* the question if one of them travelled from Berlin to Paris and the other from New York to Boston.

So the *essential* thing in order to check that $g \circ f = 1_{\{0,1\}}$ is condition (3). We have to check that $g \circ f$ acting on any element of its domain (the set $\{0,1\}$) gives the same as $1_{\{0,1\}}$. In other words, since the identity $1_{\{0,1\}}$ sends 0 to 0 and 1 to 1, what we have to check is

$$(g \circ f)(0) \stackrel{?}{=} 0 \quad \text{and} \quad (g \circ f)(1) \stackrel{?}{=} 1.$$

Now, what does $g \circ f$ mean?

OMER: First calculate f and then stick g to that.

Right. So first calculate $f(0)$, i.e. ... q, and $g(q) = 0$, so it checks: $(g \circ f)(0) = 0$. And for $(g \circ f)(1)$, first $f(1)$, i.e. r, and then $g(r) = 1$. So we really have:

3. $(g \circ f)(0) = 0$ and $(g \circ f)(1) = 1$

and we are done. The map that Katie gave us was truly a section for g.

But if you found that one, you must be able to find others. Let's see how all this can be seen directly in the pictures. We need a map $f : B \longrightarrow A$. Conditions (1) and (2) are automatically satisfied. Now we have to guarantee $(g \circ f)(0) = 0$ and $(g \circ f)(1) = 1$. The first means $g(f(0)) = 0$. But the only things that g sends to 0 are b, p, and q, so $f(0)$ has to be one of these three. Similarly, the only things that g sends to 1 are r or s, so in order that $g(f(1)) = 1$, it must be that $f(1)$ is equal to either r or s. So to find a section boils down to finding a map $f : B \longrightarrow A$ such that $f(0)$ is either b, p or q, and $f(1)$ is r or s. So how many are there?

KATIE: Six, because if 0 goes to b, 1 can either go to r or s, ...

Right. And if 0 goes to p we get two more possibilities, and two more if 0 goes to q. The pictures of all these possibilities are:

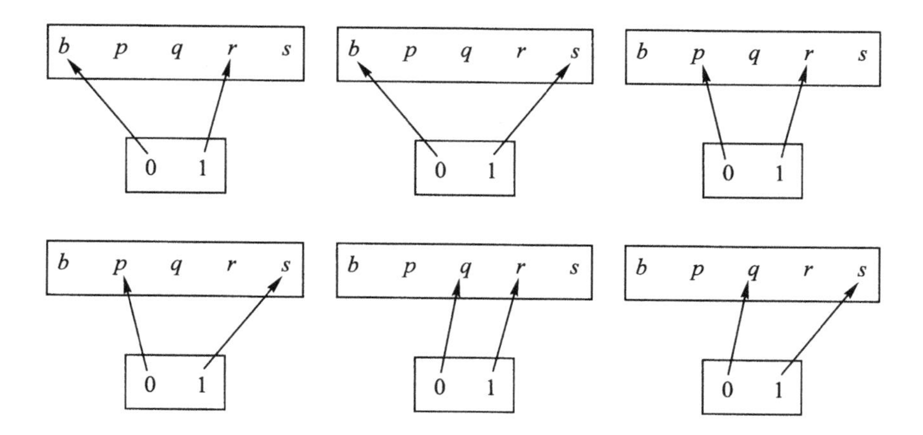

Even better would be to arrange all the possibilities systematically, something like this:

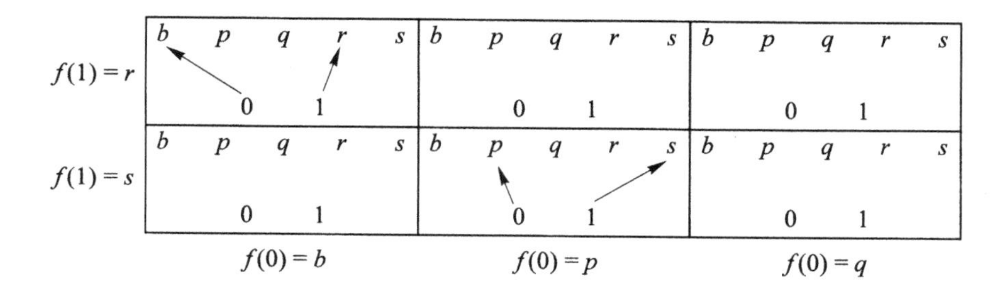

(Add the rest of the arrows yourself.)

Someone asked why these maps are called *sections*. The word 'section' here is actually short for 'cross section.' Imagine holding a cucumber vertically over the table. Consider the map that projects each point in the cucumber perpendicularly down on its shadow on the table. If you take a knife and slice through the cucumber as in the picture on the right, you have a section of that map! In general, a section of that projection map may have any funny shape, not just a straight cut.

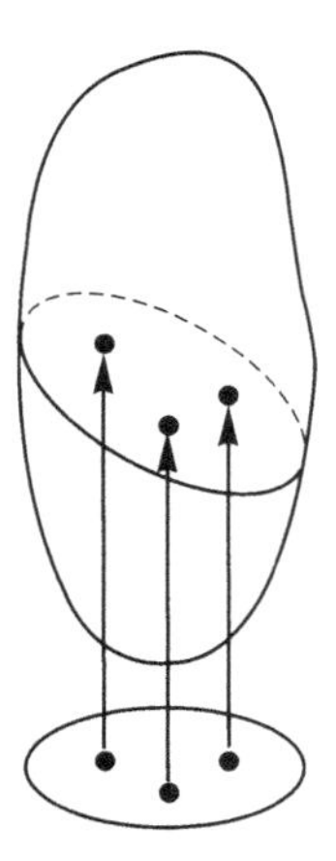

There is a similar picture for the section f that Katie gave for our map g

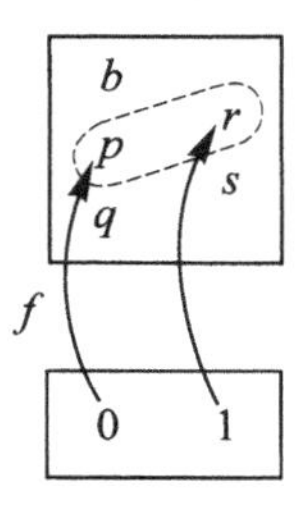

where we have put directly above 0 only the elements that are mapped by g to 0, and above 1 the elements that are mapped by g to 1. In both cases the 'cross section' is a copy of the smaller set at the bottom (the codomain of g) inside the set on top.

You should now be able to answer any question of this type. Let's see if you can. Consider the map

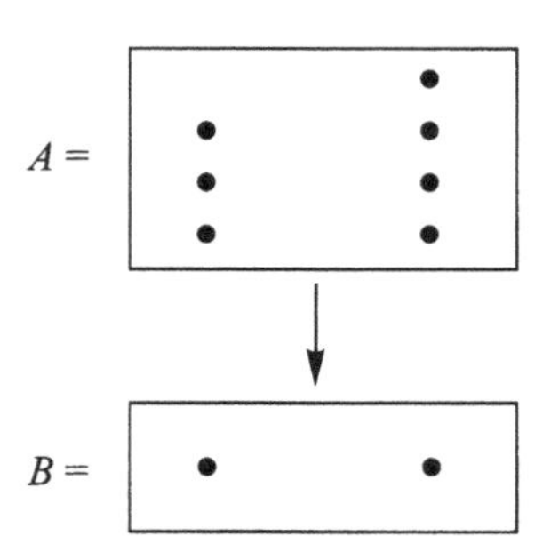

How many sections does it have?

KATIE: Twelve.

Right. Three choices for one point and four for the other .... What about

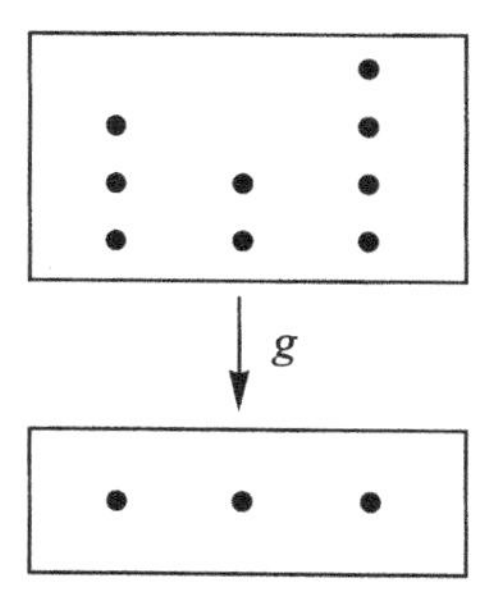

where again the map g assigns to every point in the set on top that point in the bottom set that lies directly below? How many sections does g have? Right, $24 = 3 \times 2 \times 4$. That's the formula Chad gave us earlier.

A section can also be called a **choice of representatives**. In fact, a very good example is the section of the population of the United States constituted by the congressional representatives. We have a map from the set of people in the United States to the set of all congressional districts because every person lives in some congressional district

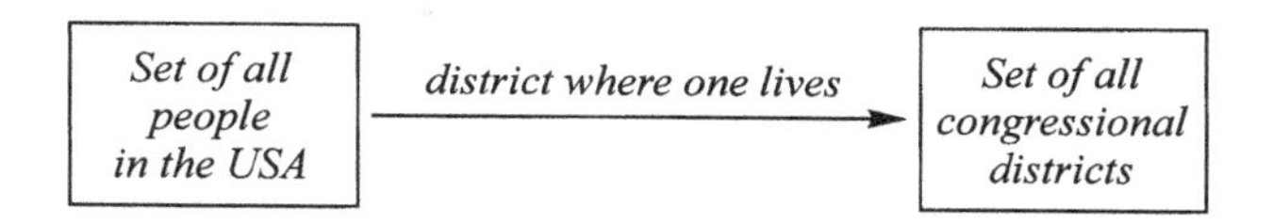

By law, a choice of congressional representatives must be done in such a way that every congressman lives in the district he represents. This is precisely to say that a choice of representatives must be a section of the map above.

We should do one more example to remind ourselves of something we noticed earlier. Suppose we have the map

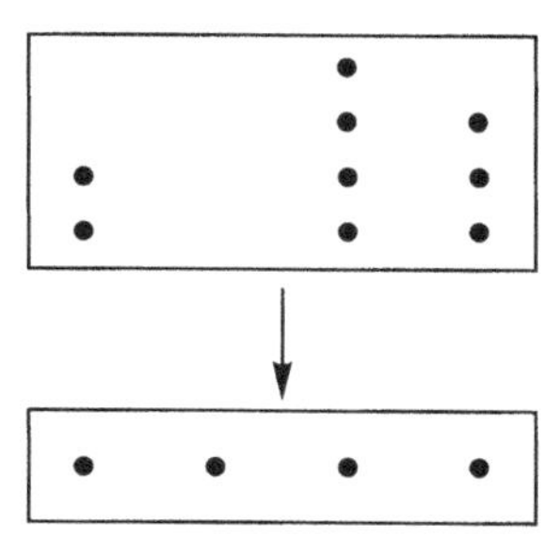

How many sections are there? Zero. Chad's formula also gives the correct answer: $3 \times 4 \times 0 \times 2 = 0$.

Our next problem is to find the number of retractions for a given map. For example we can start with the map

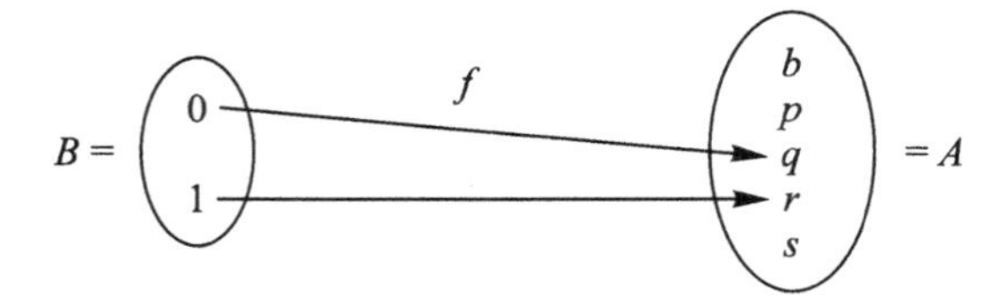

and ask how many maps $h : A \longrightarrow B$ are retractions for f. Retraction means that $h \circ f$ is the identity on .... Well, you figure it out. First you apply f, which goes from B to A, then apply h, which goes from A back to B. Therefore $h \circ f$ goes from B to B, and it must be equal to the identity on $B = \{0, 1\}$. Thus we must have $h \circ f = 1_{\{0,1\}}$. So the map h must satisfy the conditions

$$(h \circ f)(0) = 0 \qquad (h \circ f)(1) = 1$$

which are the same as

$$h(f(0)) = 0 \qquad h(f(1)) = 1$$

Just two conditions. Since we know $f(0)$ and $f(1)$ because they are given to us (they are respectively q and r), the two conditions are really

$$h(q) = 0 \qquad h(r) = 1$$

Other than that, h can send each of b, p, s to either 0 or 1.

DANILO: So, for the rest h is just like any map from $\{b, p, s\}$ to $\{0, 1\}$.

That's right. This idea can even be used to find a formula for the number of retractions when there are any at all. In this case, it shows that our map f has eight retractions, since $2^3 = 8$.

OMER: Why is it that some maps have a section and no retraction, and other maps have a retraction and no section?

The first thing one has to realize is that to say that a map $A \xrightarrow{r} B$ has a section means that there is a map coming back:

$$A \underset{s}{\overset{r}{\rightleftarrows}} B,$$

such that $r \circ s = 1_B$. According to our discussion earlier, *this implies that B is at most as big as A*. Remember that in our problems of finding sections, the big set was always on top and it was the codomain of the section. This allows us to tell whether a set is smaller than another without using numbers. If there are maps $A \underset{s}{\overset{r}{\rightleftarrows}} B$ such that $r \circ s = 1_B$, then B is smaller than (or at most as big as) A.

In fact, even before numbers were invented people knew how to tell which of two sets was smaller: you just have to pair off the elements of one set with those of the other and see which set has elements left over. This is practical too. Imagine that you are setting up the chairs for people to sit at a chamber music concert. What is the simplest way to know whether you have enough chairs? You won't start counting all the people and then counting all the chairs. You just ask everybody to sit down. If anybody remains standing you need more chairs. This is something worth remembering; the primitive notion is ISOMORPHISM; the fancy abstract notion is NUMBER.

Now let's try Exercise 8 from Article II. Prove that the composite of two maps that have sections has a section. So, suppose that we have two maps that we can compose. Let's call them k and p

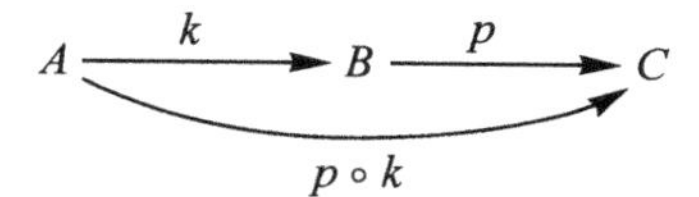

and they are supposed each to have a section. A section for p goes from C to B (let's call it s) and a section for k goes from B to A (let's call it s'). Putting all this in our diagram, we have:

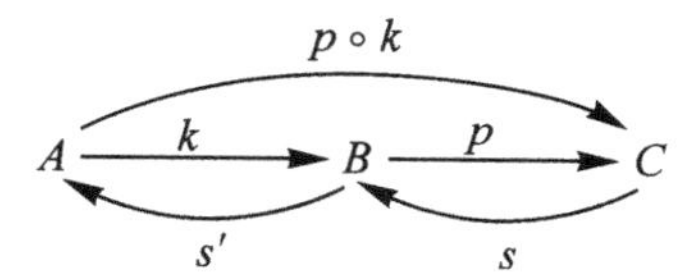

Now, what does it mean to say that s is a section of p?

OMER: Is it that s following p is the identity on B?

No, that would mean that s is a retraction of p. Remember that the domain of the section is the 'smaller' set and that the condition always involves the identity on the smaller set. So, the condition for s to be a section of p is that p following s is the identity on C, or $p \circ s = 1_C$. And what is the condition for s' to be a section for k?

ALYSIA: k following s' is the identity on B.

Right. So, what we have so far is

$$p \circ s = 1_C \quad \text{and} \quad k \circ s' = 1_B$$

What we want is a section for $p \circ k$, a map from C to A

$$A \underset{S}{\overset{p \circ k}{\rightleftarrows}} C$$

such that $(p \circ k) \circ S = 1_C$. Is there any guess as to what S could be?

OMER: Compose s and s'.

That's about the simplest thing we can try. There are other ways to go from C to A, but let's try the simplest first. So we'll try to prove that $s' \circ s$ is a section for $p \circ k$. In other words, we are faced with the question

$$(p \circ k) \circ (s' \circ s) = 1_C?$$

Any suggestions?

OMER: We can compose k with s' and substitute it with the identity on B.

Right.

$$(p \circ k) \circ (s' \circ s) = p \circ (k \circ s') \circ s = p \circ 1_B \circ s$$

And now what?

CHAD: $1_B \circ s$ is equal to s.

Very good, so we can put $p \circ 1_B \circ s = p \circ (1_B \circ s) = p \circ s$, and now we are ready to use the condition that s is a section for p, which is that $p \circ s$ is the identity on C. Therefore indeed the map $S = s' \circ s$ is a section for $p \circ k$.

Notice how similar to multiplication of numbers our calculation is. The main difference is that multiplication of numbers is commutative. Now you should try to work out the remaining exercises.

SESSION 9

Retracts and idempotents

1. Retracts and comparisons

We have seen that a reasonable notion of 'same size' is given by *isomorphism*: $A \cong B$ (read 'A is isomorphic to B') means that there is at least one invertible map (isomorphism) from A to B. For finite sets, $A \cong B$ tells us precisely that A and B have the same number of *points*, or maps from a singleton set **1**. (In other categories, we'll see that it tells us much more.) What is a good way to express that A is 'at most as big as' B? There are several answers, and we'll discuss two of them. The first is:

Definition: *$A \trianglelefteq B$ means that there is at least one map from A to B.*

This has two reasonable properties that a notion of 'smaller than' (really 'at most as big as') ought to have:

(R, for 'reflexive'):	$A \trianglelefteq A$, since there is the identity map $A \longrightarrow A$.
(T, for 'transitive'):	If $A \trianglelefteq B$ and $B \trianglelefteq C$ then $A \trianglelefteq C$, since the composite of a map from A to B and a map from B to C is a map from A to C.

For sets, the relation $A \trianglelefteq B$ doesn't tell us much, except that if A has a point, then B does too: a point $\mathbf{1} \longrightarrow A$ followed by a map $A \longrightarrow B$ gives a point of B.

Exercise 1:
(In the category of sets) Show that unless the set A has a point and B has none, then $A \trianglelefteq B$.

In other categories, this way of comparing objects can be quite interesting, but in sets it needs to be supplemented with another method. The idea comes from remembering that whenever we met two finite sets A, B and a section-retraction pair, $A \xrightarrow{s} B \xrightarrow{r} A$ with $rs = 1_A$, we saw that A was at most as big as B.

Definition: *A is a* **retract of** *B means that there are maps $A \xrightarrow{s} B \xrightarrow{r} A$ with $rs = 1_A$. (We write this as $A \leq_R B$.)*

(The notations $\triangleleft$ and $\leq_R$ are not standard, and have just been chosen to suggest 'arrow' and 'retract'. We won't use them after this session without reminding you what they mean.)

> **Exercise 2:**
> (In any category) Show that
>
> (R) $A \leq_R A$.
>
> (T) If $A \leq_R B$ and $B \leq_R C$ then $A \leq_R C$.
>
> Hint: You already proved (T) when you proved that if a composable pair of maps each has a retraction, then so does their composite.

If $A \leq_R B$, then in particular there are maps $A \longrightarrow B$ and $B \longrightarrow A$, so each of A and B is in our earlier sense at most as big as the other: $A \triangleleft B$ and $B \triangleleft A$. Our earlier sense was good for telling non-empty sets from empty, while our new sense is good for sorting out the non-empty sets. In fact, if A and B are finite non-empty sets, $A \leq_R B$ says exactly that A has at most as many points as B. (In other categories we'll see that this comparison also tells us much more, so that both our ideas for comparing relative size should be used.)

2. Idempotents as records of retracts

Suppose in sets that we have a retract $A \xrightarrow{s} B \xrightarrow{r} A$, so $rs = 1_A$. Then we have seen that the endomap e of B given by composing the maps r and s in the other order, $e = sr$, is idempotent: $ee = e$. Recall the proof?

$$ee = (sr)(sr) = s(rs)r = s1_A r = sr = e$$

This endomap e is a vestige in B of the maps s and r, but you would not expect us to be able to reconstruct s and r from their composite e. How could we even reconstruct A? Perhaps if we think of an example it will help.

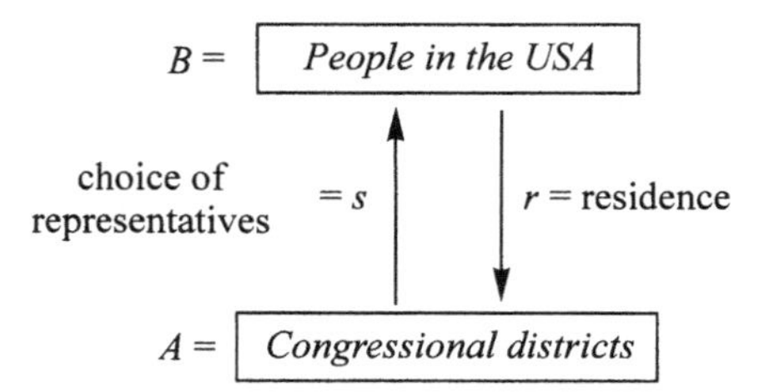

We have seen that, by law, $rs = 1_A$: the residence of the representative of each district is that district. What does the idempotent map $B \xrightarrow{e=sr} B$ look like? For example, who is e (Fatima)?

FATIMA: $e(\text{me}) = s(r(\text{me})) = s(\text{my district})$ = the honorable person or crook representing my district.

Just so. e(Fatima) is Fatima's congresswoman or congressman. This is not Fatima herself, but some people are their *own* congressional representatives. Which?

KATIE: The people who are members of the House of Representatives.

Exactly. Those people are the *fixed points* of the endomap e. That is perhaps a bit of a disappointment, not because of the quality of the representatives, but because we were hoping that from the endomap e of B we could 'reconstruct' the set A of congressional districts. We failed, but we did something just as good: we found a set (the fixed points of e, members of the House of Representatives) which is related in a very nice way to the set of congressional districts. In what way?

SHERI: It is *isomorphic* to the set of congressional districts.

Bravo! Let's see the process with pictures, too. What does the internal diagram of a typical idempotent endomap look like? For each point x, the point ex has to be a fixed point of e, since $e(ex) = (ee)x = ex$; so each point, if not already fixed by e, at least reaches a fixed point in one step. That means that the picture of an idempotent map is pretty simple. It must look something like this:

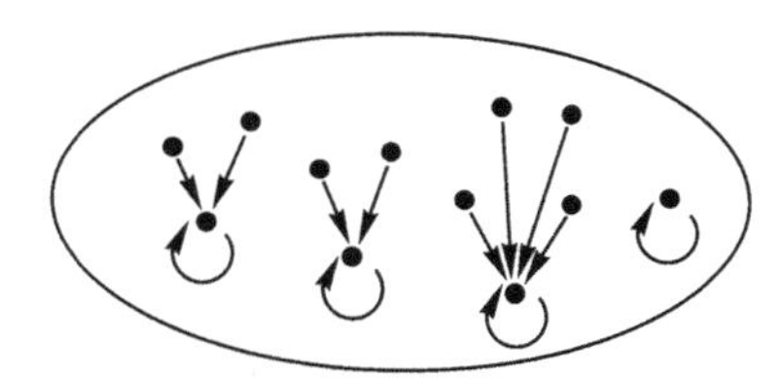

Here we have only the set B and the idempotent map e; we don't know which set A and which maps r and s gave rise to it, if indeed any did. Still, we can use the procedure we discovered with the representatives to get an A, r, and s which will do. We just copy the set of fixed points of e to serve as our A, and then the maps r and s that we should use are pretty clear. We'll draw s, and picture r as a sorting of B by A.

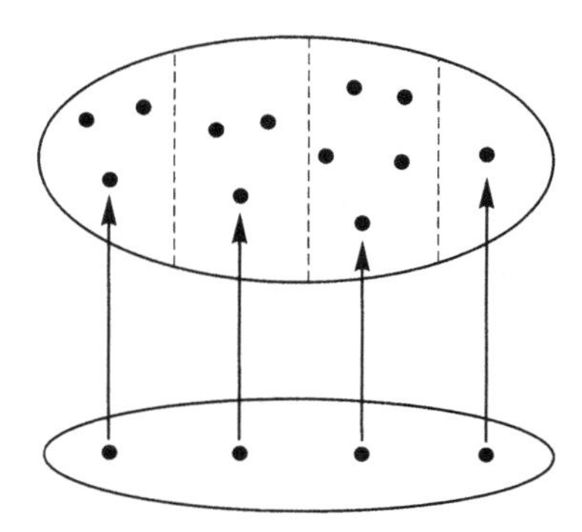

With a little practice, you can 'see' A, r, and s in the internal diagram of e. In fact we can see A either as the fixed points of e or as the sorts into which e 'clusters' B. Let's give a name to this relationship of e to A, r, and s.

Definition: *(In any category) If $B \xrightarrow{e} B$ is an idempotent map, a* **splitting of** *e consists of an object A together with two maps $A \underset{r}{\overset{s}{\rightleftarrows}} B$ with $rs = 1_A$ and $sr = e$.*

It will turn out that in many categories, a device very similar to the one we used in sets will give a splitting for any idempotent endomap. In any case, there cannot be two essentially different splittings for e, as the following exercise shows.

Exercise 3:
(In any category) Suppose that both $A \underset{r}{\overset{s}{\rightleftarrows}} B$ and $A' \underset{r'}{\overset{s'}{\rightleftarrows}} B$ split the same idempotent $B \xrightarrow{e} B$. Use these maps to construct an isomorphism $A \xrightarrow{f} A'$.

One can show that this isomorphism f is the only one 'compatible' with the maps we started from, and one can even study how to reconstruct maps between retracts from maps between the large objects B; but this should be enough to give you the crucial idea: all the essential information about A, r, and s is really contained in B and e.

Here is an example from arithmetic. Let B be the set of all 'fraction symbols' n/d with n and d whole numbers and $d \neq 0$. Different fraction symbols, like 8/6 and 4/3 may represent the same rational number. In school you were taught a 'reduction process' $B \xrightarrow{e} B$: cancel the greatest common factor in the numerator and denominator, and then if the denominator is negative change the signs of both numerator and denominator. For example, $e(6/-4) = -3/2$. This map e is idempotent, because reducing a reduced fraction doesn't change it. A rational number can now be described either as a reduced fraction (fixed point of e) or as the cluster of all fractions which reduce to that reduced fraction. In this example there is even a way to test whether two fraction symbols are in the same cluster without reducing them: $e(n/d) = e(m/c)$ exactly when $nc = md$. This is convenient, because it is easier to multiply large numbers than to find their greatest common factor. Curiously, it is easier to find the greatest common factor of two numbers, by a process called the 'Euclidean algorithm', than it is to factor either of them into its prime factors! Recent unbreakable(?) codes depend on the apparent difficulty of factoring large numbers.

3. A puzzle

If we think of B as a known set, incorporated into our 'subjective' category, what we have achieved is that the less-known set A has been captured by a description in our subjective category, namely by B and its idempotent endomap e. This seems puzzling. Why would we want to describe the smaller set A in terms of the larger set B? Normally we wouldn't, and it isn't too often done in the category of finite abstract sets (but see the examples in the section below.) One exception occurs in programming computers, where the sort of set that is most easily managed consists of all strings of zeroes and ones of a particular length, say n. There are 2^n of these, and you can see that it might be useful to represent any set A in which you are interested as a retract of a set B of n-strings, provided you can then record nicely the idempotent endomap of B. The study of how to do that gets one into 'Boolean algebra,' which is a basic topic in computer science. Still, the main use of 'describing the smaller in terms of the larger' occurs in other categories. It often happens that even though B is bigger, it is 'structurally simpler' than A.

4. Three kinds of retract problems

Let's return to the two general aspects of maps. We have seen that if a map $B \xrightarrow{r} A$ is given and we seek sections for it, a good way to picture the situation is to regard r as a 'sorting of B into A sorts':

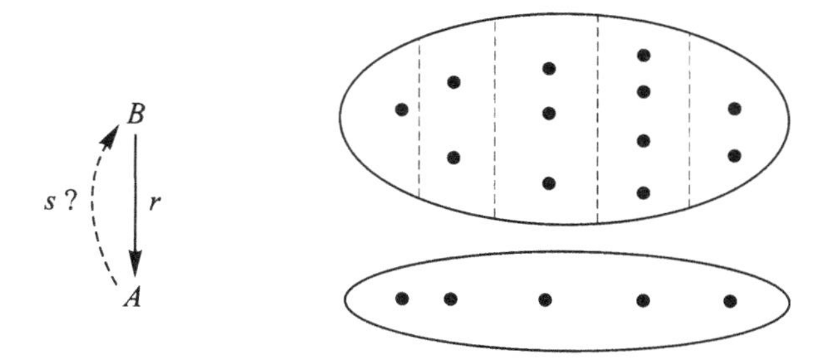

This enables us to picture a section s of r as choosing for each sort an 'example' of that sort. This might be called the 'museum director's problem'. Suppose you need to assemble an exhibit of mammals, with one mammal of each species. Then you start with the sorting map r from the set B of mammals

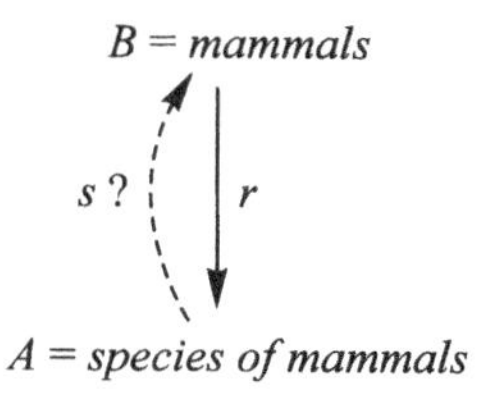

Your job is to choose a section s of r; that involves selecting one exemplary specimen for each species.

The opposite, or dual, problem is the 'bird-watcher's problem.' The bird-watcher starts with a manual giving an example of each species, a 'sampling' or 'exemplifying' map s:

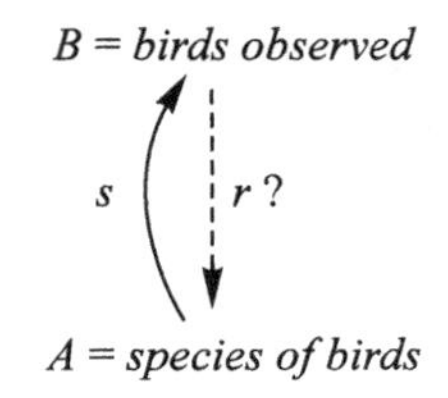

His job is to assign to each bird he sees (real or pictured) a species, and his manual at least gives him sufficient direction to ensure that $rs = 1_A$.

To the young is given the most difficult problem. The small child sees a variety of animals, and with whatever assistance can be gathered from picture-books and parents, tries to select an idempotent endomap e:

$$B = Animals \circlearrowleft^{e}$$

The map e should assign to each animal the most familiar animal it closely resembles. Having selected e, the child is asked (again with some assistance) to split this idempotent: to form the abstract idea of 'sorts of animals' (e.g. cat, dog, cow) and to master the maps:

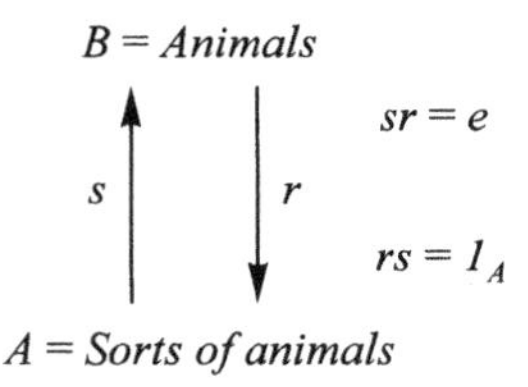

The map s assigns to each 'sort' of animal, say 'cow', the most familiar example, say Bossie; r assigns to each particular animal, say 'the big scary barking thing next door', its sort.

These three kinds of problems were described in a way that made it seem that one solution could be preferable to another. In the rarified world of abstract sets, this would not be so; the sets in our examples have additional structure.

Perhaps some abstract pictures may not be out of order, to illustrate that all three problems are solved by giving the large set B additional 'structure'.

Museum director's problem: Given $B \xrightarrow{r} A$, choose $A \xrightarrow{s} B$ satisfying $rs = 1_A$. Mental picture: View r as sorting B into A sorts:

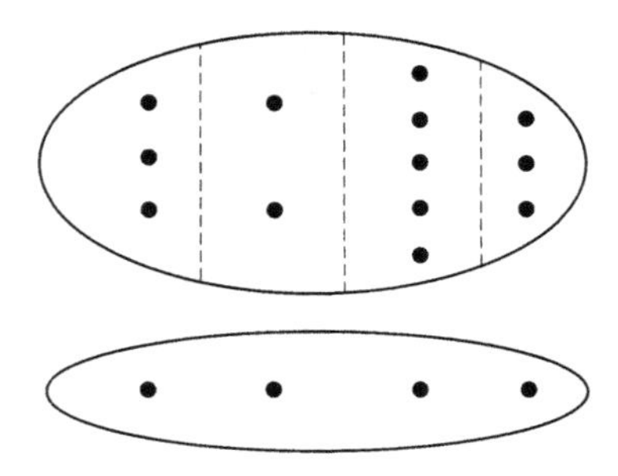

Then choose a 'cross-section':

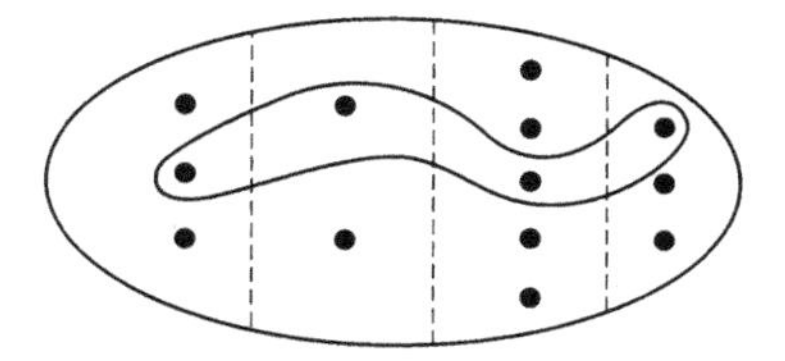

Bird-watcher's problem: Given $A \xrightarrow{s} B$, choose $B \xrightarrow{r} A$ satisfying $rs = 1_A$. Mental picture: View s as a sampling of B by A:

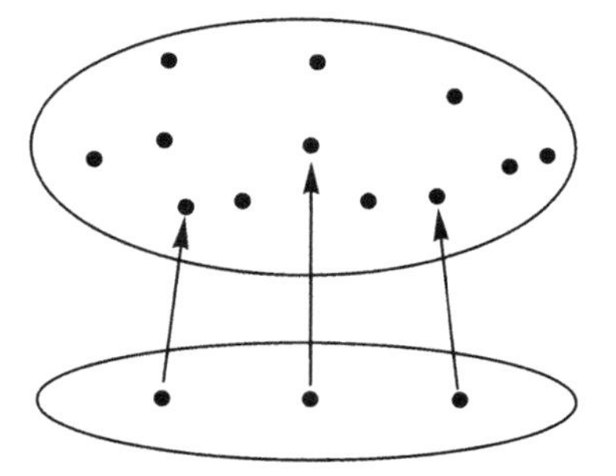

Then choose for each unidentified bird the most similar bird which is identified in the manual s:

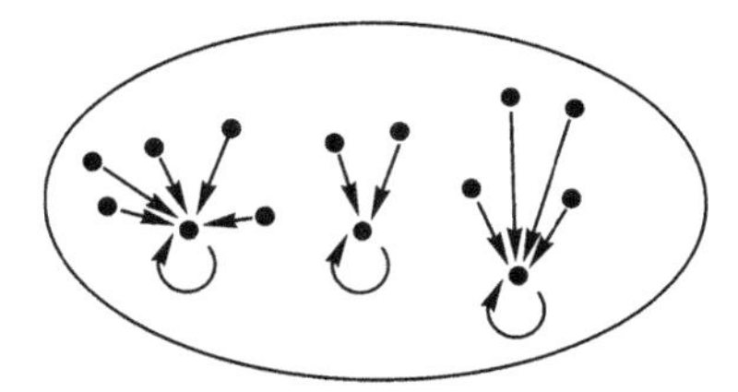

This constructs the idempotent e, clustering the birds around the sample birds, but then r is easy to find. While we're here, we should see whether we can calculate the number of solutions to the bird-watcher's problem. Suppose there were a thousand birds and only three species, so the sampling map would look like this:

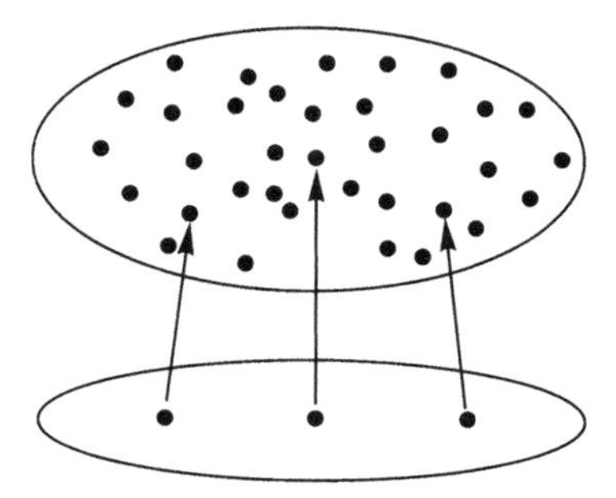

How many retractions are there for this map?

DANILO: The three birds that are sampled have to go back where they came from, but for the rest it is just any map to the three-element set of species; so there are $3^{(1000-3)}$ or 3^{997} retractions for s.

Good. You can see that Danilo's method finds the number of retractions for any map of sets that has any, i.e. any map s that 'preserves distinctness': if $x \neq y$ then $sx \neq sy$.

Child's problem: Given B, choose a map $B \xrightarrow{e} B$ satisfying $ee = e$. Having watched children for years, I remain as puzzled as ever about the selection of the idempotent endomap e associating to each animal the most familiar animal it resembles. After that's done, though, the rest of the job (splitting the idempotent) is easy:

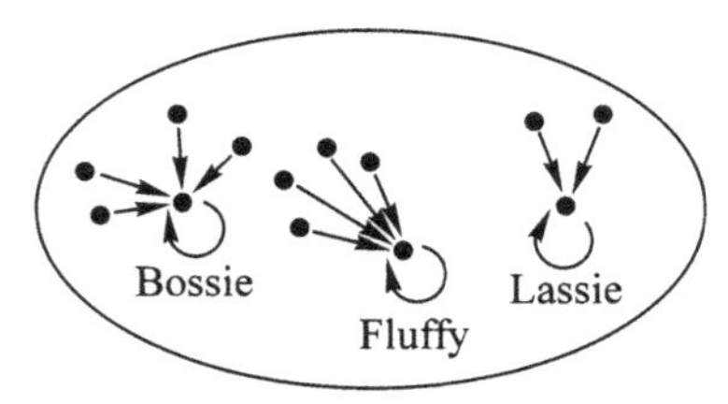

The 'sorts' are all there in the mental picture, and all that is needed is to learn the names 'cow,' 'cat,' and 'dog' for these sorts. As a picture of the actual learning process, this description is surely oversimplified, because the selection of the idempotent map and the learning of the sort-names go on concurrently.

5. Comparing infinite sets

There is one glaring omission from our account of 'same size' (isomorphism) and 'at most as big as' (retract). From our experience with finite sets, we would expect that if both $A \leq_R B$ and $B \leq_R A$ then $A \cong B$. Surprisingly, this does not follow from just the associative and identity laws: there are categories in which it is false. Its truth for (especially infinite) sets is the 'Cantor–Bernstein Theorem'. Indeed this subject pretty much began with Georg Cantor (1845–1918) who in studying the analysis of the sound wave from a violin string (or any periodic motion) into its various frequencies, found it necessary to explore the sizes of infinite sets. His striking discoveries have barely been introduced here; and we have also neglected Galileo's earlier discovery of a characteristic feature of infinite sets: a set can be isomorphic to a proper part of itself, as the isomorphism

$$\{0, 1, 2, 3, \ldots\} \xrightarrow{f} \{0, 2, 4, 6, \ldots\}$$

by $fn = 2n$ shows. Cantor's ideas, while developed for infinite sets, have proved equally useful in other categories, leading for example to 'incompleteness theorems' in logic (see Session 29). In Session 10, we will see how important it is in other categories to know, for certain special objects A and B, whether $A \leq_R B$ or not.

The expression 'each set is number' refers to the attitude toward sets in general whereby we consider relationships like $\cong$ or $\leq_R$, but don't worry about the details of the particular 'proofs,' f and f^{-1} or r and s, of these relationships. One shouldn't always neglect these; we'll see later that a given proof that A has the same number as itself (an automorphism of A) is a rich structure that needs many numbers to describe it.

Quiz

1. Give an example of two explicit sets A and B and an explicit map $A \xrightarrow{f} B$ satisfying *both*:

   (a) there is a retraction for f, *and*

   (b) there is *no* section for f.

   Then explain how you know that f satisfies (a) and (b).

2. If $C \underset{q}{\overset{p}{\rightleftarrows}} D$ satisfy $p \circ q \circ p = p$, can you conclude that

   (a) $p \circ q$ is idempotent? If so, how?

   (b) $q \circ p$ is idempotent? If so, how?

Optional questions

2*. If $C \underset{q}{\overset{p}{\rightleftarrows}} D$ satisfy $p \circ q \circ p = p$, use the given maps p and q to devise a map q' satisfying *both*:

$$p \circ q' \circ p = p$$

and

$$q' \circ p \circ q' = q'$$

(and explain how you know that your q' has these properties.)

1*. Same question as Problem 1 at top of page, except that both sets A and B are required to be *infinite* sets.

How to solve the quiz problems

I have written down the thoughts that might reasonably go through your mind in trying to solve the problems, to show how you might arrive at a solution; and then how a solution might look. You will notice that the thought-process looks long, if you write it all down, and by comparison, a solution, after you finally find it, seems brief.

Exactly how you use all this to help you learn how to solve problems is, of course, up to you. I suggest reading just a little at a time of the description of the possible thought process, and then returning to the problem to see if you are able to finish it without reading the rest. Afterwards, you can compare the way you arrived at a solution with the way this imaginary student did, and perhaps learn some new strategies to add to your techniques for thinking about problems.

Problem 1

Let's see ... I'm asked to *pick* an awful lot of stuff out of the air here. I have to pick the sets A and B, *and* the map f. My problem is that I know a lot of examples of sets, and I know a lot of examples of maps from one set to another – which should I choose?

First, I had better decide how big I need to make these sets – the smaller I can make them, the better! What do I want? $A \xrightarrow{f} B$ is supposed to be chosen so there is a *retraction* for f – let me give that retraction for f a name – maybe 'r' would be a good letter, to remind me that it is supposed to be a retraction for f. Let me add r to the picture $A \xrightarrow{f} B$. Which way must r go? That's easy – any retraction *or* section for f goes backwards from f. So my 'external diagram' is going to look like this:

$$A \underset{r}{\overset{f}{\rightleftarrows}} B$$

But *of course* I have to remember the *definition* of the phrase 'r is a retraction for f.' I *memorized the definitions*: 'r is a retraction for f' *means* $r \circ f = 1_A$. (I heeded the warning that was repeated so often in class that I had better learn by heart the difference between 'g is a retraction for f' and 'g is a section for f', since they mean *different* things!)

Now I want to pick two sets A and B and two maps r and f, arranged as in the box above; but *not just any* two maps, they have to satisfy the equation

$$\boxed{r \circ f = 1_A}$$

If I remember correctly, a retraction for a map tends to go from a bigger set to a smaller one. I ought to choose my sets so that B is at least as big as A. In fact, I think it would be safer to choose it a bit bigger. Maybe if I try taking A with no members, and B with one member, it would work?

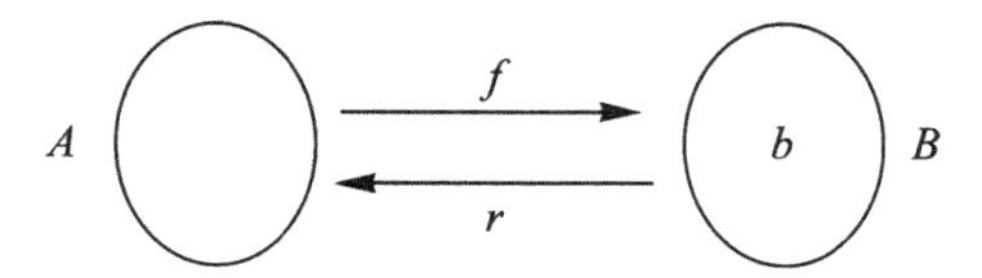

WOOPS! I can't possibly have a map r that goes from B to A, because there is no member in A to be $r(b)$!

Try again: maybe A with one member, B with two members

I still must make up my maps r and f. There is only *one* map from B to A ($1^2 = 1$), there is no choice for r; it looks like:

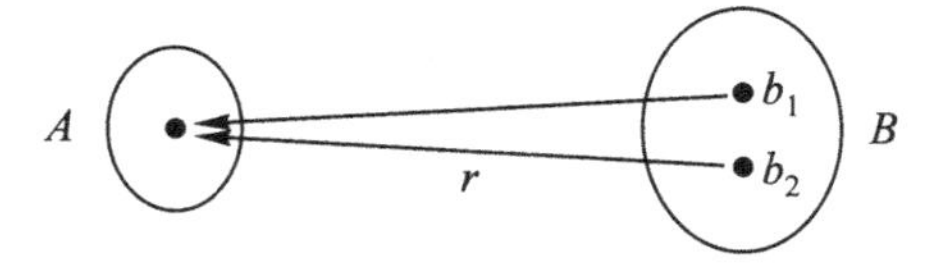

What about picking my map

$$A \xrightarrow{f} B$$

Here there are two choices, I will just pick one, since they seem to look rather alike, anyway.

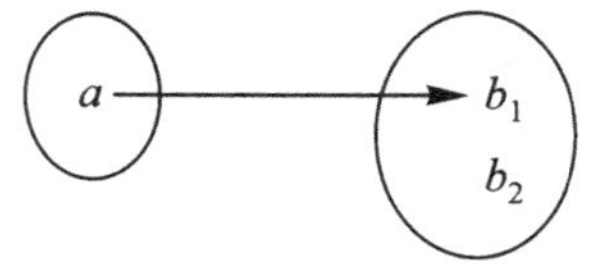

Did this work? I was supposed to choose r and f so that

$$r \circ f = 1_A$$

That means I need, for *each* member of A, that

$$r(f(\textit{that member})) = \textit{that member}$$

But there is only *one* member in A! What I need is $r(f(a)) = a$, that's all. Let's check it: $r(f(a)) = r(b_1) = a$. Yes, it's true!

(Actually, now I realize that I didn't even have to check it! There is only *one* map from A to A ($1^1 = 1$); any two maps, like $r \circ f$ and 1_A, from A to A *have to* be the same map.)

Have I finished? Let me reread the problem .... Yes, I have done everything, *except* showing that there is no section for f. How do I do that? Well, a *section for* f would be a map $B \xrightarrow{s} A$ satisfying $f \circ s = 1_B$. (Good thing I learned the definitions!) Is there a map s that satisfies that equation? Well, there is only *one map* from B to A, my s would have to be that one. And I have already named that 'r'. I need to know:

$$\text{Is } f \circ r = 1_B \text{ or not?}$$

That would say

$$f(r(b_1)) = b_1 \quad (*)$$

and

$$f(r(b_2)) = b_2 \quad (**)$$

Are these true? $f(r(b_1)) = f(a) = b_1$, $(*)$ is true; and $f(r(b_2)) = f(a) = b_1$, $(**)$ is false!

Therefore this map r is *not* a section for f; and it was the *only* map from B to A. Thus f has *no* section. Too bad! No, wait ... that's what I wanted! f has a retraction, but it does not have any section. GOOD! Maybe, just to make it prettier, and to satisfy this fussy professor who asked me to make the sets and maps 'explicit', I will give the example with 'concrete' sets, but keep the picture I drew with

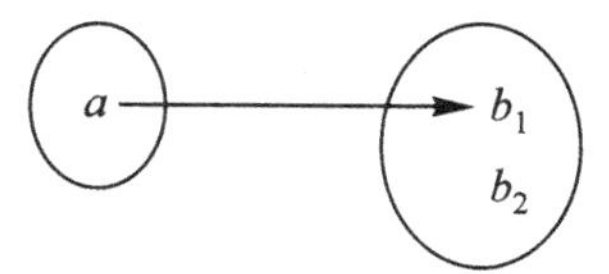

Take A to be the set whose only member is Ian; and take $B = \{Katie, Sheri\}$ and $A \xrightarrow{f} B$ to be given by $f(Ian) = Katie$.

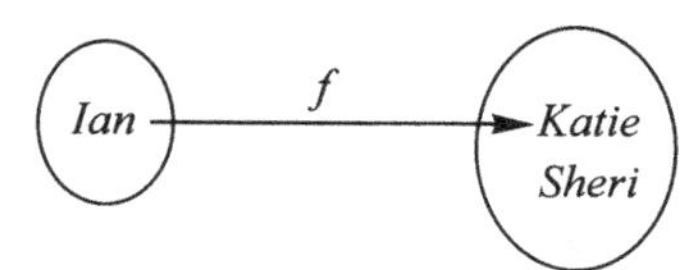

Here is how a good solution would look, written out:

I choose

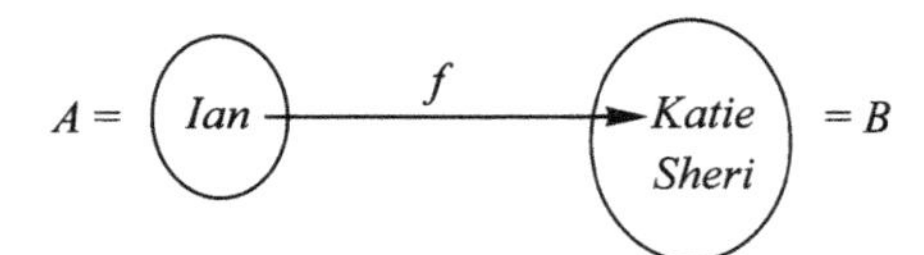

First: I claim f has a retraction r, a map $B \xrightarrow{r} A$ satisfying $r \circ f = 1_A$. What is it? r is the only map from B to A! And since $r \circ f$ and 1_A are both maps from A to A, and there is only one map from A to A, these two maps must be the same: $r \circ f = 1_A$. Here is a picture of r, if you want:

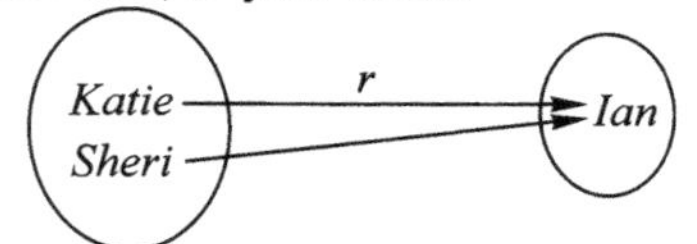

Second: I claim there is no section for f. A section for f would be a map $B \xrightarrow{s} A$ satisfying $f \circ s = 1_B$. The *Theorem on Uniqueness of Inverses* said: *If s is a section for f and r is a retraction for f then* $r = s$. So the only *possible* section for f is r! And r is *not* a section for f, since $f(r(Sheri)) = f(Ian) = Katie$, shows that $f \circ r \neq 1_{\{Katie,Sheri\}}$. QED

Note: After finding his original solution, this student found an *alternative* argument to show that this f has no section. *Either* argument would have been fine; but this one is maybe slightly better, because it employs a *general principle*: If you know that a map $A \xrightarrow{f} B$ has a retraction r, then the only possible section for f is r itself; if r is not a *section* for f, then f has no section!

Problem 2(a)

This one looks easier: I don't have to invent everything myself. What I know is that

$$C \underset{q}{\overset{p}{\rightleftarrows}} D$$

(read 'p is a map from C to D and q is a map from D to C) and that

$$p \circ q \circ p = p$$

What do I need to find out? I need to see that $p \circ q$ is *idempotent*; of course I need to know what it *means* to say that a map is 'idempotent.' Fortunately I learned that a map e is idempotent if it satisfies the equation

$$e \circ e = e$$

That seemed a bit peculiar to me at first, since usually if you have a map $A \xrightarrow{f} B$, '$f \circ f$' *doesn't make any sense*. You can follow one map by another, like $f \circ g$, only if the domains and codomains match up properly:

$$X \xrightarrow{g} Y \xrightarrow{f} Z.$$

The only case in which $f \circ f$ makes sense is if the domain and codomain of f are the *same set*, like

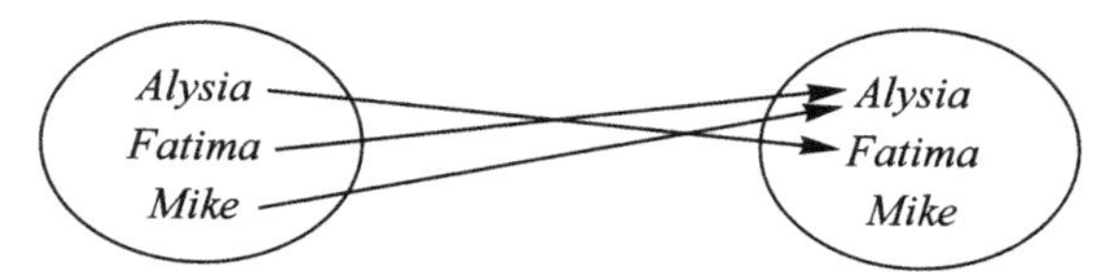

That is, f has to be – what was the word? – an ***endomap***. Only *endomaps* have a chance to be *idempotent*; and even then, *most* endomaps are *not* idempotent. Just to be sure they're not trying to trick me, I had better check: is $p \circ q$ even an *endomap*? Well, its domain is – let's see, q was done *first*, so the domain of $p \circ q$ is the domain of q, which was D. And the codomain of $p \circ q$ is the codomain of p, which was ... yes, D. Since $D \xrightarrow{p \circ q} D$ *is* an endomap, at least it has a *chance* to be idempotent. Let's write down exactly what it is that I want to show about $p \circ q$. I need to see that if you follow this complicated map by itself, you get it back again; i.e. I need to show:

$$\boxed{(p \circ q) \circ (p \circ q) \stackrel{?}{=} p \circ q}$$

What I *know* is: $p \circ q \circ p = p$.

My problem boils down to:

KNOW: $\boxed{p \circ q \circ p = p}$ $*$

WANT TO SHOW: $\boxed{(p \circ q) \circ (p \circ q) = p \circ q}$ $**$

That should be pretty easy – I have done problems like this before. Here is my solution:

$(p \circ q) \circ (p \circ q) = p \circ q \circ p \circ q$	(I can omit the parentheses)
$= (p \circ q \circ p) \circ q$	(I put parentheses back in to use $*$)
$= p \circ q$	(by $*$)

Therefore, $p \circ q$ is idempotent. QED

Now try Problem 2(b) yourself.

Composition of opposed maps

We should work through some examples of composition of maps of sets. While the algebra of composition is very simple, involving only the associative and identity laws, the understanding of how this algebra is applied is greatly aided by practice with concrete examples, first in the category of sets and later in richer categories.

Let's consider the following maps:

$$Men \underset{\text{father}}{\overset{\text{mother}}{\rightleftarrows}} Women$$

One of them assigns to each man his mother, and the other assigns to each woman her father. What is the composite $mother \circ father$, or more briefly $g \circ f$, where $g = mother$ and $f = father$? For example, let's ask Sheri: Who is $g \circ f \circ Sheri$? First you have to decide who is $f \circ Sheri$.

SHERI: My father is Mike.

And who is $g \circ Mike$?

SHERI: My father's mother was Lee.

Good, so $g \circ f \circ Sheri = Lee$. Is the map $f \circ g \circ f$ equal to the map f? How do we test whether two maps of sets are equal?

CHAD: When the same input gives the same output.

So what about these two maps?

ALYSIA: They are equal.

Really? Let's calculate both for the input

$$\mathbf{1} \xrightarrow{Alysia} Women$$

Who is $f \circ Alysia$?

ALYSIA: Rocco.

And who is $g \circ Rocco$?

ALYSIA: Dolores.

And $f \circ Dolores$?

ALYSIA: I don't remember his first name, but his family name was *R*.

All right, so $f \circ g \circ f \circ Alysia = Mr\,R$. Is $f \circ g \circ f = f$? Is $Mr\,R. = Rocco$?

ALYSIA: No, $f \circ g \circ f$ and f are different maps.

Right. There is an input for which they give different outputs, so they are different. Notice that our *test for equality of maps* $A \overset{f}{\underset{h}{\rightrightarrows}} B$ *of* sets

$$\boxed{f = h} \quad \text{if and only if} \quad \boxed{f \circ a = h \circ a \text{ for every point } \mathbf{1} \xrightarrow{a} A}$$

is equivalent to the following:

$$\boxed{f \neq h} \quad \text{if and only if} \quad \boxed{\text{for at least one } \mathbf{1} \xrightarrow{a} A \quad f \circ a \neq h \circ a}$$

Any such element a for which f is different from h is a *counterexample* proving the difference of f and h. So, Alysia, what is the counterexample proving that f and $f \circ g \circ f$ are different?

ALYSIA: Me.

Right. Because you are the member of the set of women for which we verified

$$f \circ g \circ f \circ Alysia \neq f \circ Alysia$$

In fact for these two maps $f \circ g \circ f \circ x \neq f \circ x$ for *every* woman x, since otherwise we would have the biologically impossible situation that the father of x, $y = f \circ x$, would satisfy $y = f \circ g \circ y$; he would be his mother's father. Often a composite map has a special name because of its importance. With $g = mother$ and $f = father$, the composite $g \circ f$, *mother of father*, is called 'paternal grandmother.' Notice how often it is possible to read the symbol '$\circ$' as 'of,' instead of 'following.' This is true also of the symbol '$\times$' for multiplication of numbers. We usually read it as 'times', but for fractional factors, as in $\frac{2}{3} \times 6 = 4$, we often say, 'Two-thirds *of* six is four.'

Summary/quiz on pairs of 'opposed' maps

0. 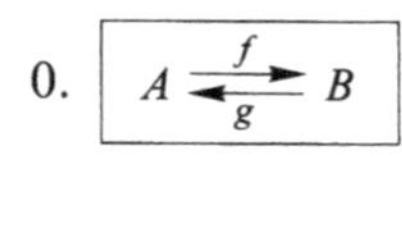

Fill in the blanks; when **vs?** occurs, cross out the false alternative

1. Given two maps f, g with domains and codomains as above, we can (sometimes **vs?** always) form the composites $g \circ f$ and $f \circ g$. All we can say about $g \circ f$ and $f \circ g$ as maps in themselves is that they are ____________.

2. If we know that g is a retraction for f, that means $g \circ f$ is actually ____________; then we can prove that $f \circ g$ is not only an ____________, but actually an ____________. The latter means that the equation ____________ is true.

3. If we even know that f is an isomorphism *and* that $g \circ f = 1_A$, then $f \circ g$ is not only an idempotent, but is ____________. If, moreover, s is a map for which $f \circ s = 1_B$, we can conclude that $s =$ ____________.

4. Going back to 0, i.e. assuming no equations, but only the domain and codomain statements about f and g, the composite $f \circ g \circ f$ (could be different from **vs?** must be the same as) f. Likewise $f \circ g \circ f \circ g$ (could be different from **vs?** must be the same as) $f \circ g$.

Summary: On the equation $p \circ j = 1_A$

If maps $A \xrightarrow{j} X \xrightarrow{p} A$ satisfy (*) $p \circ j = 1_A$, several **consequences** follow:

*In **any** category*	*In the category of **finite sets***
The endomap $X \xrightarrow{j \circ p} X$ (call it 'α' for short) satisfies $\alpha \circ \alpha = \alpha$; we say α is ***idempotent***. Written out in full, this is $(j \circ p) \circ (j \circ p) = (j \circ p)$. We will see more consequences later.	(1) p satisfies: for each member a of A, there is at least one member x of X for which $p(x) = a$; (We say p is ***surjective***.) (2) j satisfies: *if* $j(a_1) = j(a_2)$, then $a_1 = a_2$; (We say j is ***injective***.) (3) $\#A \leq \#X$, and if $\#A = 0$, then $\#X = 0$ too!

Problems involving the equation (*): (four types)

X; $?j \uparrow \downarrow p$; A	*Given* $X \xrightarrow{p} A$, *find* all $A \xrightarrow{j} X$ satisfying (*). Such a j is called a ***section*** for p.	In finite sets, j is also called a 'choice of representatives' for p. Unless p is surjective, there will be *no* sections for p. More generally, the *number* of sections for p is $\prod_{1 \xrightarrow{a} A} \#(p^{-1}a)$ ('Chad's formula').
X; $j \uparrow \downarrow p?$; A	*Given* $A \xrightarrow{j} X$, *find* all $X \xrightarrow{p} A$ satisfying (*). Such a p is called a ***retraction*** for j.	In finite sets, unless j is injective, there will be *no* retractions for j. If j is injective, the *number* of retractions for j is $(\#A)^{(\#X - \#A)}$ ('Danilo's formula').
X; $?j \uparrow \downarrow p?$; A	*Given* only X and A, *find* all p, j satisfying (*). If there is at least one such pair we say A is a ***retract*** of X (via p and j) and sometimes write '$A \leq X$'.	Unless $\#A \leq \#X$, there can be no such pairs p, j, i.e. A cannot be a retract of X. The formula for the number of pairs p, j in terms of $\#X$ and $\#A$ is rather complicated.
$X \circlearrowleft \alpha$; $?j \uparrow \downarrow p?$; $?A$	*Given* only an endomap $X \xrightarrow{\alpha} X$ *find* an A and j, p satisfying (*) *and* $j \circ p = \alpha$. Such a pair p, j is called a ***splitting*** for α. Unless α is idempotent, there cannot be a splitting for α.	In the category of finite sets, for each idempotent endomap α there is a splitting p, j. The number of elements of the desired A turns out to be the number of ***fixed points*** of α (elements x of X satisfying $\alpha(x) = x$).

Review of ‘I-words’

Identity map: For each object X there is an identity map $X \xrightarrow{1_X} X$. It satisfies $1_X f = f$ and $g 1_X = g$ whenever (the domains and codomains match, so that) the left side is defined.

Inverse, isomorphism: ‘Inverse’ is the basic word, and involves *two* maps

$$A \underset{g}{\overset{f}{\rightleftarrows}} B$$

To say that ‘g is an inverse for f’ means $fg = 1_B$ *and* $gf = 1_A$. If f has an inverse, it has only one, and we call that one f^{-1}. If f has an inverse, we say that f is an **isomorphism**.

Here is an analogy to get the grammar straight:

MAPS	PEOPLE
g is the INVERSE for f.	Ginger is the SPOUSE of Fred.
Not all maps have inverses, but a map can't have two inverses.	Not all people have spouses, but you are not allowed to have two spouses.
f is an ISOMORPHISM	Fred is MARRIED
Meaning: there is some g which is an inverse for f, in fact exactly one.	Meaning: there is some person who is a spouse for F., in fact exactly one.
f^{-1} (the inverse of f)	Fred's spouse
It is forbidden to use this as a name of a map, unless f has an inverse.	It is forbidden to use this to specify a person, unless Fred has a spouse.
$(f^{-1})^{-1} = f$	The spouse of the spouse of F. is F.
More precisely, if f has an inverse, then f^{-1} also has an inverse, namely f.	More precisely, if F. has a spouse, then that spouse also has a spouse, namely F.

Idempotent, involution: (see Article III) Both are properties that only an endomap $A \xrightarrow{f} A$ can have, since they involve $f \circ f$.
If $f \circ f = f$, we say f is (an) **idempotent**.
If $f \circ f = 1_A$, we say f is an **involution**.

Remarks: The only idempotent which has an inverse is an identity map. Every involution has an inverse, namely itself.

Test 1

1. Throughout this problem

$$A = \begin{pmatrix} \textit{Mara} \\ \textit{Aurelio} \\ \textit{Andrea} \end{pmatrix}$$

(a) Find an *invertible* map $A \xrightarrow{f} A$, different from the identity map 1_A.

(b) Find an *idempotent* map $A \xrightarrow{e} A$, different from the identity map 1_A.

(Draw the 'special internal diagrams' of your maps f and e – the diagrams that are available only for endomaps.)

(c) Find another set B and two maps

$$B \underset{r}{\overset{s}{\rightleftarrows}} A$$

for which $r \circ s = 1_B$ and $s \circ r = e$.

(Draw the internal diagrams of r and s. In this part, e is still the map you chose in part (b).)

2. $\mathbb{R}$ is the set of all real numbers, and $\mathbb{R} \xrightarrow{f} \mathbb{R}$ is the map given by the explicit formula $f(x) = 4x - 7$ for each input x. show that f has an inverse map. To do this, give an explicit formula for the inverse map g, and then show that

(a) $(g \circ f)(x) = x$ for each real number x, and that

(b) $(f \circ g)(x) = x$ for each real number x.

SESSION 10

Brouwer's theorems

1. Balls, spheres, fixed points, and retractions

The Dutch mathematician L.E.J. Brouwer (1881–1966) proved some remarkable theorems about 'continuous' maps between familiar objects: circle, disk, solid ball, etc. The setting for these was the 'category of topological spaces and continuous maps.' For our purposes it is unnecessary to have any precise description of this category; we will instead eventually list certain facts which we will call 'axioms' and deduce conclusions from these axioms. Naturally, the axioms will not be selected at random, but will reflect our experience with 'cohesive sets' (sets in which it makes sense to speak of closeness of points) and 'continuous maps.' (Roughly, a map f is continuous if $f(p)$ doesn't instantaneously jump from one position to a far away position as we gradually move p. We met this concept in discussing Galileo's idea of a continuous motion of a particle, i.e. a continuous map from an interval of time into space.) There is even an advantage in not specifying our category precisely: our reasoning will apply to any category in which the axioms are true, and there are, in fact, many such categories ('topological spaces', 'smooth spaces', etc.).

We begin by stating Brouwer's theorems and by trying to see whether our intuition about continuous maps makes them seem plausible. First we describe the *Brouwer fixed point theorems.*

(1) *Let I be a line segment, including its endpoints (I for Interval) and suppose that $f\colon I \longrightarrow I$ is a continuous endomap. Then this map must have a fixed point: a point x in I for which $f(x) = x$.*

Example: Suppose that I is an interval of time, and that R is an interval of road, say the highway from Buffalo to Rochester. Suppose that two cars drive on this road. The first car drives at a constant speed from Buffalo to Rochester, so its motion is described by $I \xrightarrow{u} R$ (u for 'uniform' motion). Meanwhile, the second car starts anywhere along the road and just travels aimlessly along, perhaps occasionally parking for a while, then retracing its path for a while, and ending its journey at any point along the road. Let's denote the motion of this second car by $I \xrightarrow{m} R$. Now u is an invertible map, so we get $R \xrightarrow{u^{-1}} I$, and let $I \xrightarrow{f} I$ be the composite $f = u^{-1} \circ m$. Brouwer's theorem tells us that there must be some time t in I at which $f(t) = t$; that is, $u^{-1}mt = t$; so $mt = ut$, which says there is some time t at which the two cars are at the same point on the road. This seems not very surprising; if the first car drives from Buffalo to Rochester and the second car is always on the road, then of course the first car must at some time meet the second.

The next theorem is similar, but about a disk instead of an interval, and I find it much less obvious.

(2) *Let D be a closed disk (the plane figure consisting of all the points inside or on a circle), and f a continuous endomap of D. Then f has a fixed point.*

Example: Rotating the disk by a certain angle gives a continuous endomap of the disk; f could be the process 'turn 90 degrees.'

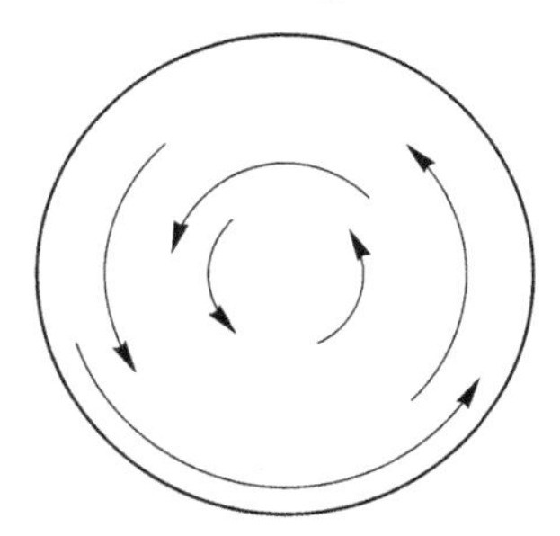

ALYSIA: What about the center?

Exactly! That is a fixed point. For this map it was easy to see that it has a fixed point, but for other maps it may not be so easy; yet the theorem says that as long as f is continuous, it will have at least one fixed point. This theorem seems to me much more surprising than the previous one.

Example: Suppose my disk is a portion of the Washington DC area, say the part inside or on the circular beltway. I also bring a map of the region, drawn on a piece of paper P. My map is thus a continuous map $D \xrightarrow{m} P$. If I am so callous as to crumple up the map and throw it out of the car window, so that it lands inside the beltway, I get an additional continuous map $P \xrightarrow{p} D$ (p for 'projection'), assigning to each point on the crumpled paper the point on the ground directly under it. Brouwer's theorem, applied to the map $f = p \circ m$: $D \longrightarrow D$, tells me that some point x inside the beltway is directly under the point $m(x)$ that represents x on the map. Do you find that surprising? I did when I first heard it. You can try the experiment, but please pick up the map afterward.

If it occurred to you that a perfect map would show every detail of the area, even including a picture of the discarded map, congratulate yourself. You have discovered the idea behind *Banach's fixed point theorem* for 'contraction' maps. You only have to go a step further: the discarded map has a small picture of the discarded map, and that picture has a smaller picture which has a smaller picture .... These pictures gradually close in on the one and only fixed point for our endomap. This beautifully simple idea only works for an endomap which shrinks distances, though. Brouwer's theorem applies to *every* continuous endomap of the disk.

Example: Here is a map to which Brouwer's theorem applies and Banach's doesn't. Suppose D is a disk-shaped room in a doll's house, and F is a larger-than-life floor plan of that room; we crumple F and discard it on D as before. The composite map $D \xrightarrow{m} F \xrightarrow{p} D$ this time will not shrink all distances, so the Banach idea doesn't apply. (In fact $p \circ m$ may have many fixed points, but they are not so easy to locate. It often happens that if a problem has only one solution, it's easy to find it; but if there are many solutions, it's hard to find even one of them.) The next theorem is about .... Any guess?

FATIMA: A ball?

Exactly! A solid ball. It says the following:

(3) *Any continuous endomap of a solid ball has a fixed point.*

To imagine an endomap, think of deforming the ball in any arbitrary way, but without tearing it.

DANILO: Something like folding dough?

Yes, but without breaking it into separate pieces. I find it easier to imagine this endomap of the ball if I first have *two* 'objects,' a wad W of dough and a ball-shaped region B in space. Then I can use two maps from W to B: a 'uniform' placement $u : W \longrightarrow B$ in which the wad W exactly fills the region B, and the new placement after kneading the dough, $p : W \longrightarrow B$. Now u is invertible, and the endomap we want is pu^{-1}. It assigns to each point in the region the new location of the point in the dough that was originally there; it's a sort of 'change of address' map.

Now we describe the sequence of theorems known as *Brouwer retraction theorems.*

(I) *Consider the inclusion map $j : E \longrightarrow I$ of the two-point set E as boundary of the interval I. There is no continuous map which is a retraction for j.*

Recall that this means there is no continuous map $r : I \longrightarrow E$ such that $r \circ j = 1_E$.

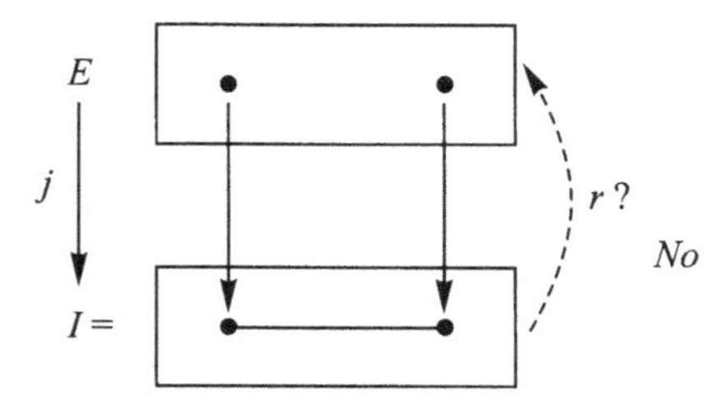

In other words, it is not possible to map the interval continuously to its two endpoints and leave the endpoints in place. Isn't this reasonable? Isn't it pretty obvious that one cannot put one part of the interval on one of its endpoints and another part of it on the other without tearing it?

The next retraction theorem is about the disk and its boundary.

(II) *Consider the inclusion map $j : C \longrightarrow D$ of the circle C as boundary of the disk D into the disk. There is no continuous map which is a retraction for j.*

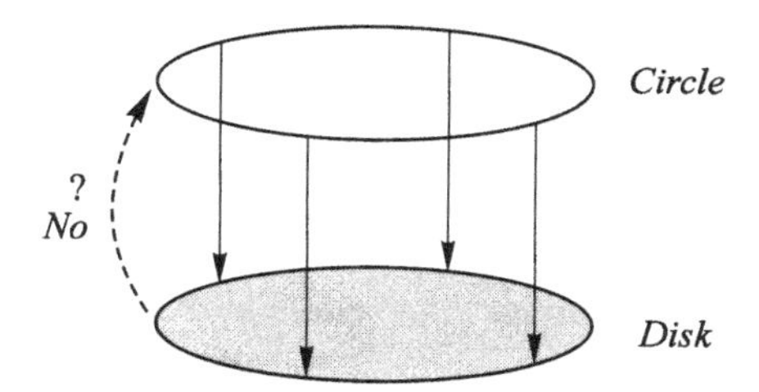

Again, this should seem quite reasonable. Suppose we have a drum made of a very flexible stretchable sheet. To get a retraction for the inclusion of the boundary we might imagine taking the sheet and squeezing it into the rim but without moving its boundary. One would think that this is not possible without puncturing or tearing the sheet. What this retraction theorem says is that this thought is correct.

The third retraction theorem is, as you can imagine, about the ball and its boundary (the sphere).

(III) *Consider the inclusion $j : S \longrightarrow B$ of the sphere S as boundary of the ball B into the ball. There is no continuous map which is a retraction for j.*

Now, here is the point about all these theorems: (1) and (I) are actually equivalent theorems, and so are the Theorems (2) and (II), and also Theorems (3) and (III). In other words, after proving the retraction theorems, which seem so reasonable, Brouwer could easily get as a consequence the fixed point theorems (which seem much less intuitive). We shall illustrate this by showing how Brouwer proved that (II) implies (2), and we'll leave the other cases for you to think about.

Let's write clearly what Brouwer promised to show:

> *If there is no continuous retraction of the disk to its boundary then every continuous map from the disk to itself has a fixed point.*

However, Brouwer did not prove this directly. Instead of this he proved the following:

> *Given a continuous endomap of the disk with no fixed points, one can construct a continuous retraction of the disk to its boundary.*

This is an example of the *contrapositive* form of a logical statement. The contrapositive form of 'A implies B' is '*not* B implies *not* A,' which conveys exactly the same information as 'A implies B,' just expressed in a different way. Below is an example of how it is used.

2. Digression on the contrapositive rule

A friend of mine, Meeghan, has many uncles. All of Meeghan's uncles are doctors. In Meeghan's world

$$\textit{uncle} \xrightarrow{(\textit{implies})} \textit{doctor} \qquad \text{(PARTICULAR SITUATION)}$$

I went to her wedding and met some of them. There I had an interesting discussion with an intelligent man who I thought was another uncle, but in the course of the conversation he said that he was a mechanic. So I thought

$$\textit{mechanic} \xrightarrow{(\textit{implies})} \textit{not doctor} \qquad \text{(GENERAL KNOWLEDGE about our society)}$$

$$\textit{not doctor} \xrightarrow{(\textit{implies})} \textit{not Meeghan's uncle} \qquad \text{(CONTRAPOSITIVE of what is known of the particular situation)}$$

Therefore this man is not one of Meeghan's uncles.

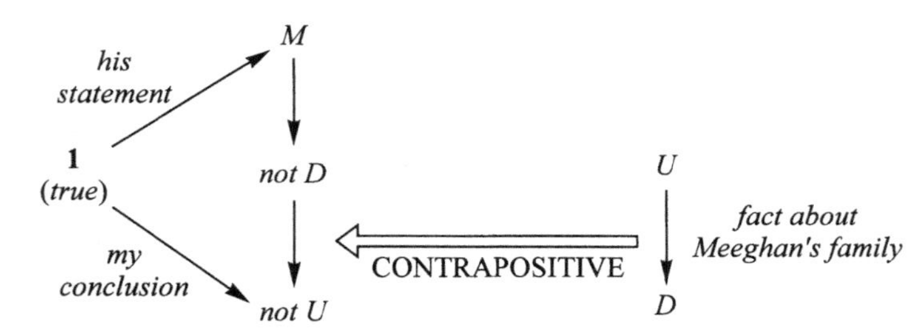

3. Brouwer's proof

We return to Brouwer's theorems. To prove that the non-existence of a retraction implies that every continuous endomap has a fixed point, all we need to do is to assume that there is a continuous endomap of the disk which does not have any fixed point, and to build from it a continuous retraction for the inclusion of the circle into the disk.

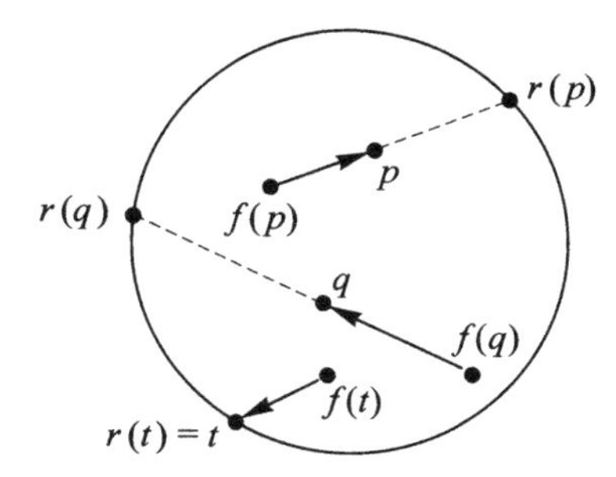

So, let $j : C \longrightarrow D$ be the inclusion map of the circle into the disk as its boundary, and let's assume that we have an endomap of the disk, $f: D \longrightarrow D$, *which does not have any fixed point*. This means that for every point x in the disk D, $f(x) \neq x$.

From this we are going to build a retraction for j, i.e. a map $r: D \longrightarrow C$ such that $r \circ j$ is the identity on the circle. The key to the construction is the assumed property of f, namely that for every point x in the disk, $f(x)$ is different from x. Draw an arrow with its tail at $f(x)$ and its head at x. This arrow will 'point to' some point $r(x)$ on the boundary. When x was already a point on the boundary, $r(x)$ is x itself, so that r is a retraction for j, i.e. $rj = 1_C$.

Two things are worth noting: first, that sometimes something that looks impossible or hard to prove may be easily deduced from something that looks much more reasonable and is, in fact, easier to prove; and second, that to know that a map has no retraction often has very powerful consequences.

The reasoning leading to the proof of Brouwer's fixed point theorem can be summarized in the following diagram:

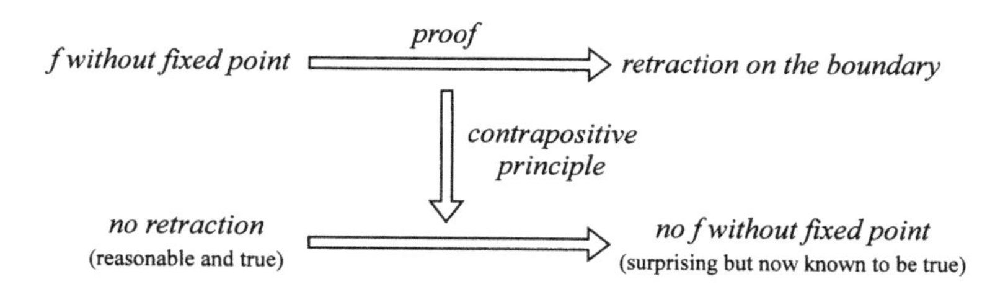

DANILO: Your conclusion sounds peculiar. Instead of 'every f has a fixed point,' you get 'there is no f without fixed point.'

You are right. We need to use another principle of logic, that *not*(*not A*) implies A, to reach 'every f has a fixed point.' Brouwer himself seriously questioned this rule of logic; and we will later see that there are examples of useful categories in whose 'internal' logic this rule does not hold. (This 'logical' difficulty turns out to be connected with the difficulty of actually locating a fixed point for f, if f is not a 'contraction map'.)

4. Relation between fixed point and retraction theorems

Exercise 1:
Let $j : C \longrightarrow D$ be, as before, the inclusion of the circle into the disk. Suppose that we have two continuous maps $D \overset{f}{\underset{g}{\rightrightarrows}} D$, and that g satisfies $g \circ j = j$. Use the retraction theorem to show that there must be a point x in the disk at which $f(x) = g(x)$. (Hint: The fixed point theorem is the special case $g = 1_D$, so try to generalize the argument we used in that special case.)

I mentioned earlier that each retraction theorem is equivalent to a fixed point theorem. That means that not only can we deduce the fixed point theorem from the retraction theorem, as we did, but we can also deduce the retraction theorem from the fixed point theorem. This is easier, and doesn't require a clever geometrical construction. Here is how it goes.

Exercise 2:
Suppose that A is a 'retract' of X, i.e. there are maps $A \overset{s}{\underset{r}{\rightleftarrows}} X$ with $r \circ s = 1_A$. Suppose also that X has the fixed point property for maps from T, i.e. for every endomap $X \xrightarrow{f} X$, there is a map $T \xrightarrow{x} X$ for which $fx = x$. Show that A also has the fixed point property for maps from T. (Hint: The proof should work in any category, so it should only use the algebra of composition of maps.)

Now you can apply Exercise 2 to the cases: T is $\mathbf{1}$ (any one-point space), X is the interval, the disk, or the ball, and A is its boundary (two points, circle, or sphere.) Notice that in each of these cases, there is an obvious 'antipodal' endomap a of A, sending each point to the diametrically opposite point; and a has no fixed point.

Exercise 3:
Use the result of the preceding exercise, and the fact that the antipodal map has no fixed point, to deduce each retraction theorem from the corresponding fixed point theorem.

In solving these exercises, you will notice that you have done more than was required. For example, from the fixed point theorem for the disk, you will have concluded not only that the inclusion map $C \longrightarrow D$ has no retraction, but also that C is not a retract of D (by *any* pair of maps.) In fact, the argument even shows that *none* of E, C, S, is a retract of *any* of I, D, B.

You will probably have noticed that the same reasoning is used in all dimensions; for instance, Exercise 1 applies to the interval or ball as well as the disk. In the next section we state things for the 'ball' case, but draw the pictures for the 'disk' case.

5. How to understand a proof: The objectification and 'mapification' of concepts

You may have felt that none of our reasoning about Brouwer's theorems was valid, since we still have no precise notion of 'continuous map.' What we wish to do next is to extract those properties which are needed for our reasoning, and see that our conclusions are valid in any category in which these properties (which we will call Axiom 1 and Axiom 2) hold.

Brouwer introduces in his proof, besides the sphere S and ball B and the inclusion map $S \xrightarrow{j} B$, several new *concepts*:

1. *arrows* in B:

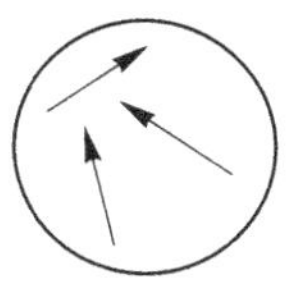

2. each arrow has a *head*, in B:

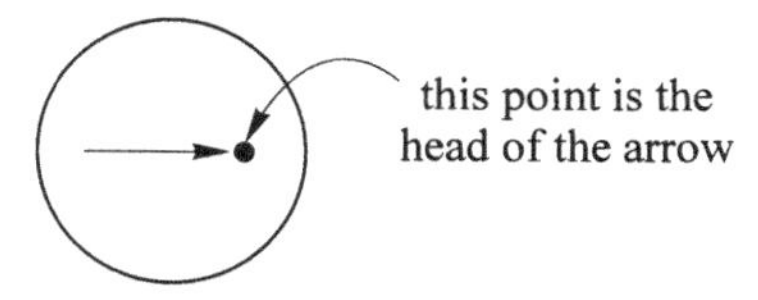

3. each arrow in B *points* to a point in S:

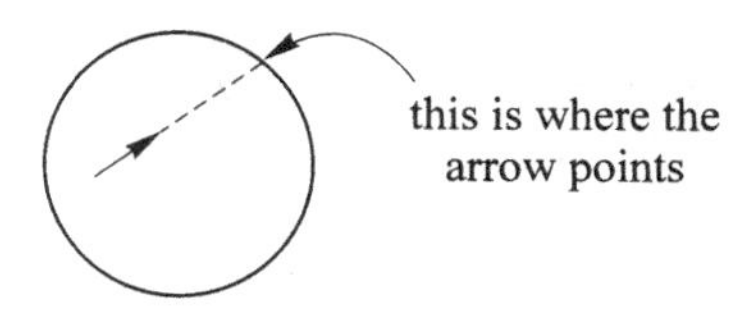

To analyze his proof, then, we must bring these concepts into our category $\mathcal{C}$. This means that we will need:

1. an *object* A (whose points are the *arrows* in B);
2. a *map* $A \xrightarrow{h} B$ (assigning to each arrow its *head*); and
3. a *map* $A \xrightarrow{p} S$ (telling where each arrow *points*).

(Remember that a map in $\mathcal{C}$ *means* a 'continuous' map, so that any map obtained by composing maps in $\mathcal{C}$ will automatically be continuous.)

Now we have three objects and three maps:

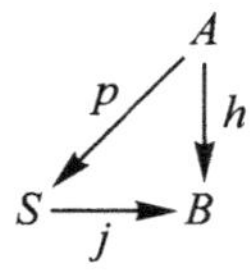

and we can begin to ask: what special properties of these (now 'objectified') concepts are used in Brouwer's proof?

First, we observe that if an arrow has its head on the boundary, then its head *is* the place to which it points:

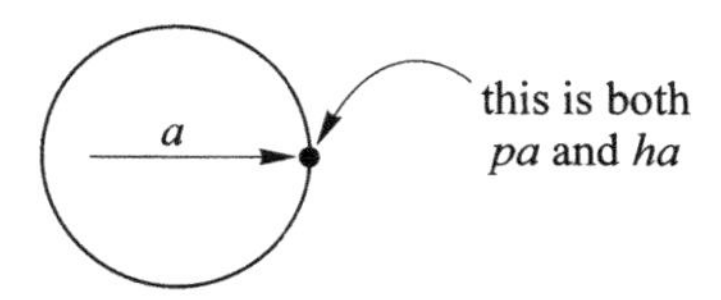

We will bring this into our category, by noting that a map $T \xrightarrow{a} A$ is a (smooth) 'listing' of arrows: $T \xrightarrow{a} A$.

Axiom 1: *If T is any object in $\mathcal{C}$, and $T \xrightarrow{a} A$ and $T \xrightarrow{s} S$ are maps satisfying $h\,a = j\,s$, then $p\,a = s$.*

The diagram below shows all the maps involved.

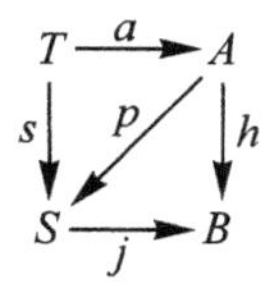

(Instead of just one arrow, we imagine a 'parameterized family' of arrows, one for each point in a 'parameter space' or 'test object' T; i.e. a map $T \xrightarrow{a} A$. The rest of the translation process leading to Axiom 1 just requires taking care to notice that p of an arrow is in S, while h of an arrow is in B; so to compare them we need to use the inclusion map $S \xrightarrow{j} B$.)

Already from Axiom 1, we can carry out part of Brouwer's argument:

Theorem 1: *If $B \xrightarrow{\alpha} A$ satisfies $h\alpha j = j$, then $p\alpha$ is a retraction for j.*

Proof: Put $T = S$, $s = 1_S$, and $a = \alpha j$ in Axiom 1.

Corollary: *If $h\alpha = 1_B$, then $p\alpha$ is a retraction for j.*

Second, we notice that if two points of B are *different*, there is an arrow from the first to the second; in fact each arrow in A should be thought of as having its head and tail *distinct*, otherwise it wouldn't 'point to' a definite place on the boundary S. We use the method of test objects again, with the idea that for each t, αt is the arrow from ft to gt.

Axiom 2: *If T is any object in $\mathcal{C}$, and $T \overset{f}{\underset{g}{\rightrightarrows}} B$ are any maps, then either there is a point $\mathbf{1} \xrightarrow{t} T$ with $ft = gt$, or there is a map $T \xrightarrow{\alpha} A$ with $h\alpha = g$.*

Now we can finish his argument:

Theorem 2: *Suppose we have maps*

$$B \overset{f}{\underset{g}{\rightrightarrows}} B$$

and $gj = j$, then either there is a point $\mathbf{1} \xrightarrow{b} B$ with $fb = gb$, or there is a retraction for $S \xrightarrow{j} B$.

Proof: Take $T = B$ in Axiom 2. We get: either there is a point $\mathbf{1} \xrightarrow{b} B$ with $fb = gb$, or there is a map $B \xrightarrow{\alpha} A$ with $h\alpha = g$; but then $h\alpha j = gj = j$, so Theorem 1 says that $p\alpha$ is a retraction for j.

If we take $g = 1_B$ in Theorem 2, we get a corollary.

Corollary: *If $B \xrightarrow{f} B$, then either there is a fixed point for f or there is a retraction for $S \xrightarrow{j} B$.*

(We gave, in Theorem 2, the more general version of Brouwer's theorem; the corollary is the original version.)

We will see later that in many categories $\mathcal{C}$, an object T may be large, and still have no 'points' $\mathbf{1} \xrightarrow{t} T$. In such a category, we should notice that we really didn't use the full strength of Axioms 1 and 2 in our proofs. It was enough to have Axiom 1 just for $T = S$, and Axiom 2 for $T = B$.

The main thing to study, though, is the way in which by objectifying certain concepts as maps in a category, the combining of concepts becomes *composition* of maps! Then we can condense a complicated argument into simple calculations using the associative law. Several hundred years ago, Hooke, Leibniz, and other great scientists foresaw the possibility of a 'philosophical algebra' which would have such features. This section has been quite condensed, and it may take effort to master it. You will need to go back to our previous discussion of Brouwer's proof, and carefully compare it with this version. Such a study will be helpful because this example is a model for the method of 'thinking categorically.'

6. The eye of the storm

Imagine a fluid (liquid or gas) moving in a spherical container. (If you want a two-dimensional example, you can imagine water swirling in a teacup and observe the surface current, say by imagining tiny boats drifting.) Right now, each point in our ball is moving, and we draw an arrow with tail at that point to represent its velocity. That is, the length of the arrow is proportional to the speed of the point, and the arrow points in the direction of travel. Could it be that every point is moving with non-zero speed, or must there be at least one instantaneous 'eye of the storm?'

To answer this, we take a slightly different arrow-object A than we imagined before. Its points are to be the possible velocity arrows of particles moving in our ball with non-zero speed. These arrows are less constrained than in our previous arrow-object, since the head of the arrow may be outside the ball; the only restriction is that if a point is on the surface of the ball, its velocity arrow cannot 'point outward' – at worst it is tangent to the sphere. Here is a picture for dimension 2:

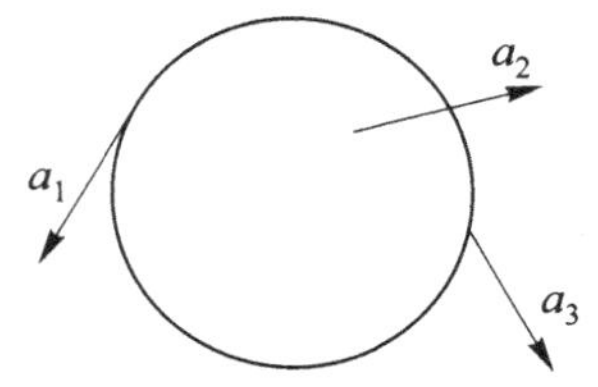

The arrows a_1 and a_2 are allowed as points in A, but a_3 is forbidden. Now we'll suppose that every point is moving, so we get a map $B \xrightarrow{\alpha} A$, assigning to each point of B the 'velocity arrow' at that point. For the map $A \xrightarrow{h} B$, we take the map assigning to each arrow its 'home.' (Remember that an arrow is supposed to represent the velocity of a moving point, so the tail of the arrow is the current home of the point.) Finally, for the map $A \xrightarrow{p} S$, we assign to each arrow its imaginary 'place of birth.' (It is customary to name winds in this way, as if a wind arriving from the north had always blown in one direction, and came from the farthest point that it could.)

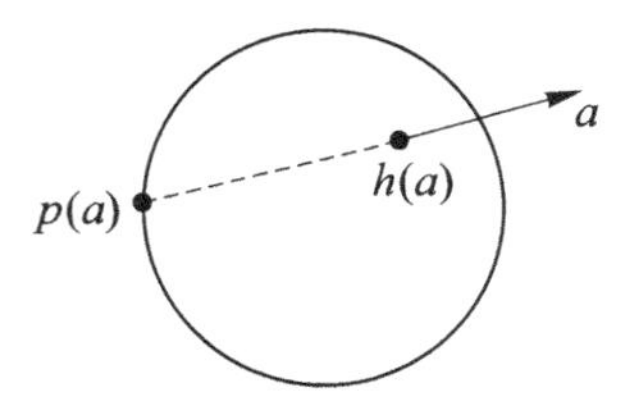

Axiom 1 says that if the moving point is on the sphere, then its 'place of birth' is its current location:

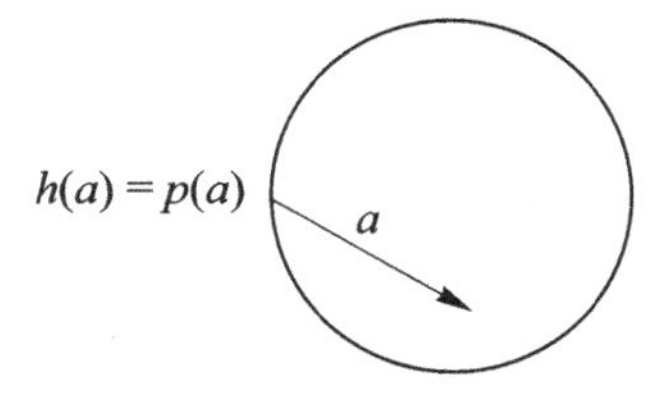

That is, the dot in the picture above is both $h(a)$ (as a point in the ball) and $p(a)$ (as a point on the sphere). Now you can work out for yourself that the corollary to Theorem 1 tells us that if there were a storm with no instantaneous 'eye,' there would be a retraction for the inclusion of the sphere into the ball.

7. Using maps to formulate guesses

Let's return to the one-dimensional case, the two cars traveling on the highway.

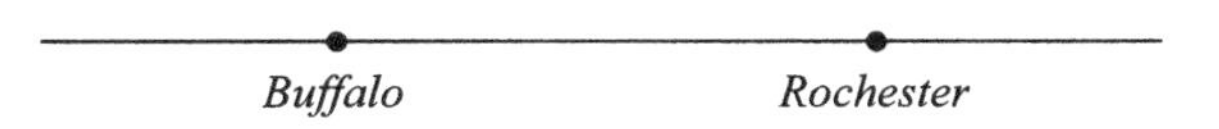

Actually, the highway extends beyond these two cities. Suppose I drive along the road, starting in Buffalo and ending in Rochester; you start and finish at the same times, starting and finishing anywhere between Buffalo and Rochester. During our travels, we're allowed to go anywhere along the highway we want, even west of Buffalo or east of Rochester. Are you convinced that at some time we must meet? Why?

Notice that there are now three objects involved: I, an interval; E, its endpoints; and R, the long road. (You can imagine R as the whole line if you want.) We also have two 'inclusion maps':

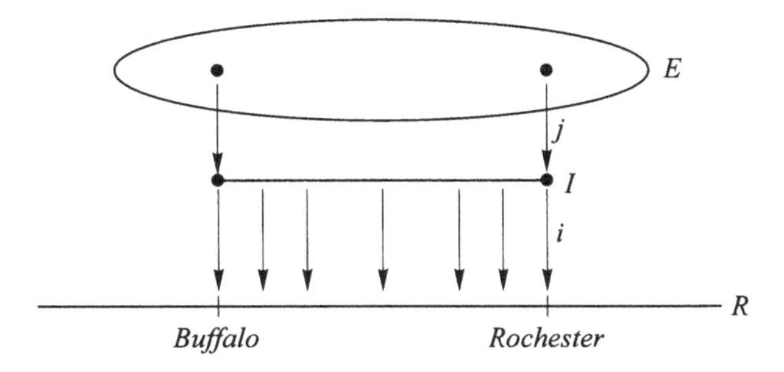

My travel gives an additional map: $I \xrightarrow{m} R$, and your travel gives another: $I \xrightarrow{y} R$. The relations among these four maps are investigated in the exercises below.

Exercise 4:

(a) Express the restrictions given above on my travel and yours by equations involving composition of maps, introducing other objects and maps as needed.

(b) Formulate the conclusion that at some time we meet, in terms of composition of maps. (You will need to introduce the object **1**.)

(c) Guess a stronger version of Brouwer's fixed point theorem in two dimensions, by replacing E, I, and R by the circle, disk, and plane. (You can do it in three dimensions too, if you want.)

(d) Try to test your guess in (c); e.g. try to invent maps for which your conjectured theorem is not true.

PART III

Categories of structured sets

We use maps to express extra 'structure' on sets, leading to graphs, dynamical systems, and other examples of 'types of structure.' We then investigate 'structure-preserving' maps.

ARTICLE III

Examples of categories

Directed graphs and other structures

We recall from Session 10:

1. Given an endomap of the ball with no fixed point, we can construct a retraction of the ball to its boundary.
2. Brouwer proved that no such retraction is possible.

We deduced by pure logic:

3. Every endomap of a ball has a fixed point.

We saw further that:

4. The sphere and the ball cannot be isomorphic (since the sphere does have a fixed point free endomap, for example, its antipodal map.)

It is critical that the category which we were discussing is not the category of abstract sets and arbitrary functions; it must rather be some

category of 'cohesive' objects and 'continuous' maps

Precisely which category of this type does not matter for our purposes; it only matters that spheres and balls are related by certain maps with certain properties, which we specified in detail in the case of (1). Crucially, (2) does not hold in the category of sets and functions on which most of our earlier discussions centered. Nor does (3) or (4): for example, the sphere, ball, and circle are all isomorphic in the category of sets! There are, in fact, many categories having the needed properties:

the category of topological spaces and continuous maps
the category of smooth spaces and smooth maps, etc.

These categories differ substantially from each other, and to specify these categories precisely and discuss their differences is a task better left to more advanced works.

We propose instead to define and study certain simpler categories which are of great interest in their own right, and which exhibit many of the features of cohesive categories. All our examples will illustrate a basic method: to make precise some imprecisely-known category, we can try to model it by *structures in* the category of abstract sets. These structures are always expressed by some configuration of given maps. When these categories arise later, we will treat them more slowly and in more detail; our main aim here is to give a rapid preview of some possible notions of structure, and especially to provide a first introduction to the powerful idea of *structure-preserving map*. The exercises in this article, like our previous exercises, involve only the application of the associative law to given definitions.

1. The category $\mathbf{S}^{\circlearrowright}$ of endomaps of sets

An important example already implicitly alluded to is the category in which an object is a set equipped with a specified endomap. Before defining it, let us denote by $\mathbf{S}$ the category of sets and maps which we have been discussing up to now. The maps in $\mathbf{S}$ are to be thought of as 'arbitrary'; that is, *any* conceivable process or scheme, which has only the property that to every point of a specified domain it gives a unique value in a specified codomain, counts as a map in $\mathbf{S}$. As a consequence $\mathbf{S}$ itself cannot really assign any property to distinguish one point of a set A from another point of A, though the number of points of A is an isomorphism-invariant in $\mathbf{S}$. Most of the interesting examples we discussed, such as the time-line, are only partly captured by $\mathbf{S}$, since, for example the order of time involves further 'structure'; however, all the examples have their shadow in $\mathbf{S}$, and already the calculations we can do in $\mathbf{S}$ (using composition, the forthcoming products, etc.) shed some light on the real examples. By ascending to the consideration of categories of objects more richly structured than those in $\mathbf{S}$, we can hope to see a much sharper image above the shadow, and to shed much more light on the examples by the same kind of categorical calculations. Schematically, the program is

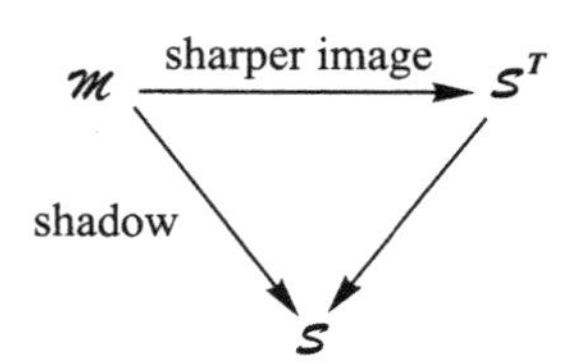

where $\mathcal{M}$ denotes an imprecise but real category, $\boldsymbol{T}$ denotes a specific chosen notion of structure, and $\mathbf{S}^{\boldsymbol{T}}$ denotes the category of structures of kind $\boldsymbol{T}$ which can be built in $\mathbf{S}$; the arrows denote the appropriate kind of maps between categories, known as *functors*, which we will discuss later.

Now we return to the category in which an object is an endomap of a set. A suggestive notation for it is $\mathbf{S}^{\circlearrowright}$. An object of $\mathbf{S}^{\circlearrowright}$ is any set X equipped with an endomap α. But the most important thing about a category is its maps and how they compose – what are the maps of $\mathbf{S}^{\circlearrowright}$? They are maps which 'respect the given structure,' i.e. a map

$$\boxed{X^{\circlearrowright \alpha}} \xrightarrow{\quad f \quad} \boxed{Y^{\circlearrowright \beta}}$$

between two objects of $\mathbf{S}^{\circlearrowright}$ is an $\mathbf{S}$-map $X \xrightarrow{f} Y$ which, moreover, satisfies

$$\boxed{f \circ \alpha = \beta \circ f}$$

After doing several exercises you will see that this equation really is the most appropriate expression of the idea that f preserves the given structure, i.e. that f is a way of mirroring the structure of α in the structure of β.

> **Exercise 1:**
> Show that if both f as above and also
>
> $$\boxed{Y^{\circlearrowright\beta}} \xrightarrow{\;g\;} \boxed{Z^{\circlearrowright\gamma}}$$
>
> are maps in $\mathbf{S}^{\circlearrowright}$, then the composite $g \circ f$ in $\mathbf{S}$ actually defines another map in $\mathbf{S}^{\circlearrowright}$. Hint: What should the domain and the codomain (in the sense of $\mathbf{S}^{\circlearrowright}$) of this third map be? Transfer the definition (given for the case f) to the cases g and $g \circ f$; then calculate that the equations satisfied by g and f imply the desired equation for $g \circ f$.

An object of $\mathbf{S}^{\circlearrowright}$ actually has all the structure suggested by our internal picture of an endomap α:

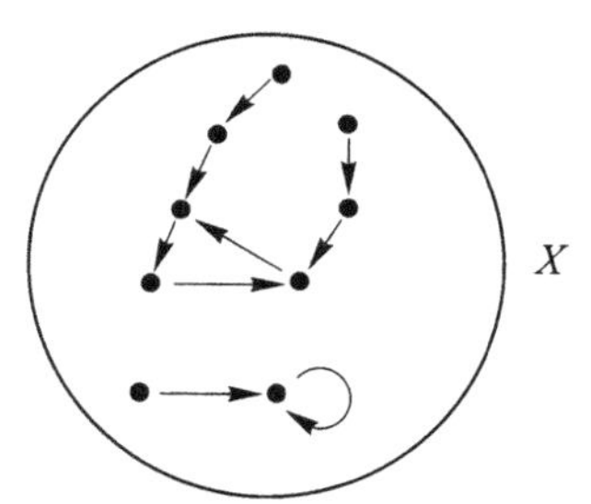

This is because an isomorphism in $\mathbf{S}^{\circlearrowright}$ has an inverse which is *also* a map in $\mathbf{S}^{\circlearrowright}$. It can thus be shown that if between two objects of $\mathbf{S}^{\circlearrowright}$ there exists an isomorphism of $\mathbf{S}^{\circlearrowright}$, then not only do the two sets have the same total number of points (as already mere $\mathbf{S}$-isomorphism would imply), but also equal numbers of fixed points, the same number of cycles of length seven, equal numbers of points that move four steps before stopping, equal numbers of points which move two steps before entering a cycle of length three, etc. and, moreover, equal numbers of components, etc. This array of numbers (which we may learn to organize) describes the kind of structure inherent in an object of $\mathbf{S}^{\circlearrowright}$.

2. Typical applications of $\mathbf{S}^{\circlearrowright}$

Objects of $\mathbf{S}^{\circlearrowright}$ arise frequently as *dynamical systems* or *automata*. The idea is that X is the set of possible *states*, either of a natural system or of a machine, and that the given endomap α represents the evolution of states, either the natural evolution in one unit of time of the system left to itself, or the change of internal state that will occur as a result of pressing a button (or other control) α on the outside of the machine once. If the system happens to be in state x now, then after one unit of time or one activation of the control, it will be in state $\alpha(x)$. After two units of time or pressing the button twice, it will be in the state.

$$\alpha(\alpha(x)) = (\alpha \circ \alpha)(x)$$

Similarly, $\alpha^3 = \alpha \circ \alpha \circ \alpha$ effects the three-step evolution, etc. Questions which could be asked about a particular object of $\mathcal{S}^{\circlearrowright}$ thus include the question of *accessibility*:

> Given a state x, is it possible to get into that state, i.e. does there exist a state x' for which $\alpha(x') = x$?

as well as the question of *convergence to equilibrium*:

> Given a state x, is it possible by activating α enough times (or waiting long enough, in the natural system view) to arrive at a state which no longer changes, i.e. for some n, $\alpha^{n+1}(x) = \alpha^n(x)$?

3. Two subcategories of $\mathcal{S}^{\circlearrowright}$

By putting restrictions on the kind of endomaps allowed, we obtain subcategories

$$\mathcal{S}^{\circlearrowright} \supset \mathcal{S}^{e} \qquad \mathcal{S}^{\circlearrowright} \supset \mathcal{S}^{\circlearrowleft}$$

where $\mathcal{S}^e$ means the category whose objects are all *idempotent* endomaps of sets and $\mathcal{S}^{\circlearrowleft}$ means the category whose objects are all *invertible* endomaps of sets (also known as automorphisms of sets or just as *permutations*); in both of these categories the definition of *map* between objects is the same as that stated above for $\mathcal{S}^{\circlearrowright}$. The numerical (or other) description of the detailed structure of a typical object in one of these two subcategories may be regarded as a specialization (somewhat less complicated) of the description for $\mathcal{S}^{\circlearrowright}$. But, as categories in their own right, these three are strikingly different, as we will see.

4. Categories of endomaps

If $\mathcal{C}$ is any category, we can build $\mathcal{C}^{\circlearrowright}$ from $\mathcal{C}$ in the same way we built $\mathcal{S}^{\circlearrowright}$ from $\mathcal{S}$. An object is an endomap in $\mathcal{C}$ and a map is a $\mathcal{C}$-map satisfying the same equation as before. There are many full subcategories of $\mathcal{C}^{\circlearrowright}$ (the category whose objects are endomaps in $\mathcal{C}$ and whose maps are 'equivariant' maps), for example

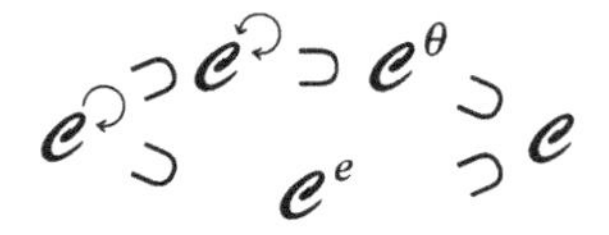

meaning:

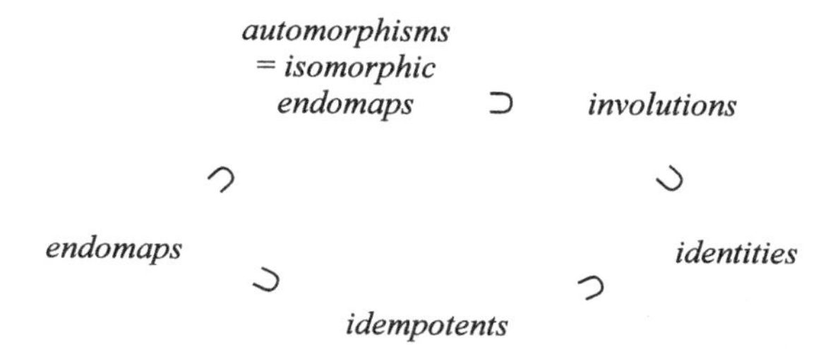

where θ is an *involution* of A if and only if $\theta \circ \theta = 1_A$. Note that an involution is automatically an automorphism (i.e. an endomap which is also an isomorphism) for it has an obvious inverse: θ^{-1} is θ itself if θ is an involution.

Every object A of $\mathcal{C}$ has only *one* identity map, but may have many idempotents and many involutions, some automorphisms which are not involutions, and some endomaps which are neither idempotents nor automorphisms.

Question: Could an endomap be both an automorphism and idempotent? Yes, 1_A is obviously both. Are there any others? Well, suppose we know both

$$\alpha \circ \alpha = \alpha$$
$$\alpha \circ \beta = 1_A$$
$$\beta \circ \alpha = 1_A$$

that is, that α is idempotent and also has a (two-sided) inverse β. Then

$$1_A = \alpha \circ \beta = (\alpha \circ \alpha) \circ \beta = \alpha \circ (\alpha \circ \beta) = \alpha \circ 1_A = \alpha$$

in other words, the *only* idempotent automorphism is the identity. From the proof we see in fact that the only idempotent which has even a section is 1_A.

Exercise 2:
What can you prove about an idempotent which has a retraction?

When $\mathcal{C} = \mathcal{S}$, what do the internal pictures of such special endomaps look like? If $\theta \circ \theta = 1_A$, then the internal diagram of θ must look like this

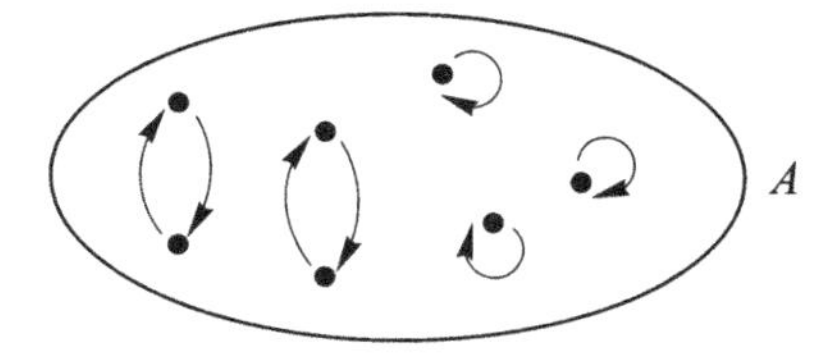

that is, a certain number of '2-cycles' and a certain number of *fixed points* (x for which $\theta(x) = x$).

Exercise 3:
A finite set A has an even number of elements iff (i.e. if and only if) there is an involution on A with *no fixed points*; A has an odd number of elements iff there is an involution on A with just *one* fixed point. Here we rely on known ideas about numbers – but these properties can be used as a *definition* of oddness or evenness that can be verified without counting if the structure of a real situation suggests an involution. The map 'mate of' in a group A of socks is an obvious example.

Let us exemplify the above types of endomaps on the set

$$\mathbb{Z} = \{\ldots -3, -2, -1, 0, 1, 2, 3, \ldots\}$$

of all (positive and negative) whole numbers, considered as an object of $\mathcal{S}$.

Exercise 4:
If $\alpha(x) = -x$ is considered as an endomap of $\mathbb{Z}$, is α an involution or an idempotent? What are its fixed points?

Exercise 5:
Same questions as above, if instead $\alpha(x) = |x|$, the absolute value.

Exercise 6:
If α is the endomap of $\mathbb{Z}$, defined by the formula $\alpha(x) = x + 3$, is α an automorphism? If so, write the formula for its inverse.

Exercise 7:
Same questions for $\alpha(x) = 5x$.

There are many other subcategories of $\mathcal{C}^{\circlearrowright}$, for example, the one whose objects are all the endomaps α in $\mathcal{C}$ which satisfy

$$\alpha \circ \alpha \circ \alpha = \alpha$$

Exercise 8:
Show that both $\mathcal{C}^e$, $\mathcal{C}^\theta$ are subcategories of the category above, i.e. that either an idempotent or an involution will satisfy $\alpha^3 = \alpha$.

Exercise 9:
In $\mathcal{S}$, consider the endomap α of a three-element set defined by the internal picture

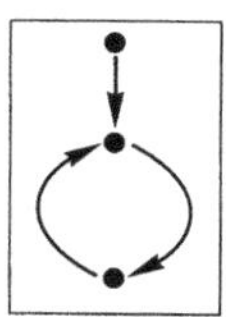

Show that it satisfies $\alpha^3 = \alpha$, but that it is *not* idempotent and that it is *not* an involution.

5. Irreflexive graphs

There is another important category of structures of which $\mathcal{S}^{\circlearrowright}$ itself may be considered to be a subcategory. We refer to the category $\mathcal{S}^{\downdownarrows}$ of (irreflexive directed multi-) *graphs*. An *object* of this category is any pair of sets equipped with a parallel pair of maps, as in this diagram:

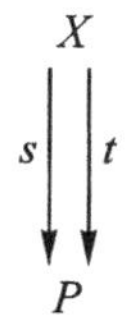

where X is called the set of *arrows* and P the set of *dots* of the graph. If x is an 'arrow' (element of X), then $s(x)$ is called the *source* of x, and $t(x)$ is called the *target* of x. The terminology refers to the fact that any graph has an internal picture of the type

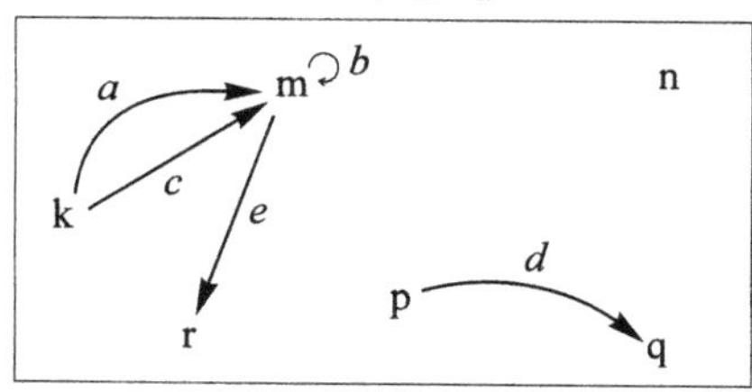

Here X has five elements $(a, b, \ldots)$ and P has six (k, m, ...) and $s(a) = \text{k}$, $t(e) = \text{r}$, $t(d) = \text{q}$, etc.

Exercise 10:
Complete the specification of the two maps

$$X \xrightarrow{s} P \quad \text{and} \quad X \xrightarrow{t} P$$

which express the source and target relations of the graph pictured above. Is there any element of X at which s and t take the same value in P? Is there any element to which t assigns the value k?

The maps in $\mathbf{S}^{\downdownarrows}$ are again defined so as to respect the graph structure. That is, a map

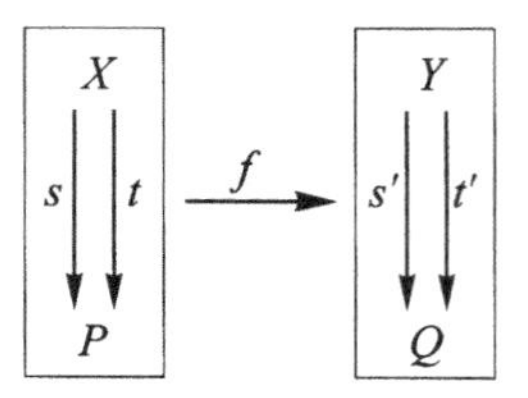

in $\mathbf{S}^{\downdownarrows}$ is defined to be any *pair* of $\mathbf{S}$-maps $X \xrightarrow{f_A} Y$, $P \xrightarrow{f_D} Q$ for which *both* equations

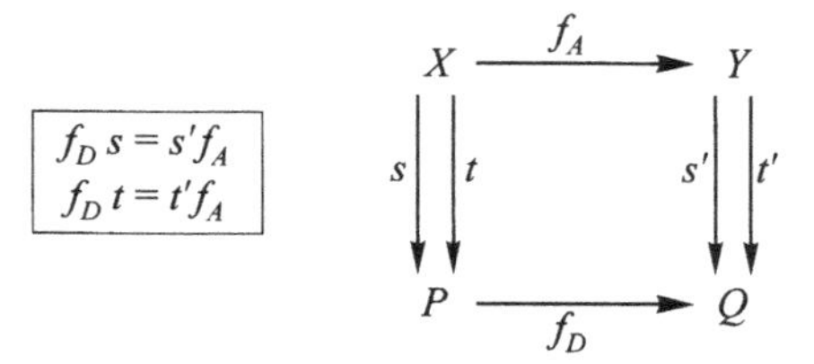

are valid in $\mathbf{S}$. We say briefly that 'f preserves the source and target relations' of the graphs. (The subscripts A, D are merely to suggest the part of the map f that operates on arrows and the part that operates on dots.)

Exercise 11:
If f is as above and if

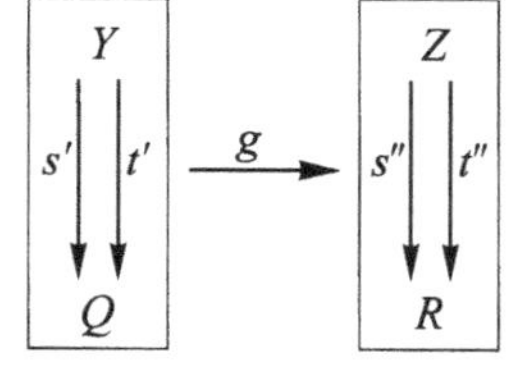

is another map of graphs, show that the pair $g_A \circ f_A$, $g_D \circ f_D$ of $\mathbf{S}$-composites is also an $\mathbf{S}^{\downdownarrows}$-map.

Graphs have many important applications – we might consider dots as towns, and arrows as possible roads; or dots as nouns, and arrows as transitive verbs with specified subject and object. Electrical wiring diagrams, information flow diagrams, etc. are often considered explicitly as graphs, i.e. as objects in $\mathbf{S}^{\downdownarrows}$, and many important relationships between graphs are expressed in terms of maps in $\mathbf{S}^{\downdownarrows}$.

Among the many numerical properties of graphs which remain unchanged by isomorphism are, not only the total number of arrows and of dots, but also the number of loops and the number of components.

6. Endomaps as special graphs

Why did we say that $\mathcal{S}^{\circlearrowleft}$ 'may be considered as a subcategory of' $\mathcal{S}^{\downdownarrows}$? Any such statement involves a specific way I of inserting

$$\mathcal{S}^{\downdownarrows} \xleftarrow{I} \mathcal{S}^{\circlearrowleft}$$

which in this case is the following: Given any set $X^{\circlearrowleft\alpha}$ equipped with an endomap, we may consider

$$X \underset{\alpha}{\overset{1_X}{\downdownarrows}} X$$

as a special kind of graph, in which the number of arrows is the same as the number of dots, and in which more precisely the source of the arrow named x is the dot also named x, but the target of the arrow named x is the dot named $\alpha(x)$. Now you see the method in our madness: the internal picture of an endomap is a special case of the internal picture of a graph!

We said that the *category* $\mathcal{S}^{\circlearrowleft}$ could be considered as a *subcategory* of $\mathcal{S}^{\downdownarrows}$. Since a big part of a category is its maps, this means that our insertion idea must apply to maps as well. Indeed, if

$$\boxed{X^{\circlearrowleft\alpha}} \xrightarrow{f} \boxed{Y^{\circlearrowleft\beta}}$$

in $\mathcal{S}^{\circlearrowleft}$, then it is easy to see that

$$\begin{array}{ccc} X & \xrightarrow{f} & Y \\ {\scriptstyle 1_X}\downarrow\downarrow{\scriptstyle \alpha} & & {\scriptstyle 1_Y}\downarrow\downarrow{\scriptstyle \beta} \\ X & \xrightarrow[f]{} & Y \end{array}$$

satisfies the two equations required of a map in $\mathcal{S}^{\downdownarrows}$.

Exercise 12:
If we denote the result of the above process by $I(f)$, then $I(g \circ f) = I(g) \circ I(f)$ so that our insertion I preserves the fundamental operation of categories.

Exercise 13:
(Fullness) Show that if we are given any $\mathcal{S}^{\downarrow\downarrow}$-morphism

$$\begin{array}{ccc} X & \xrightarrow{f_A} & Y \\ {\scriptstyle 1_X}\downarrow\downarrow{\scriptstyle \alpha} & & {\scriptstyle 1_Y}\downarrow\downarrow{\scriptstyle \beta} \\ X & \xrightarrow[f_D]{} & Y \end{array}$$

between the special graphs that come via I from endomaps of sets, then it follows that $f_A = f_D$, so that the map itself comes via I from a map in $\mathcal{S}^{\circlearrowright}$.

Considering I as understood, we see that our examples are related as

$$\mathcal{S}^{\downarrow\downarrow} \supset \mathcal{S}^{\circlearrowright} \begin{array}{l} \supset \mathcal{S}^{e} \\ \supset \mathcal{S}^{\circlearrowright} \end{array}$$

7. The simpler category $\mathcal{S}^{\downarrow}$: Objects are just maps of sets

A different subcategory of $\mathcal{S}^{\downarrow\downarrow}$ is $\mathcal{S}^{\downarrow}$, in which an object is an arbitrary *single* map between two sets, and a map is a 'commutative square of maps' in $\mathcal{S}$. Here the intended inclusion involves considering those graphs for which the source and target structures are the *same* map (i.e. graphs all of whose arrows are loops). Since an endomap is a special case of a map, there is also an obvious insertion J of $\mathcal{S}^{\circlearrowright}$ into $\mathcal{S}^{\downarrow}$; but, crucially, it does *not* satisfy 'fullness.' There are maps $J(X^{\circlearrowright\alpha}) \longrightarrow J(Y^{\circlearrowright\beta})$ in $\mathcal{S}^{\downarrow}$ which do not come via J from maps $X^{\circlearrowright\alpha} \longrightarrow Y^{\circlearrowright\beta}$ in $\mathcal{S}^{\circlearrowright}$.

Exercise 14:
Give an example of $\mathcal{S}$ of two endomaps and two maps as in

$$\begin{array}{ccc} X & \xrightarrow{f_A} & Y \\ {\scriptstyle \alpha}\downarrow & & \downarrow{\scriptstyle \beta} \\ X & \xrightarrow[f_D]{} & Y \end{array}$$

which satisfy the equation $f_D \circ \alpha = \beta \circ f_A$, but for which $f_A \neq f_D$.

Since it is easy to give many examples as in the last exercise, we may say that the structure preserved by an $\mathcal{S}^{\downarrow}$-map

$$J(X^{\circlearrowright\alpha}) \longrightarrow J(Y^{\circlearrowright\beta})$$

is much 'looser' than the structure preserved by an actual $\mathcal{S}^{\circlearrowleft}$-map

$$\boxed{X^{\circlearrowleft\alpha}} \longrightarrow \boxed{Y^{\circlearrowleft\beta}}$$

This remains true even for isomorphisms, so that the rich structure which $\mathcal{S}^{\circlearrowleft}$ sees in an endomap is degraded by considering it as just a map (that happens to be an endomap) to the much simpler questions: How many points are in the set and how many α-stacks of each possible size are there?

8. Reflexive graphs

A final very important example is *reflexive graphs*: these may be considered as graphs with a third structural map i

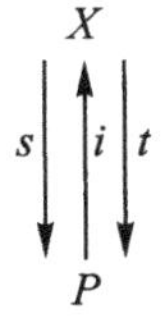

$$X \underset{t}{\overset{s}{\rightrightarrows}} P, \quad i: P \to X \qquad \boxed{\begin{matrix} s\,i = 1_P \\ t\,i = 1_P \end{matrix}}$$

of which both source and target are retractions; or equivalently, i is a given common section of both the source map and the target map. The following exercise asks you to prove certain consequences of these equations.

Exercise 15:
In a reflexive graph, the two endomaps $e_1 = is$, $e_0 = it$ of the set of arrows are not only idempotent, but even satisfy *four* equations:

$$e_k e_j = e_j \quad \text{for} \quad k,j = 0,1$$

Of course, maps of reflexive graphs are required to respect not only source and target, but also the extra ingredient i. You should formulate the definition of map of reflexive graphs before beginning Exercise 16.

Exercise 16:
Show that if f_A, f_D in $\mathcal{S}$ constitute a map of reflexive graphs, then f_D is determined by f_A and the internal structure of the two graphs.

Exercise 17:
Consider a structure involving two sets and four maps as in

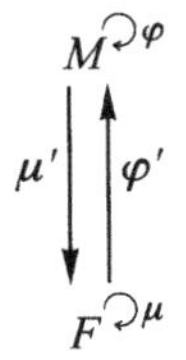

$$M^{\circlearrowleft\varphi} \underset{\varphi'}{\overset{\mu'}{\rightleftarrows}} F^{\circlearrowleft\mu} \qquad \text{(no equations required)}$$

(for example M = *males*, F = *females*, φ and φ' are *father*, and μ and μ' are *mother*). Devise a rational definition of *map* between such structures in order to make them into a category.

9. Summary of the examples and their general significance

In the diagram below, the horseshoe symbols indicate full insertions. Note that J followed by the inclusion to $\mathbf{S}^{\downarrow\downarrow}$ is *not* the same as the insertion I. The relation U between reflexive and irreflexive graphs is not a full insertion, but a *forgetful* functor (it just neglects the structural ingredient i); similarly for V.

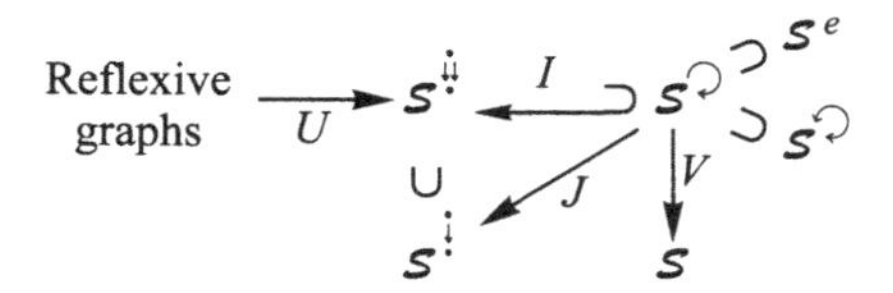

In all of the examples, the general kind of 'structure' involved can be more precisely described. Each example involves a 'category' (species or mode) of *cohesive* or *active* sets. As opposed to the abstract sets $\mathbf{S}$, which have zero internal cohesion or internal motion, these 'sets' have specific ways of internally sticking together and/or internally moving, and the *maps* in these categories permit comparing and studying these objects without tearing or interrupting them. By applying specified forgetful functors, we can also study how the objects compare if we imagine permitting (partial) tearing or interrupting to specified degrees.

10. Retractions and injectivity

When does a map a have a retraction? An important necessary condition is that it should be *injective*. Recall the definition:

Definition:
We say that a map $X \xrightarrow{a} Y$ is **injective** *iff for any maps $T \xrightarrow{x_1} X$ and $T \xrightarrow{x_2} X$ (in the same category) if $ax_1 = ax_2$ then $x_1 = x_2$ (or, in contrapositive form, 'the map a does not destroy distinctions,' i.e. if $x_1 \neq x_2$ in the diagram below, then $ax_1 \neq ax_2$ as well).*

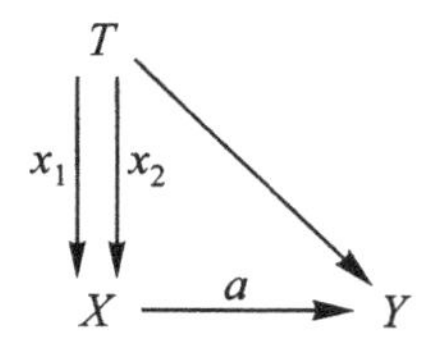

Exercise 18:
If a has a retraction, then a is injective. (Assume $pa = 1_X$ and $ax_1 = ax_2$; then try to show by calculation that $x_1 = x_2$.)

In the category $\boldsymbol{S}$ of abstract sets and arbitrary maps, the converse of the above exercise is almost true: If $X \xrightarrow{a} Y$ is any injective map in $\boldsymbol{S}$ for which $X \neq 0$, then there exist maps $Y \xrightarrow{p} X$ for which $pa = 1_X$, as we have seen before. However, it is very important that this converse is *not* true in most categories. For example, in a category of continuous maps, the inclusion a of a circle X as the boundary of a disk Y does *not* have a retraction; any of the $\boldsymbol{S}$-retractions of p of the underlying sets of points would have to 'tear' the disk, i.e. would not underlie a *continuous* retraction of the spaces. We now consider an example of the same phenomenon in $\boldsymbol{S}^{\circlearrowleft}$.

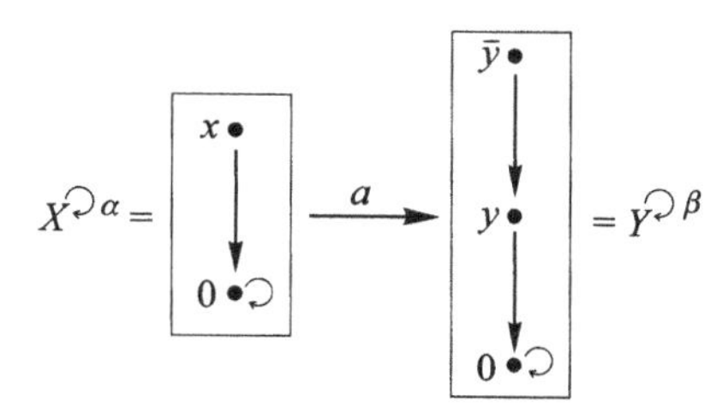

Let $ax = y$ and $a0 = 0$, with X, Y, α, and β as pictured above.

Exercise 19:
Show that a is a map $\boxed{X^{\circlearrowleft\alpha}} \xrightarrow{a} \boxed{Y^{\circlearrowleft\beta}}$ in $\boldsymbol{S}^{\circlearrowleft}$.

Exercise 20:
Show that a is *injective*.

Exercise 21:
Show that, as a map $X \xrightarrow{a} Y$ in $\boldsymbol{S}$, a has exactly two retractions p.

Exercise 22:
Show that neither of the maps p found in the preceding exercise is *a map* $\boxed{Y^{\circlearrowleft\beta}} \longrightarrow \boxed{X^{\circlearrowleft\alpha}}$ in $\boldsymbol{S}^{\circlearrowleft}$. Hence a has no retractions in $\boldsymbol{S}^{\circlearrowleft}$.

Exercise 23:
How many of the eight $\boldsymbol{S}$-maps $Y \longrightarrow X$ (if any) are actually $\boldsymbol{S}^{\circlearrowleft}$-maps?

$$\boxed{Y^{\circlearrowleft\beta}} \longrightarrow \boxed{X^{\circlearrowleft\alpha}}$$

Exercise 24:
Show that our map a does not have any retractions, even when considered (via the insertion J in Section 7 of this article) as being a map in the 'looser' category $\boldsymbol{S}^{\downarrow}$.

Exercise 25:
Show that for any two graphs and any $\mathcal{S}^{\downarrow\downarrow}$-map between them

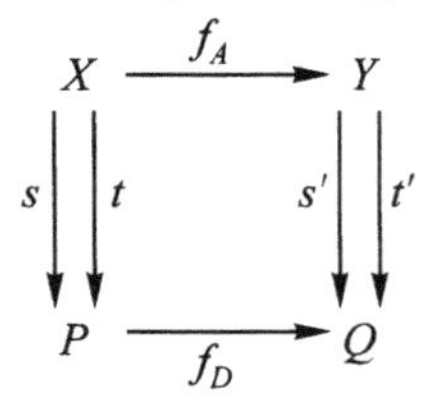

the equation $f_D \circ s = f_D \circ t$ can only be true when f_A maps every arrow in X to a *loop* (relative to s', t') in Y.

To say that

$$\mathbb{Z}^{\circlearrowright 5\times()}$$

is an automorphism would be wrong, since $\mathbb{Z}$ doesn't have fractions in it. On the other hand, there would be a germ of truth in the statement, because if $\mathbb{Q}$ denotes the set of rational numbers, then:

Exercise 26:
There is an 'inclusion' map $\mathbb{Z} \xrightarrow{f} \mathbb{Q}$ in $\mathcal{S}$ for which

1. $\boxed{\mathbb{Z}^{\circlearrowright 5\times()}} \xrightarrow{f} \boxed{\mathbb{Q}^{\circlearrowright 5\times()}}$ is a map in $\mathcal{S}^{\circlearrowright}$, and
2. $\mathbb{Q}^{\circlearrowright 5\times()}$ is an automorphism, and
3. f is *injective*.

Find the f and prove the three statements.

Exercise 27:
Consider our standard idempotent

$$X^{\circlearrowright\alpha} = \boxed{\begin{matrix}\bullet\\ \downarrow\\ \bullet\circlearrowright\end{matrix}}$$

and let $Y^{\circlearrowright\beta}$ be any *automorphism*. Show that any $\mathcal{S}^{\circlearrowright}$-map $X^{\circlearrowright\alpha} \xrightarrow{f} Y^{\circlearrowright\beta}$ must be non-injective, i.e. must map *both* elements of X to the *same* (fixed) point of β in Y.

Exercise 28:
If $X^{\circlearrowright\alpha}$ is any object of $\mathcal{S}^{\circlearrowright}$ for which there exists an *injective* $\mathcal{S}^{\circlearrowright}$-map f to some $Y^{\circlearrowright\beta}$ where β is in the subcategory of *automorphisms*, then α itself must be injective.

11. Types of structure

A type of structure can be specified by giving the following ingredients:

1. a set of names (perhaps more than one or two) for the objects we expect as components of each single structure of the type;
2. another set of names for the crucial structural maps that must be specified to determine any single structure of the type; and
3. the specification of which structural component object is required to be the domain and codomain of each structural map, but in terms of the abstract names.

Each concrete structure of the type is required to have its structural maps conform to the abstract specification. For example, discrete dynamical systems have one component object of 'states' and one structural map, the 'dynamic', whereas graphs have two component objects 'arrows' and 'dots' and two structural maps 'source' and 'target.' Reflexive graphs have three structural maps. Our discussion of kinship systems involves also component sets and structural maps. (Note that an abstract specification of a type of structure can itself be considered a *graph* – see Session 17.)

The pattern for defining the notion of map in any category of concrete structures is now explicitly the same for all abstract types. Namely, suppose X and Y are two structures of a given type, modeled in sets. Then for each component name A in the type, there are given sets $X(A)$ and $Y(A)$, so that a map $X \xrightarrow{f} Y$ is required to involve, for each such A, a map of sets $X(A) \xrightarrow{f_A} Y(A)$; but these maps are required to *preserve all the structure* in order to be considered to constitute together a single map of structures. Namely, for each structural map-name α in the type, X has specified a map

$$\alpha_X : X(A) \longrightarrow X(B)$$

where A, B are the source and target of α in the type, and also Y has specified a structural map

$$\alpha_Y : Y(A) \longrightarrow Y(B)$$

with the same name α and the same A, B; thus the natural meaning of the statement that 'f preserves α' is that

$$\alpha_Y f_A = f_B \alpha_X \qquad \begin{array}{ccc} X(A) & \xrightarrow{f_A} & Y(A) \\ {\scriptstyle \alpha_X}\downarrow & & \downarrow{\scriptstyle \alpha_Y} \\ X(B) & \xrightarrow[f_B]{} & Y(B) \end{array}$$

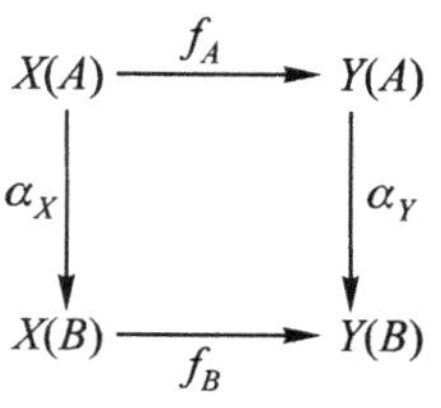

in the background category of sets. To be a map of structures, f is required to preserve *all* the structural maps as named by the type of structure. Thus a map in a category of structures has as many component maps as there are component-object names in the type, and is required to satisfy one preservation-equation for each structural-map name in the type.

There are many more categories than just those given by abstract types of structure; however, those can be construed as full subcategories of the latter, so that the notion of map does not change. Such full subcategories are determined by putting restrictive conditions on the diagram that constitutes an object, the simplest sort of such condition just being a composition-equation that the structural maps are required to satisfy. For example, a dynamical system might be required to be an *involution*, or a 'preferred loop' structure in a graph might be required to have source and target both equal to the identity on dots in order to have a *reflexive* graph, etc.

An abstract structure type often arises from a particular example as follows. Suppose $\mathcal{A}$ is a small family of objects and maps in a category $\mathcal{X}$, with the domain and codomain of any map in $\mathcal{A}$ being in $\mathcal{A}$. Let each object A in $\mathcal{A}$ be considered as the name of 'A-shaped figures' and each map α in $\mathcal{A}$ be considered as a name α^* of structural map. The domain of α^* is the codomain of α, and the codomain of α^* is the domain of α. Then every object X of $\mathcal{X}$ gives rise to an $\mathcal{A}$-structure whose A-th component set is the set of all $\mathcal{X}$-maps $A \longrightarrow X$ and wherein for each $B \xrightarrow{\alpha} A$ the structural map on these figures has for all x

$$\alpha^*_X(x) = x \circ \alpha$$

Exercise 29:
Every map $X \xrightarrow{f} Y$ in $\mathcal{X}$ gives rise to a map in the category of $\mathcal{A}$-structures, by the associative law.

For example, the abstract notion of *graph structure* can be identified with the concrete diagram $\mathcal{A}$ of graphs below

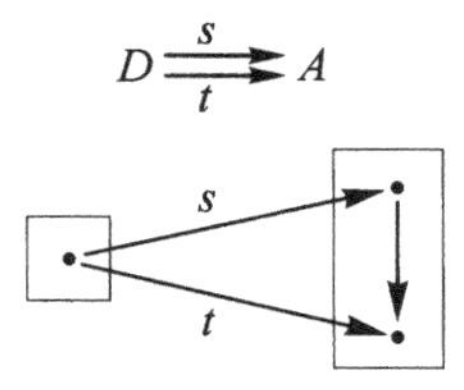

because for any graph X the arrows in X may be identified with the graph maps $A \xrightarrow{x} X$, the dots in X with the graph maps $D \longrightarrow X$, and the source of x is then $s^*_X(x) = x \circ s$; for any graph map $X \xrightarrow{f} Y$, the associativity $f(xs) = (fx)s$ then substantiates, inside the category itself, the fact that f preserves sources:

$$s^*_Y \circ f = f \circ s^*_X$$

Any instance of a structure 'opposite' to a given type (e.g. the type 'graph') in any category $\mathcal{C}$ gives rise to an interpretation of $\mathcal{C}$ into the category of 'sets with structure' of the given type. For example if $\mathcal{C}$ is some category of cohesive spaces, we might take in place of the objects D and A the objects $\mathbf{1}$ and S, a one-point space and an object representing the space of a room. In addition, we need two selected points in the room, $\mathbf{1} \xrightarrow{s} S$ and $\mathbf{1} \xrightarrow{t} S$. Once these data are fixed, each object in the category $\mathcal{C}$ gets an 'interpretation' as a graph. For example, if T is the temperature line, a dot of the 'temperature graph' is a point of T (a map $\mathbf{1} \longrightarrow T$), and an arrow of the graph is a 'temperature field' in this room (a map $S \longrightarrow T$). The 'source' of a temperature field is the temperature at the point s in the room; the 'target' is the temperature at t.

Exercise 30:
If S, s, t is a given bipointed object as above in a category $\mathcal{C}$, then for each object X of $\mathcal{C}$, the graph of 'X fields' on S is actually a *reflexive* graph, and for each map $X \xrightarrow{f} Y$ in $\mathcal{C}$, the induced maps on sets constitute a map of reflexive graphs.

12. Guide

Several useful examples of categories have been constructed by a common method, and we have begun to explore some ways in which these categories do and do not resemble the category of sets. Extended discussion of these and other categories is given in Sessions 11–18, along with a sample test after Session 17.

SESSION 11

Ascending to categories of richer structures

1. A category of richer structures: Endomaps of sets

A simple example of a 'type of structure' $\boldsymbol{T}$, that is used to construct a category richer than $\boldsymbol{\mathcal{S}}$, is the idea of a single endomap, $\boldsymbol{T} = \boxed{\bullet\circlearrowleft}$. A structure of that type in $\boldsymbol{\mathcal{S}}$ is just a given set *with a given endomap*, and the resulting category of sets-with-an-endomap is denoted accordingly by $\boldsymbol{\mathcal{S}}^{\circlearrowleft}$. If you remember that an endomap of a set has a special type of internal diagram, you will see why an endomap of a set can be considered to be a particular type of structure on that set. For example, a typical endomap looks something like this:

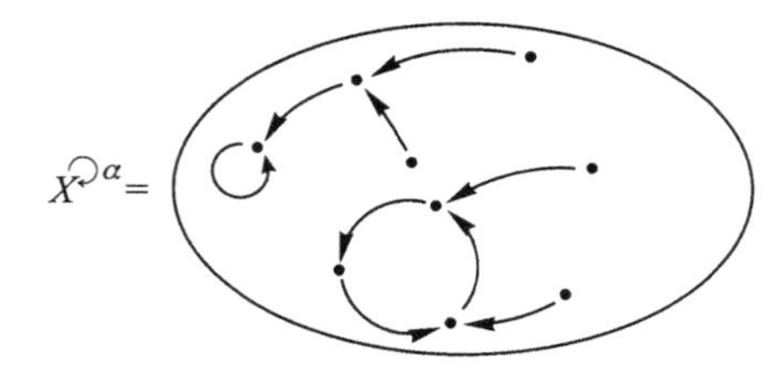

(Remember that the internal diagram of an endomap has exactly one arrow *leaving* each dot, but no special condition on how many arrive at each dot.) This really looks like a set with some 'structure.' This set X together with this particular endomap α is an example of an *object* of the category $\boldsymbol{\mathcal{S}}^{\circlearrowleft}$, denoted by $X^{\circlearrowleft\alpha}$.

Besides objects, we must also have *maps* in the category $\boldsymbol{\mathcal{S}}^{\circlearrowleft}$. Given two sets-with-endomap, say $X^{\circlearrowleft\alpha}$ and $Y^{\circlearrowleft\beta}$, the appropriate maps between them are set maps $X \longrightarrow Y$, which are not allowed to be arbitrary, since they should 'be consistent' with the structures given by the endomaps α and β. A few moments of reflection will suggest that the appropriate restriction on a map of sets $f : X \longrightarrow Y$ in order to 'preserve' or 'be consistent with' the structures given by the endomaps α and β is that the equation

$$f \circ \alpha = \beta \circ f$$

must be satisfied by f.

Definition:

$$\boxed{X^{\circlearrowleft\alpha} \xrightarrow{f} Y^{\circlearrowleft\beta} \text{ in } \mathcal{S}^{\circlearrowleft}} \quad \textit{means} \quad \boxed{\begin{array}{c} X \xrightarrow{f} Y \text{ in } \mathcal{S} \\ \textit{and} \\ f \circ \alpha = \beta \circ f \end{array}}$$

As an example let's try to find a map in the category $\mathcal{S}^{\circlearrowright}$ from a one-element set to the $X^{\circlearrowright\alpha}$ pictured before. Of course, before we can attempt this we must say what endomap of the one-element set we mean to use. The endomap has to be always specified beforehand for every object of $\mathcal{S}^{\circlearrowright}$. However, a one-element set has only one endomap (its identity, of course), so that the only object of $\mathcal{S}^{\circlearrowright}$ that one can mean is the one pictured below.

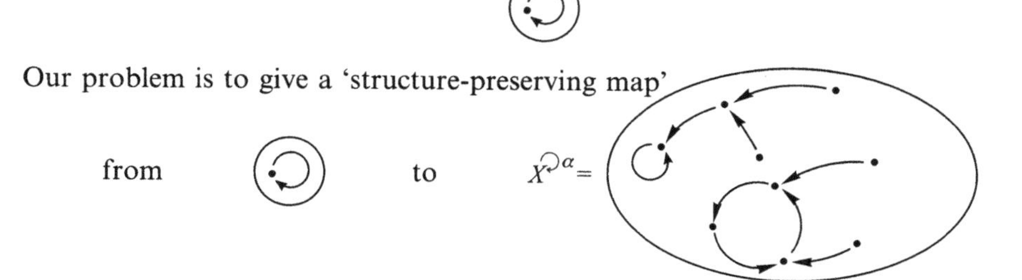

Our problem is to give a 'structure-preserving map'

You might guess that a structure-preserving map does not 'alter' the loop, and that it can only map to another loop. Since there is only one loop in the codomain, this guess suggests that the only map is

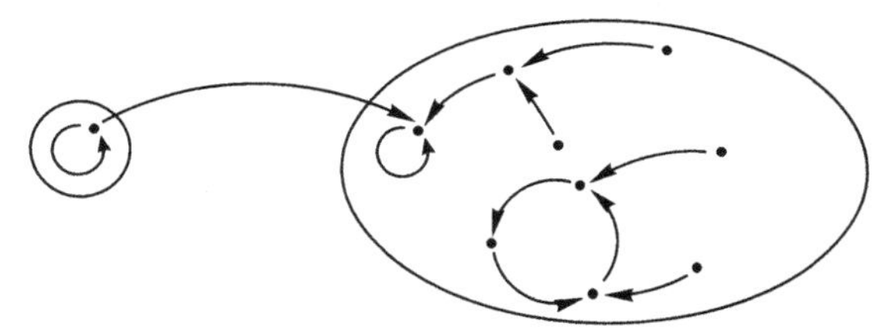

You should verify that indeed this is the only map satisfying the defining property of the maps in $\mathcal{S}^{\circlearrowright}$.

Suppose that we ask for maps

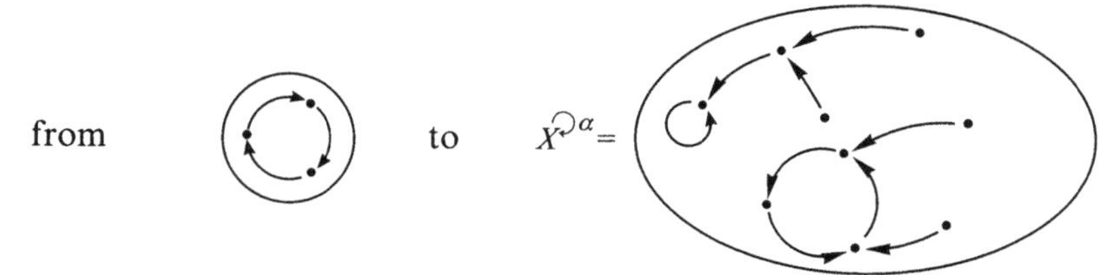

Exercise 1:
How many maps can you find? (There are fewer than seven.)

We have been referring to 'the category $\mathcal{S}^{\circlearrowright}$', but we haven't finished saying what it is. You should look back at the definition of 'category' to see what we need to do to specify a particular category. We have decided what the objects are, and what the maps are, but we have not yet specified the other two data: composition of maps and identity maps. I think any reasonable person, though, would come up with the following choice for the composite of

$$X^{\circlearrowleft\alpha} \xrightarrow{f} Y^{\circlearrowleft\beta} \xrightarrow{g} Z^{\circlearrowleft\gamma}$$

namely to define it to be the composite as maps of sets; i.e.

$$X^{\circlearrowleft\alpha} \xrightarrow{g \circ f} Z^{\circlearrowleft\gamma}$$

Warning! It is conceivable that this reasonable choice might not be allowed; perhaps $g \circ f$ is not a map in the category $\boldsymbol{S}^{\circlearrowleft}$. We must check that

$$(g \circ f) \circ \alpha \stackrel{?}{=} \gamma \circ (g \circ f)$$

This is Exercise 1 in Article III. All we know is that g and f are maps in $\boldsymbol{S}^{\circlearrowleft}$ ('structure-preserving' maps), i.e. $f \circ \alpha = \beta \circ f$ and $g \circ \beta = \gamma \circ g$.

Can you see a way to deduce the equation we need from the two equations we have?

FATIMA: Use the associative law.

Right, the associative law and substitution,

$$(g \circ f) \circ \alpha = g \circ (f \circ \alpha) = g \circ (\beta \circ f) = (g \circ \beta) \circ f = (\gamma \circ g) \circ f$$
$$= \gamma \circ (g \circ f)$$

Later, when we want to shorten the writing, we can leave out the parentheses, and even leave out the circles, and just write

$$gf\alpha = g\beta f = \gamma gf$$

the first equality because $f\alpha = \beta f$, and the second because $g\beta = \gamma g$. For now, though, it is probably better to make the use of the associative law more explicit, since it is the most important fact about composition of maps.

We still need to select the identity map for each object $X^{\circlearrowleft\alpha}$; and it seems the only reasonable choice is to take $X^{\circlearrowleft\alpha} \xrightarrow{1_X} X^{\circlearrowleft\alpha}$, the identity that X had (as an unstructured set). Of course, we need to check that this is a map in $\boldsymbol{S}^{\circlearrowleft}$, that is

$$1_X \circ \alpha \stackrel{?}{=} \alpha \circ 1_X$$

Can you see how to do it?

EVERYBODY: Yes, these are both equal to α.

Good. Now we have all the *data* to specify a category: objects, maps, composition, and identity maps. We still must check that the associative and identity laws are true. But fortunately these verifications are easy, because the composition and the identity maps were chosen to be those from $\boldsymbol{S}$, and in $\boldsymbol{S}$ we already know these rules are true.

Now that we know that we have a category we can consider the notion of isomorphism. In the category of sets an isomorphism meant that two sets had the same number of points, but in this category of sets-with-an-endomap, isomorphism means much more. It means that the structure of the endomaps is the same. In particular, the two endomaps must have the same number of fixed points, the same number of cycles of length 2, the same number of cycles of length 3, etc. and more.

This completes what I wanted to say for now about this new example of a category. There are many other examples of categories of structures, but note that, paradoxically, these structures are all built from sets, which can be considered to have no structure. Some people interpret this by saying that sets are the foundation of mathematics. What this really reveals is that although an abstract set is completely described by a single number, the set has the potentiality to carry all sorts of structure with the help of maps.

2. Two subcategories: Idempotents and automorphisms

$\mathcal{S}^{\circlearrowright}$ is the category of endomaps of sets. If we put a restriction on the endomaps we will obtain a subcategory. Two examples of this are the following:

1. The category $\mathcal{S}^e$ of sets with an endomap *which is idempotent*. Then a set-with-an-endomap $X^{\circlearrowright\alpha}$ is an object in $\mathcal{S}^e$ if and only if $\alpha \circ \alpha = \alpha$. The picture of an object in $\mathcal{S}^e$ looks like this:

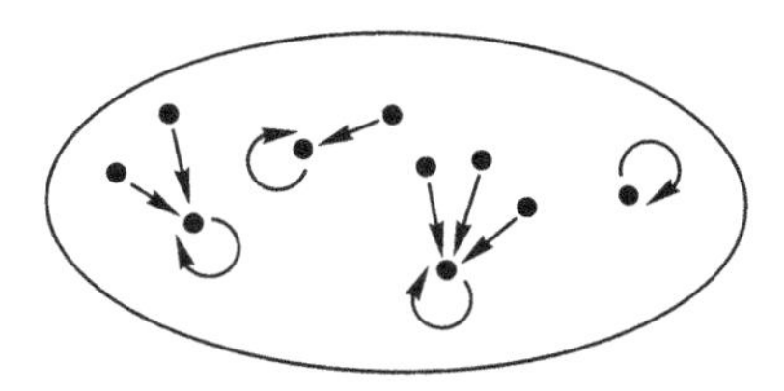

Every point is either a fixed point or reaches a fixed point in one step. (In particular there are no cycles of length two or more.) An isomorphism in $\mathcal{S}^e$ means 'correspondence between fixed points and correspondence between branches at corresponding fixed points.'

2. The category $\mathcal{S}^{\circlearrowleft}$ of sets with an endomap *which is invertible*. $X^{\circlearrowright\alpha}$ is an object in $\mathcal{S}^{\circlearrowleft}$ if and only if the endomap α has an inverse, i.e. a map β such that $\alpha \circ \beta = 1_X$ and $\beta \circ \alpha = 1_X$. The endomap can have cycles of any length, but no branches, so that pictures of the objects in $\mathcal{S}^{\circlearrowleft}$ look like this:

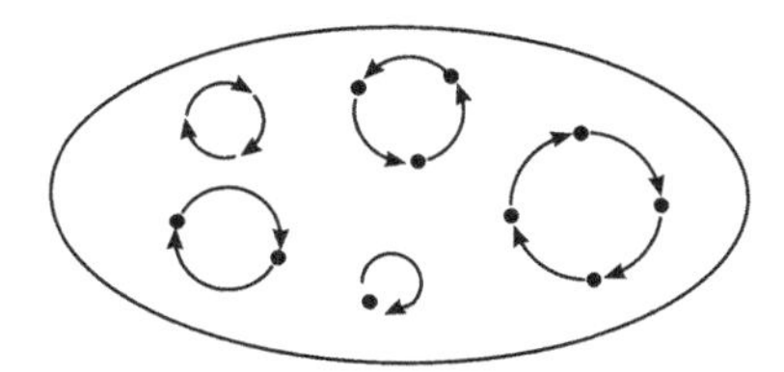

Recall that an invertible endomap, i.e. an endomap which is also an isomorphism, is called an *automorphism*. An automorphism of a finite set is also known as a *permutation* of the set.

3. The category of graphs

Besides these two categories which are subcategories of $\mathcal{S}^{\circlearrowright}$ we can give an example of a category, denoted $\mathcal{S}^{\downdownarrows}$, of which $\mathcal{S}^{\circlearrowright}$ is a subcategory. An *object* in $\mathcal{S}^{\downdownarrows}$ is a pair of maps with the same domain and with the same codomain. Thus an object in $\mathcal{S}^{\downdownarrows}$ consists of two sets X, Y, and two maps s and t (called 'source' and 'target') from one to the other:

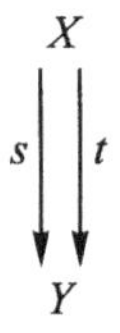

Such a thing is called a *graph*. (Specifically, since there are many kinds of graphs that are used, these are 'irreflexive directed multigraphs.') To depict a graph, we draw a dot for each element of Y and then we join the dots with arrows in the following way: for each element x of X we draw an arrow from the dot sx to the dot tx. The result will be something like this:

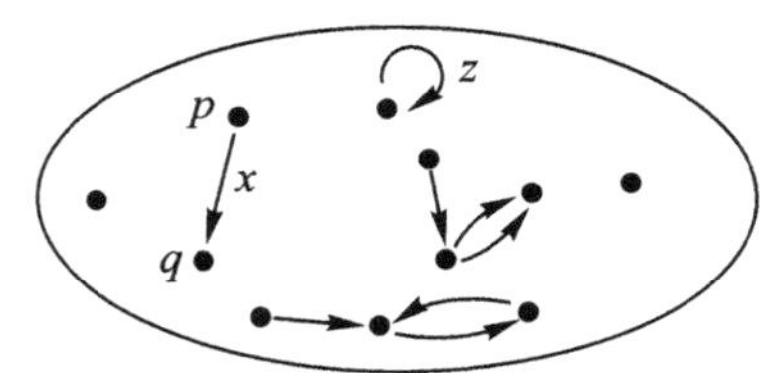

where the dots are the elements of Y and the arrows are the elements of X. If X has an element z such that $sz = tz$ then we draw z as a loop. For any object in $\mathcal{S}^{\downdownarrows}$ we can draw such a picture and each picture of this kind represents a pair of maps with the same domain and same codomain.

Now everybody should ask. What should we mean by a *map* in this category,

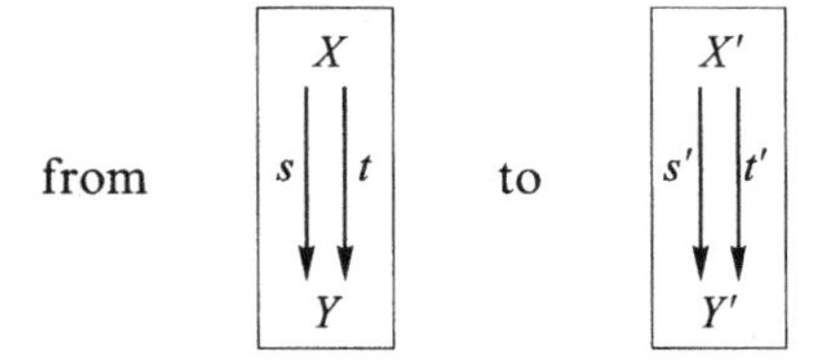

The idea is that it should be a map that 'preserves the structure' of the graph. Now, the structure of the graph consists of dots, arrows, and the source and target relations between them. Thus a map in this category should map dots to dots and arrows to arrows, in such a way that if one arrow is sent to another, then the source-dot of the first arrow must be sent to the source-dot of the second (and the similar restriction for targets.) If you think for a while what all this means you will see that we should define:

Definition:

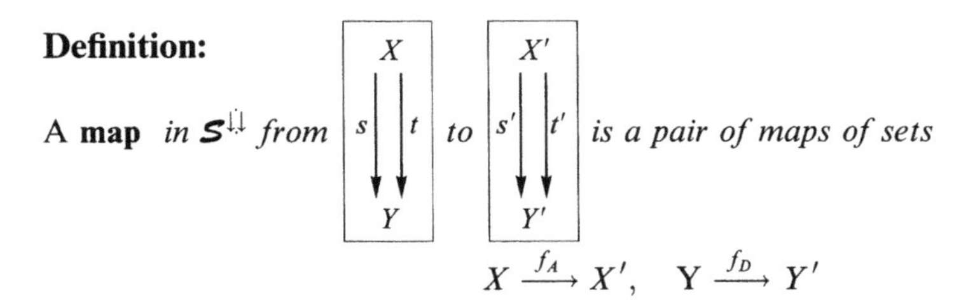

$$X \xrightarrow{f_A} X', \quad Y \xrightarrow{f_D} Y'$$

such that

$$f_D \circ s = s' \circ f_A \quad \text{and} \quad f_D \circ t = t' \circ f_A$$

These equations can be remembered by drawing this diagram:

$$\begin{array}{ccc} X & \xrightarrow{f_A} & X' \\ s \downarrow\downarrow t & & s' \downarrow\downarrow t' \\ Y & \xrightarrow[f_D]{} & Y' \end{array}$$

Given two maps in $\mathcal{S}^{\downarrow\downarrow}$,

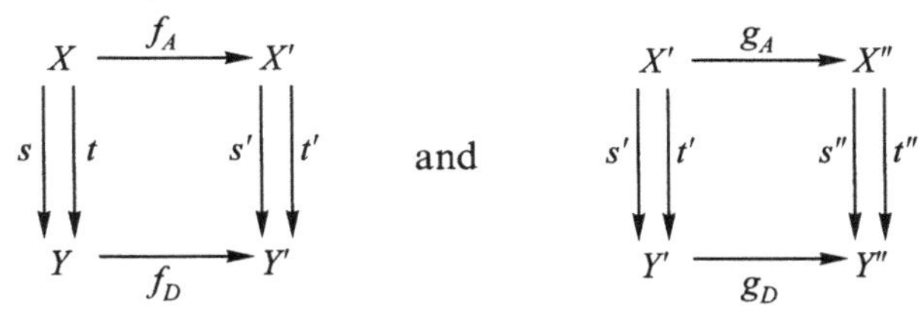

we can obtain the two composites $g_A \circ f_A$ and $g_D \circ f_D$ and form the diagram

$$\begin{array}{ccc} X & \xrightarrow{g_A \circ f_A} & X'' \\ s \downarrow\downarrow t & & s'' \downarrow\downarrow t'' \\ Y & \xrightarrow[g_D \circ f_D]{} & Y'' \end{array}$$

and we *define* this to be the composite of the two maps. Is it a map? We need to verify the equations

$$(g_D \circ f_D) \circ s = s'' \circ (g_A \circ f_A) \quad \text{and} \quad (g_D \circ f_D) \circ t = t'' \circ (g_A \circ f_A).$$

You should do this, and also define identity maps and check the associative and identity laws.

Two graphs are isomorphic if we can exactly match arrows of one to arrows of the other and dots of one to dots of the other, taking care that if two arrows are matched then so are their source-dots and so are their target-dots. The exercises below illustrate this. This category has many applications, e.g. in electrical engineering, transport problems, and even in linguistics, since graphs appear in all these subjects, be it as electric circuits, road systems between towns, or as nouns and verbs relating the nouns.

In what sense is it meant that $\boldsymbol{S}^{\circlearrowleft}$ is a subcategory of $\boldsymbol{S}^{\downarrow\downarrow}$? It means that there is a specific procedure by which the objects and the maps in $\boldsymbol{S}^{\circlearrowleft}$ can be viewed as graphs and maps of graphs. This procedure is suggested by our picture for an endomap, which is also a picture for a graph. But one can ask: What is the pair of maps that corresponds to an endomap in the passage from $\boldsymbol{S}^{\circlearrowleft}$ to $\boldsymbol{S}^{\downarrow\downarrow}$? The answer is the following:

$$X^{\circlearrowleft\alpha} \quad \text{corresponds to} \quad X \underset{\alpha}{\overset{1_X}{\rightrightarrows}} X$$

and in this correspondence

the picture 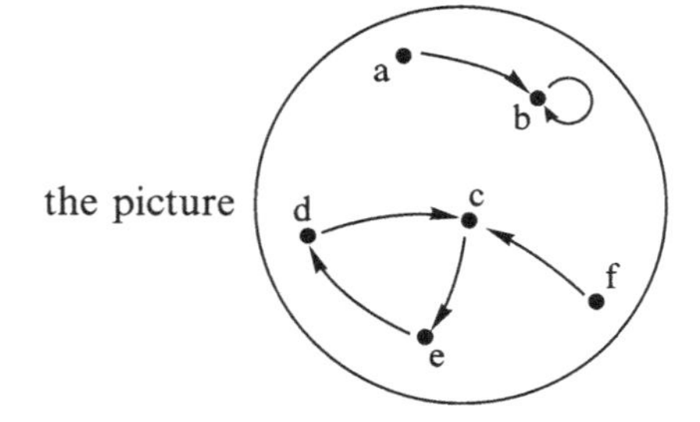

is to be regarded as 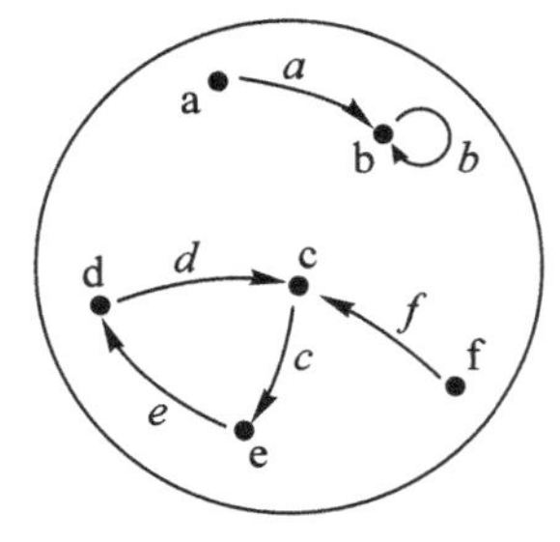

The next four exercises concern isomorphisms in $\boldsymbol{S}^{\circlearrowleft}$.

Exercise 2:

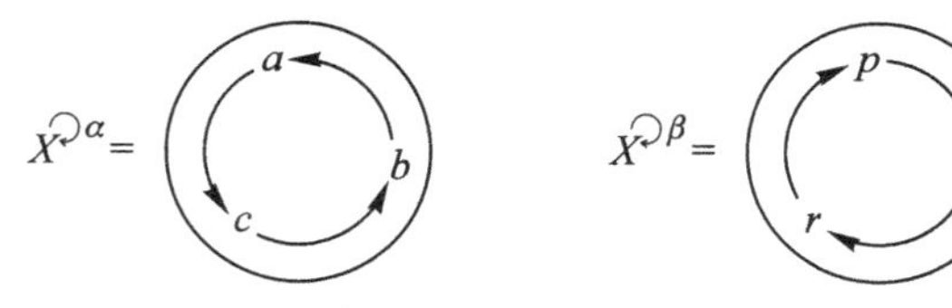

Find an isomorphism from $X^{\circlearrowleft\alpha}$ to $Y^{\circlearrowleft\beta}$. How many such isomorphisms are there?

Hint: You need to find $X \xrightarrow{f} Y$ such that $f\alpha = \beta f$, and check that f has an inverse $Y \xrightarrow{f^{-1}} X$ (meaning $f^{-1}f = 1_X$ and $ff^{-1} = 1_Y$). Then you'll still need to check that f^{-1} is a map in $\boldsymbol{S}^{\circlearrowleft}$ (meaning $f^{-1}\beta = \alpha f^{-1}$), but see Exercise 4, below.

Exercise 3:
Prove that there is no isomorphism (in $\boldsymbol{\mathcal{S}}^{\circlearrowleft}$)

from $X^{\circlearrowleft \alpha} =$ to $Y^{\circlearrowleft \beta} =$

Hint: In fact, more is true: there is no map (in $\boldsymbol{\mathcal{S}}^{\circlearrowleft}$) from $X^{\circlearrowleft \alpha}$ to $Y^{\circlearrowleft \beta}$.

Exercise 4:
Suppose $A^{\circlearrowleft \alpha} \underset{f^{-1}}{\overset{f}{\longrightarrow}} B^{\circlearrowleft \beta}$ is a map in $\boldsymbol{\mathcal{S}}^{\circlearrowleft}$, and that as a map of sets, $A \xrightarrow{f} B$ has an inverse $B \xrightarrow{f^{-1}} A$. Show that f^{-1} is automatically a map in $\boldsymbol{\mathcal{S}}^{\circlearrowleft}$.

Putting Exercises 3 and 4 together, we see that if two sets-with-endomap, $A^{\circlearrowleft \alpha}$ and $B^{\circlearrowleft \beta}$ have A isomorphic to B as sets, we *cannot* conclude that $A^{\circlearrowleft \alpha}$ is isomorphic to $B^{\circlearrowleft \beta}$. Nevertheless, if we are given a map in $\boldsymbol{\mathcal{S}}^{\circlearrowleft}$, $A^{\circlearrowleft \alpha} \xrightarrow{f} B^{\circlearrowleft \beta}$ which is an isomorphism of A and B (as sets), then it is also an isomorphism of $A^{\circlearrowleft \alpha}$ and $B^{\circlearrowleft \beta}$ (as sets-with-endomap).

Exercise 5:
$\mathbb{Z} = \{\ldots, -2, -1, 0, 1, 2, 3, \ldots\}$ is the set of integers, and $\mathbb{Z}^{\circlearrowleft \alpha}$ and $\mathbb{Z}^{\circlearrowleft \beta}$ are the maps which add 2 and 3: $\alpha(n) = n + 2$, $\beta(n) = n + 3$. Is $\mathbb{Z}^{\circlearrowleft \alpha}$ isomorphic to $\mathbb{Z}^{\circlearrowleft \beta}$? (If so, find an isomorphism $\mathbb{Z}^{\circlearrowleft \alpha} \xrightarrow{f} \mathbb{Z}^{\circlearrowleft \beta}$; if not, explain how you know they are not isomorphic.)

The next two exercises concern isomorphisms in $\boldsymbol{\mathcal{S}}^{\downdownarrows}$.

Exercise 6:
Each of the following graphs is isomorphic to exactly one of the others. Which?

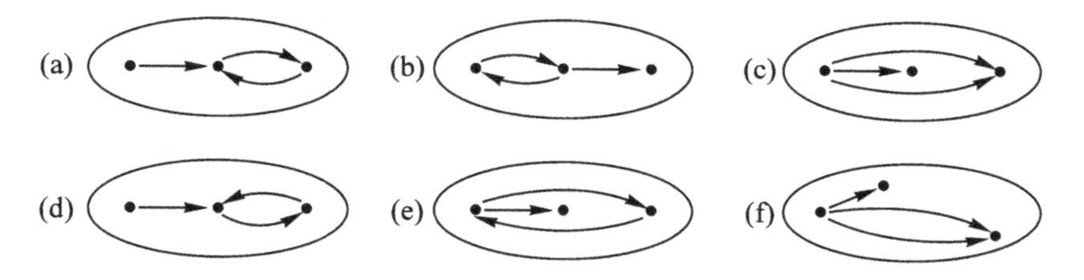

Exercise 7:
If these two graphs are isomorphic, find an isomorphism between them; if they are not isomorphic, explain how you know they are not.

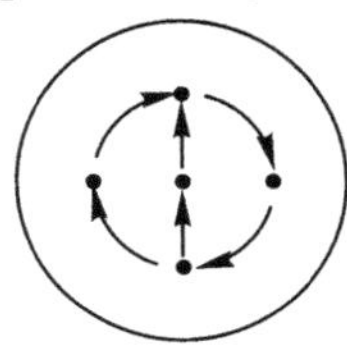
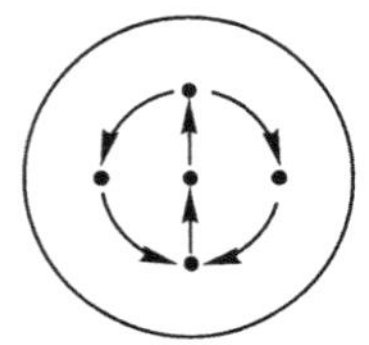

Exercise 8:
(Impossible journeys) J is the graph

G is any graph, and b and e are dots of G.

(*a*) Suppose that $G \xrightarrow{f} J$ is a map of graphs with $fb = 0$ and $fe = 1$. Show that there is no path in G that begins at b and ends at e.

(*b*) Conversely, suppose that there is no path in G that begins at b and ends at e. Show that there is a map $G \xrightarrow{f} J$ with $fb = 0$ and $fe = 1$.

You can think of the dots as cities and the arrows as available airline flights, or the dots as states of a physical system and the arrows as simple processes for getting from one state to another, if you want. Here is an example:

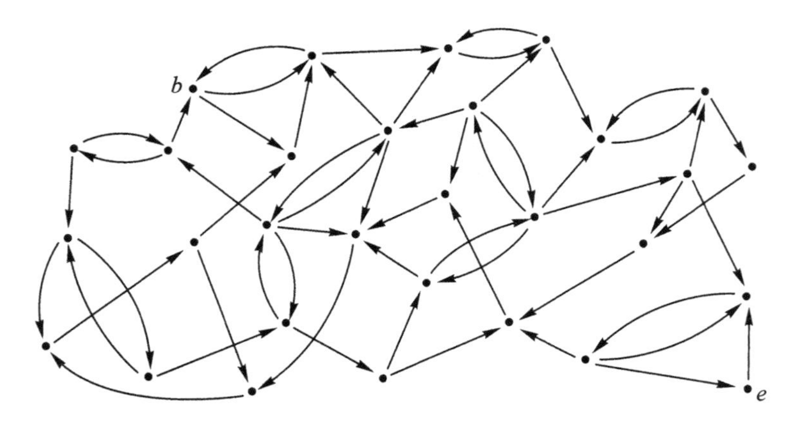

Can one get from b to e? Is there a map $G \xrightarrow{f} J$ with $fb = 0$ and $fe = 1$?

SESSION 12

Categories of diagrams

1. Dynamical systems or automata

The practical use of the category $\mathbf{S}^{\circlearrowright}$, studied last session, is suggested by two names which have been given it: the category of dynamical systems, or the category of automata. Remember that an object in $\mathbf{S}^{\circlearrowright}$ is a set equipped with an endomap, $X^{\circlearrowright\alpha}$, and that a map from $X^{\circlearrowright\alpha}$ to $Y^{\circlearrowright\beta}$ is a map of sets from X to Y, $f : X \longrightarrow Y$, such that $f \circ \alpha = \beta \circ f$. This equation can be remembered by drawing the diagram of all the maps involved:

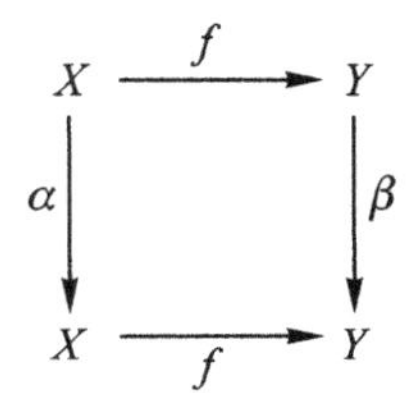

In the dynamical system view, we have the set X of all the different possible states of the system, and the endomap α of X which takes each state x to the state in which the system will be one unit of time later. If instead we think of an object of $\mathbf{S}^{\circlearrowright}$ as an automaton or machine, X is the set of all possible states in which the machine can be, and α gives for each state, the state in which the machine will be if one 'pushes the button' once. Composing α with itself, $\alpha \circ \alpha = \alpha^2$ gives the operation of 'pushing the button twice.' A simple example of such a system is a push button that turns a lamp on and off. In this machine the set of states has only two elements, and the endomap interchanges them, so that this automaton can be pictured like this:

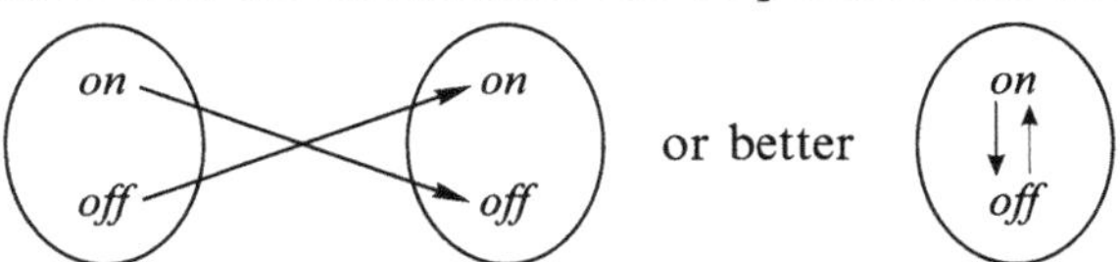

If $X^{\circlearrowright\alpha}$ and $Y^{\circlearrowright\beta}$ are two dynamical systems then a map from $X^{\circlearrowright\alpha}$ to $Y^{\circlearrowright\beta}$ sends a state x of the first system to a state which transforms under the dynamics β 'in the same way' that x transforms under the dynamics α. Exercise 1 gives an example.

Exercise 1:
Suppose that $x' = \alpha^3(x)$ and that $X^{\circlearrowright\alpha} \xrightarrow{f} Y^{\circlearrowright\beta}$ is a map in $\mathbf{S}^{\circlearrowright}$. Let $y = f(x)$ and $y' = \beta^3(y)$. Prove that $f(x') = y'$.

Exercise 2 shows a general characteristic of finite dynamical systems, and Exercise 3 gives an idea of how kinship patterns can be formulated in an appropriate category.

Exercise 2:
'With age comes stability.' In a finite dynamical system, every state eventually settles into a cycle.

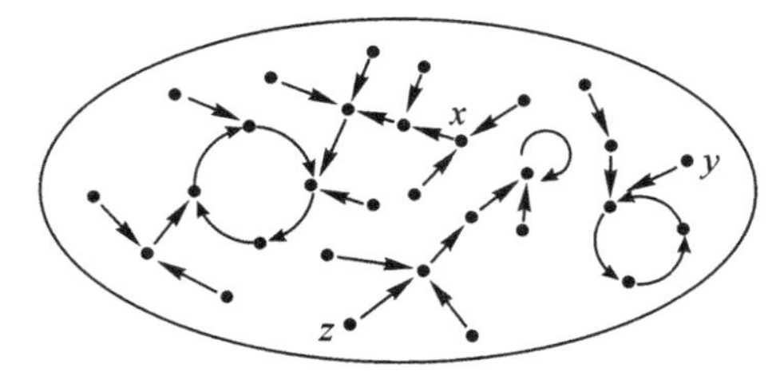

For two units of time, x is living on the fringes, but after that he settles into an organized periodic behaviour, repeating the same routine every four units of time. What about y and z? Don't take the title seriously; humans can change the system! This sort of thing applies to light bulbs, though. If a particular light bulb can only be lit four times before burning out, after which pressing the on–off button has no effect, draw the automaton modeling its behavior.

2. Family trees

The study of family trees begins with the set of all people and two *endo*maps, $f = father$ and $m = mother$. This suggests a new category, in which an object is a set with a specified pair of endomaps. In keeping with our general scheme of notation, we should denote this category by $\mathbf{S}^{\circlearrowright\cdot\circlearrowleft}$. Of course we must say what are to be the maps in this category, but I hope by now that you can see what the reasonable notion of 'structure-preserving map' is. Since our notion of structure this time involves one set and two structural maps, a map in $\mathbf{S}^{\circlearrowright\cdot\circlearrowleft}$ should be one map of sets satisfying two equations; you should figure out precisely what they are. You will notice that this category contains many objects that cannot reasonably be interpreted as a set of people with 'father' and 'mother' maps; for example, a 'person' can be its own 'mother', or even its own 'mother' and its own 'father'. In Exercise 3 you will see that these strange objects are still very useful for sorting other objects. Just as the *set* of all people can be sorted into genders by a map into the set {female, male}, we can sort the *object* of all people by a map into a certain 'gender object' in our category $\mathbf{S}^{\circlearrowright\cdot\circlearrowleft}$.

Exercise 3:

(a) Suppose $\mathbf{P} = {}^{m}\circlearrowright \mathbf{P} \circlearrowleft^{f}$ is the set P of all people together with the endomaps $m = \textit{mother}$ and $f = \textit{father}$. Show that 'gender' is a map in the category $\mathcal{S}^{\circlearrowright\cdot\circlearrowleft}$ from **P** to the object

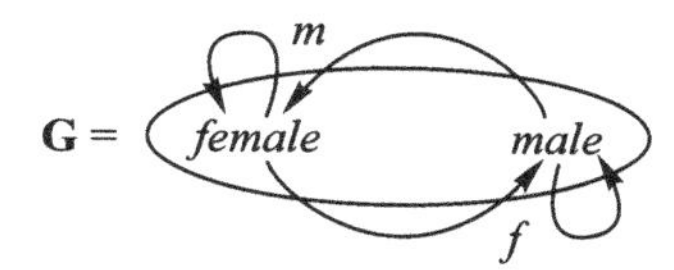

(b) In a certain society, all the people have always been divided into two 'clans,' the Wolf-clan and the Bear-clan. Marriages within a clan are forbidden, so that a Wolf may not marry a Wolf. A child's clan is the same as that of its mother. Show that the sorting of people into clans is actually a map in $\mathcal{S}^{\circlearrowright\cdot\circlearrowleft}$ from **P** to the object

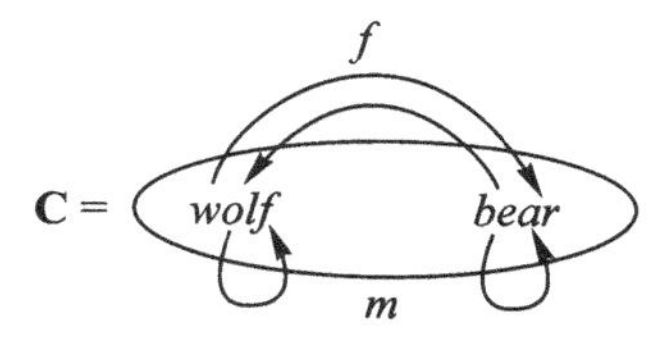

(c) Find appropriate 'father' and 'mother' maps to make

$\mathbf{G} \times \mathbf{C} =$ *he-wolf* *he-bear* *she-wolf* *she-bear*

into an object in $\mathcal{S}^{\circlearrowright\cdot\circlearrowleft}$ so that 'clan' and 'gender' can be combined into a single map $\mathbf{P} \longrightarrow \mathbf{G} \times \mathbf{C}$. (Later, when we have the precise definition of multiplication of objects in categories, you will see that $\mathbf{G} \times \mathbf{C}$ really *is* the product of **G** and **C**.)

3. Dynamical systems revisited

Some of the recent exercises use only the associative and identity laws, and so the results are valid in any category. In spite of this greater generality, these are the easiest problems; they must be, since they use so little. Other exercises are designed to give you a feel for the idea of 'structure-preserving map'; these will gradually acquire more significance as we study more examples.

As suggested in Section 4 of Article III, we can construct new categories from any category $\mathcal{C}$ in the same way that we constructed new categories from $\mathcal{S}$. Let $\mathcal{C}$ be any category whatsoever, and now write X or Y to stand for any object of $\mathcal{C}$, so that an arrow $X \longrightarrow Y$ means a map of the category $\mathcal{C}$.

Just as we invented the category $\mathcal{S}^{\circlearrowright}$, we can make a new category $\mathcal{C}^{\circlearrowright}$ in which an object is to be a '$\mathcal{C}$-object-equipped-with-an-endomap.' That is, an object of $\mathcal{C}^{\circlearrowright}$ is of the form $X^{\circlearrowright\alpha}$, where X is an object of $\mathcal{C}$ and α is an endomap of this object in $\mathcal{C}$. Now we want to complete the specification of the category $\mathcal{C}^{\circlearrowright}$ and to check that we have satisfied the associative and identity laws (and, of course, the 'bookkeeping' laws about domains and codomains).

What do we need to do to complete our specification of $\mathcal{C}^{\circlearrowright}$? We must decide what are the maps in $\mathcal{C}^{\circlearrowright}$, what is the composite of two maps, and what are the identity maps. Just as we did with $\mathcal{S}^{\circlearrowright}$, we decide:

1. a map $X^{\circlearrowright\alpha} \xrightarrow{f} Y^{\circlearrowright\beta}$ will be a map $X \xrightarrow{f} Y$ in $\mathcal{C}$ satisfying $f \circ \alpha = \beta \circ f$;
2. the composite of $X^{\circlearrowright\alpha} \xrightarrow{f} Y^{\circlearrowright\beta} \xrightarrow{g} Z^{\circlearrowright\gamma}$ will just be the composite in $\mathcal{C}$, $X \xrightarrow{g \circ f} Z$; and
3. the identity map on $X^{\circlearrowright\alpha}$ will just be the identity map on X in $\mathcal{C}$, $X \xrightarrow{1_X} X$.

What must we check to be sure we have specified a category? We must check, first, that if f and g are maps in $\mathcal{C}^{\circlearrowright}$ (i.e. $f \circ \alpha = \beta \circ f$ and $g \circ \beta = \gamma \circ g$) then the composite map $g \circ f$ in (2) really is a map in $\mathcal{C}^{\circlearrowright}$ (i.e. $\gamma \circ (g \circ f) = (g \circ f) \circ \alpha$.) This just uses the associative law in $\mathcal{C}$.

$$\gamma(gf) = (\gamma g)f = (g\beta)f = g(\beta f) = g(f\alpha) = (gf)\alpha$$

Do you see the justification for each step? By now we can even shorten things by leaving out the parentheses, thus taking the associative law in $\mathcal{C}$ for granted, and just write

$$\gamma gf = g\beta f = gf\alpha$$

It is helpful, in order to picture the equations and guide the calculations, to draw the following diagrams:

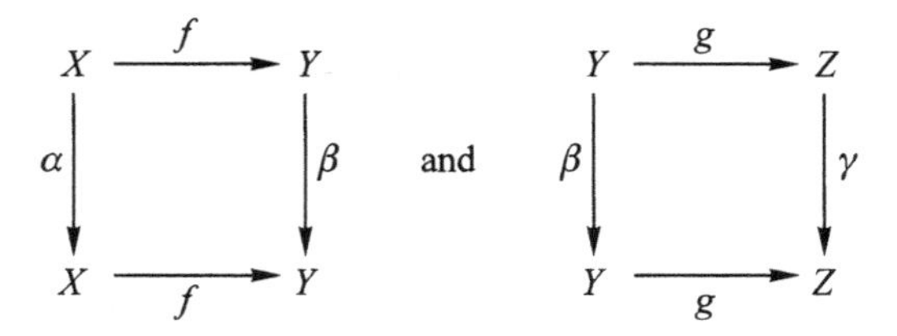

or combine them into the single diagram

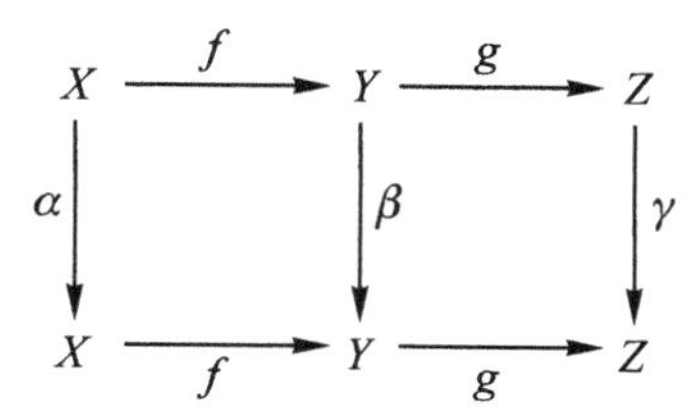

This enables us to follow the calculation pictorially:

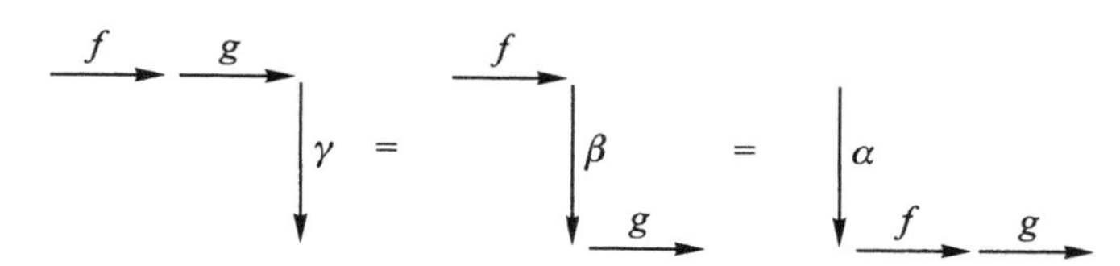

That is, we use the fact that the two ways of getting from northwest to southeast in each of the two small squares are the same, to guide us in seeing how to prove that the two outer routes from northwest to southeast in the large rectangle are the same.

Of course, we still have to check that our supposed identity maps in $\mathcal{C}^{\circlearrowright}$ really are maps in $\mathcal{C}^{\circlearrowright}$: that means, for any $X^{\circlearrowright\alpha}$ in $\mathcal{C}^{\circlearrowright}$, 1_X satisfies $\alpha \circ 1_X = 1_X \circ \alpha$; but that's easy, since both sides are α.

Finally we must check the associative and identity laws in $\mathcal{C}^{\circlearrowright}$. However, I say these laws are obvious for $\mathcal{C}^{\circlearrowright}$? Why? How is the composition defined in $\mathcal{C}^{\circlearrowright}$?

OMER: By the composition in $\mathcal{C}$.

Right. And if you check you'll see that the identity and associative laws for $\mathcal{C}^{\circlearrowright}$ are therefore direct consequences of those for $\mathcal{C}$.

Just as in the case of the category $\mathcal{S}^{\circlearrowright}$ of endomaps of sets, for $\mathcal{C}^{\circlearrowright}$ also we can form certain subcategories:

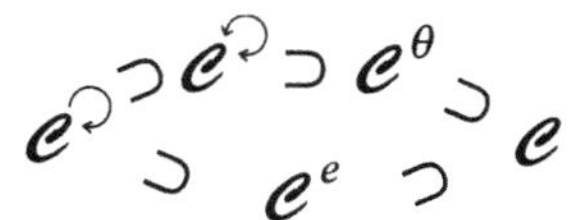

$\mathcal{C}^{e}$ consists of those endomaps of $\mathcal{C}$ which are idempotent;
$\mathcal{C}^{\circlearrowleft}$ consists of those endomaps of $\mathcal{C}$ which are invertible;
$\mathcal{C}^{\theta}$ consists of those endomaps of $\mathcal{C}$ which are not only invertible, but are their own inverse.

DANILO: About the category $\mathcal{C}$. Is $\mathcal{C}$ just less specified than $\mathcal{S}$?

Yes. $\mathcal{C}$ can be any category whatsoever, so that all that we say about $\mathcal{C}$ is necessarily true for all categories. For example, $\mathcal{C}$ can be $\mathcal{S}$ itself, or it can be $\mathcal{S}^{\circlearrowright}$, or $\mathcal{S}^{\downarrow\downarrow}$, or any other category.

SESSION 13

Monoids

In general, in order to specify a category completely I must specify what are the *objects*, what are the *maps*, which object is the *domain of each map*, which object is the *codomain of each map*, which map is the *identity of each object*, and which map is the *composite* of any two 'composable' maps – six things to be specified. Of course, this cannot be done in any arbitrary way. Recall these laws that must be satisfied:

bookkeeping laws
associative law
identity laws

Here is a special case. Suppose we have only one object, which we call '$*$'. This means that all the maps in the category are endomaps (of this unique object). Nevertheless there may be many maps in this category. Suppose that as maps I take all natural numbers: 0 is a map, 1 is a map, 2 is a map, and so on. They all are maps from $*$ to $*$, so that we can write,

$$* \xrightarrow{0} *, \quad * \xrightarrow{1} *, \quad * \xrightarrow{2} *, \quad * \xrightarrow{3} *, \quad \text{etc.}$$

What shall we take as the composition in this category? There are many possibilities, but the one that I will choose is just multiplication. In other words, the composite of two maps in this category – two numbers – is their product: $n \circ m = n \times m$. Because there is only one object, the bookkeeping laws are automatically satisfied.

Now we must specify the identity map of the only object, $*$. What number should we declare to be 1_*? Now 1_* is supposed to satisfy $1_* \circ n = n$ and $n \circ 1_* = n$ for every number n, and according to our definition of composition these equations just mean that 1_* is a number which multiplied by any other number n gives n. Therefore the only choice is clear: the identity of $*$ must be the number one: $1_* = 1$.

Definition: *A category with exactly one object is called a* **monoid**.

Such a category seems to be a little strange in the sense that the object seems featureless. However, there are ways of interpreting any such category in sets, so that the object takes on a certain life. Let's call the category we defined above $\mathcal{M}$ for multiplication. An interpretation will be denoted this way:

$$\mathcal{M} \longrightarrow \mathcal{S}$$

One interpretation 'interprets' the only object $*$ of $\mathcal{M}$ as the set $\mathbb{N}$ of natural numbers, and each map in $\mathcal{M}$ (a natural number) is interpreted as a map from the set of natural numbers to itself

$$\mathbb{N} \xrightarrow{f_n} \mathbb{N}$$

defined by

$$f_n(x) = n \times x$$

Thus $f_3(x) = 3x$, $f_5(x) = 5x$, and so on. According to this what map is f_1? How do you evaluate this map? By multiplying by 1, right? So what is f_1?

ALYSIA: The identity?

That's right, $f_1 = 1_{\mathbb{N}}$. Also, the composite of two maps $f_n \circ f_m$ is the map which evaluated at a number x gives

$$\begin{aligned}(f_n \circ f_m)(x) &= f_n(f_m(x)) = n \times (m \times x) = (n \times m) \times x = (nm) \times x \\ &= f_{nm}(x)\end{aligned}$$

so we conclude that

$$f_n \circ f_m = f_{nm}$$

This shows that this interpretation preserves the structure of the category, because the objects go to objects, the maps go to maps, the composition is preserved, and the identity maps go to identity maps. Such a 'structure-preserving' interpretation of one category into another is called a **functor** (from the first category to the second). Actually a functor is required to preserve also the notions of 'domain' and 'codomain,' but in our example this is automatic since all the maps have the same domain and codomain.

Such a functor also sheds light on the sense in which we can use the symbol of raising to minus one as a vast generalization of 'inverse.' If we change the example slightly, taking rational numbers instead of natural numbers as the maps in $\mathcal{M}$, we'll find that $(f_3)^{-1} = f_{(3^{-1})}$. The inverse map of a map in the list of interpretations is also an example of the maps in the list, so if a 'named' map is invertible, the inverse can also be named. In the example above, f_3 is invertible and its inverse is named by the inverse of 3.

But if the maps in $\mathcal{M}$ consist only of the natural numbers, and $*$ is interpreted as the set of rational numbers, then f_3 has an inverse, but now it is not named since there is no natural number inverse of 3.

DANILO: I can see that f_3 has a retraction, but why does it have a section?

Well, the commutativity of multiplication of numbers implies that $f_n \circ f_m = f_m \circ f_n$.

Are all the maps in the list invertible?

OMER: No, f_0 is not.

Right. $f_0(x) = 0 \times x = 0$, so many different numbers are mapped to 0; f_0 isn't invertible.

Let's now introduce another category which we can call $\mathcal{N}$. This is also a monoid, so that it will only have one object, denoted again $*$. The maps will be again numbers, but now the composition will be addition instead of multiplication. What number should be the identity of $*$? The condition that 1_* is required to satisfy means that 'adding 1_* to any number n gives n.' Therefore 1_* must be 0. To give a functor $\mathcal{N} \longrightarrow \mathcal{S}$ means that we interpret $*$ as some set S and each map $* \xrightarrow{n} *$ in $\mathcal{N}$ as an endomap $S \xrightarrow{g_n} S$ of the set S, in such a way that $g_0 = 1_S$ and $g_n \circ g_m = g_{n+m}$. We might take S to be a set of numbers and define

$$g_n(x) = n + x$$

In particular, take $S = \mathbb{N}$ (the natural numbers) so that for example the number 2 (as a map in $\mathcal{N}$) is interpreted as the map g_2 (in $\mathcal{S}$) whose internal picture is

$$g_2 = \boxed{\begin{array}{l} 0 \to 2 \to 4 \to 6 \to \ldots \\ \quad 1 \to 3 \to 5 \to \ldots \end{array}}$$

Now we have to check that $g_{n+m}(x) = g_n(g_m(x))$, which is similar to the previous case:

$$g_n(g_m(x)) = n + (m + x) = (n + m) + x = g_{n+m}(x)$$

All the above suggests the 'standard example' of interpretation of a monoid in sets, in which the object of the monoid is interpreted as the set of maps of the monoid itself. In this way we get a standard functor from any monoid to the category of sets.

There are many functors from $\mathcal{N}$ to sets other than the standard one. Suppose I take a set X together with an endomap α, and I interpret $*$ as X and send each map n of $\mathcal{N}$ (a natural number) to the composite of α with itself n times, i.e. α^n, and in order to preserve identities, I send the number 0 to the identity map on X. In this way we get a functor from $\mathcal{N}$ to sets, $h : \mathcal{N} \longrightarrow \mathcal{S}$ which can be summarized this way:

1. $h(*) = X$,
2. $h(n) = \alpha^n$, and
3. $h(0) = 1_X$. (It is reasonable, for an endomap α of an object X, to define the symbol α^0 to mean 1_X; then (3) becomes a special case of (2).)

Then it is clear that $h(n + m) = h(n) \circ h(m)$.

In this way, whenever we specify a set-with-endomap $X^{\circlearrowright \alpha}$ we obtain a functorial interpretation of $\mathcal{N}$ in sets. This suggests that another reasonable name for $\mathcal{S}^{\circlearrowright}$ would be $\mathcal{S}^{\mathcal{N}}$ to suggest that an object is a functor from $\mathcal{N}$ to $\mathcal{S}$. This was the category of dynamical systems (more appropriately called 'discrete-time dynamical systems'). A discrete-time dynamical system is just a functor from this monoid $\mathcal{N}$ to the category of sets. What would be a 'continuous-time dynamical system'?

DANILO: Just replace the natural numbers with the real numbers?

Right. Allow all real numbers as maps in the monoid. Then, giving a functor from the new monoid $\mathcal{R}$ to sets amounts to giving a set X and an endomap $X \circlearrowleft^{\alpha_t}$ for each real number t. To preserve composition, we must ensure that $\alpha_0 = 1_X$, and that $\alpha_{s+t} = \alpha_s \circ \alpha_t$. We can think of X as the set of 'states' of a physical system which, if it is in the state x at a certain time, then t units of time later it will be in the state $\alpha_t(x)$.

SESSION 14

Maps preserve positive properties

Here are some exercises on the meaning of maps in $\mathbf{S}^{\circlearrowleft}$. The idea of these exercises is to see that the condition (in $\mathbf{S}$)

$$f \circ \alpha = \beta \circ f$$

which we took as the definition of 'map'

$$X^{\circlearrowleft\alpha} \xrightarrow{f} Y^{\circlearrowleft\beta}$$

really is the appropriate notion of 'map of dynamical systems.' Assume in these problems that f, α, and β satisfy the above relation.

Exercise 1:
Let x_1 and x_2 be two points of X and define $y_1 = f(x_1)$ and $y_2 = f(x_2)$. If

$$\alpha(x_1) = \alpha(x_2) \text{ in } X$$

(i.e. pushing the button α we arrive at the same state whether the initial state was x_1 or x_2) then show that

$$\beta(y_1) = \beta(y_2) \text{ in } Y$$

(the 'same' statement with button β on the machine $Y^{\circlearrowleft\beta}$ with regard to its two states y_1 and y_2).

Exercise 2:
If instead we know that

$$x_2 = \alpha^5(x_1) \text{ in } X$$

(i.e. that starting from state x_1, five pushes of the button α will bring X to the state x_2), show that the 'same' statement is true of the states y_1 and y_2 in $Y^{\circlearrowleft\beta}$; i.e.

$$y_2 = \beta^5(y_1) \text{ in } Y$$

Exercise 3:
If $\alpha(x) = x$ (i.e. x is an 'equilibrium state' or 'fixed point' of α), then the 'same' is true of $y = f(x)$ in $Y^{\circlearrowleft\beta}$.

Exercise 4:
Give an example in which x is not a fixed point of α but $y = f(x)$ is a fixed point of β. This illustrates that although certain important properties are 'preserved' by f they are not necessarily 'reflected.' Hint: The simplest conceivable example of $Y^{\circlearrowright\beta}$ will do.

Exercise 5:
Show that if $\alpha^4(x) = x$, then the 'same' is true of $y = f(x)$. (Same idea as Exercise 2.) But give an example where $\alpha^4(x) = x$ and $\alpha^2(x) \neq x$, while $\beta^2(y) = y$ and $\beta(y) \neq y$. This illustrates that while f preserves the property of being in a small cycle, the size of the cycle may decrease.

Now we are going to work out some of these exercises. In the first one we have a map $X^{\circlearrowright\alpha} \xrightarrow{f} Y^{\circlearrowright\beta}$ in $\boldsymbol{S}^{\circlearrowright}$, i.e. satisfying the condition $f \circ \alpha = \beta \circ f$, and we have two points x_1 and x_2 of X such that $\alpha(x_1)$ equals $\alpha(x_2)$, so that *part* of the internal diagram of α is:

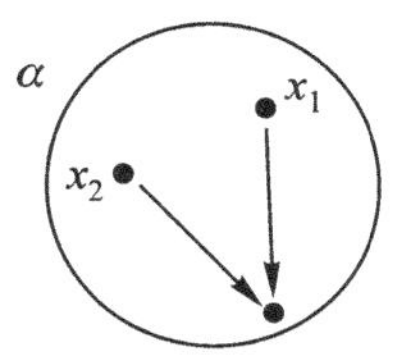

The problem is to prove that the points y_1 and y_2 obtained by applying f to x_1 and x_2 – i.e. $y_1 = f(x_1)$ and $y_2 = f(x_2)$ – satisfy the 'same property' as x_1 and x_2. In other words, the problem is to prove that $\beta(y_1) = \beta(y_2)$.

Solution: By direct substitution and use of the associative law it is immediate that $\beta(y_1)$ is equal to $f(\alpha(x_1))$ and that $\beta(y_2)$ is equal to $f(\alpha(x_2))$. For example,

$$\beta(y_1) = \beta(f(x_1)) = (\beta \circ f)(x_1) = (f \circ \alpha)(x_1) = f(\alpha(x_1))$$

and replacing the subscript 1 by 2 proves the other equality. But we know that $f(\alpha(x_1)) = f(\alpha(x_2))$ because by hypothesis $\alpha(x_1) = \alpha(x_2)$. Therefore we can conclude that $\beta(y_1) = \beta(y_2)$.

The idea of the exercise is to learn that a map f for which

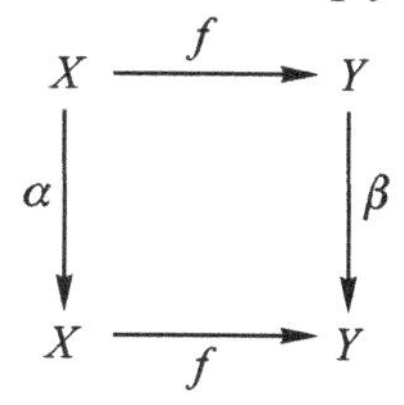

commutes ($f \circ \alpha = \beta \circ f$) *preserves the property* of two points 'reaching the same point in one step.' Most of these exercises are of the same nature: to prove that some relation among points in the domain also holds among their images in the codomain.

Note: The phrase 'point of an object' will later be given a precise definition, useful in many categories. We will then find that for a dynamical system $X^{\circlearrowright\alpha}$, there are *two* relevant notions of point: (1) a point of the *set* X, $\mathbf{1} \longrightarrow X$, and (2) a point of the *dynamical system* $X^{\circlearrowright\alpha}$, which will turn out to amount to a very special kind of point of X, a *fixed* point, an x for which $\alpha(x) = x$. In this session we discuss only *points of* X, not points of $X^{\circlearrowright\alpha}$; we mention this distinction now only to avoid confusion if you reread this session after you have learned the notion *point of* $X^{\circlearrowright\alpha}$.

Now let's look at Exercise 2. We assume that we have two points x_1, x_2 in X such that $\alpha^5(x_1) = x_2$. If $X^{\circlearrowright\alpha}$ is a machine, we can intepret that property as saying that starting in the state x_1 and pressing the button five times we end up in the state x_2. If $X^{\circlearrowright\alpha} \xrightarrow{f} Y^{\circlearrowright\beta}$ is a map in $\boldsymbol{S}^{\circlearrowright}$ and $y_1 = f(x_1)$ and $y_2 = f(x_2)$, the problem is to prove that $\beta^5(y_1) = y_2$. Said otherwise, this problem amounts to showing that if $f \circ \alpha = \beta \circ f$ then $f \circ \alpha^5 = \beta^5 \circ f$.

DANILO: So, just substitute α^5 and β^5 for α and β.

No, it is not so immediate. The fact that $f \circ \alpha = \beta \circ f$ is true for particular maps α and β doesn't allow us to substitute any maps for α and β. One has to prove that it works for α^5 and β^5, by using the associative law. It is true that this will imply that f is also a map for the new dynamics determined by the endomaps α^5 and β^5, but it has to be proved, it can't be assumed. The proof consists in applying the associative law several times:

$$\begin{aligned}
f \circ \alpha^5 &= f \circ (\alpha \circ \alpha^4) = (f \circ \alpha) \circ \alpha^4 = (\beta \circ f) \circ \alpha^4 = \beta \circ (f \circ \alpha^4) \\
&= \beta \circ ((f \circ \alpha) \circ \alpha^3) = \beta \circ ((\beta \circ f) \circ \alpha^3) = (\beta \circ (\beta \circ f)) \circ \alpha^3 \\
&= (\beta^2 \circ f) \circ \alpha^3 = \beta^2 \circ (f \circ \alpha^3) \\
&= \ldots \\
&= \ldots \\
&= \beta^4 \circ (\beta \circ f) = \beta^5 \circ f
\end{aligned}$$

Now we can write:

$$\begin{aligned}
\beta^5(y_1) &= \beta^5(f(x_1)) = (\beta^5 \circ f)(x_1) = (f \circ \alpha^5)(x_1) = f(\alpha^5(x_1)) \\
&= f(x_2) = y_2
\end{aligned}$$

Exercise 3 concerns the fixed points of an endomap $X^{\circlearrowright\alpha}$, and it consists in proving that a map $X^{\circlearrowright\alpha} \xrightarrow{f} Y^{\circlearrowright\beta}$ (meaning, as always, $f \circ \alpha = \beta \circ f$) takes every fixed point of $X^{\circlearrowright\alpha}$ to a fixed point of $Y^{\circlearrowright\beta}$. A fixed point of α is an element x of X such that $\alpha(x) = x$. Such points can be thought of as 'equilibrium states' of the dynamics determined by α. For example, the endomap

$$\bullet \rightleftarrows \bullet$$

has no fixed points, while for the endomap

$$\bullet \to \bullet \to \bullet \circlearrowright$$

only one of the three points is a fixed point.

To do the exercise we assume $\alpha(x) = x$ and $y = f(x)$ and prove $\beta(y) = y$.

OMER: Just substitute and apply associativity.

Right. So, the proof is: $\beta y = \beta f x = f \alpha x = f x = y$.

OMER: You left out the little circles and you are missing the parentheses.

Yes. I leave that for you to fill in. After some practice with the associative law, you realize that you wouldn't make any mistake by omitting these, and that the proofs become much shorter by doing so. For example, the last proof took four steps. If you write all the parentheses it would take six steps and many more symbols. For now I suggest that in your exercises you do the proofs both ways, to ensure that you understand perfectly well the justification for every step.

1. Positive properties versus negative properties

One property an element x of $X^{\circlearrowright\alpha}$ may have is *to be a value of* α. This means that there exists an element $\bar{x}$ such that $x = \alpha(\bar{x})$. This property of x can be called 'accessibility'; it says that there is an 'access' (the $\bar{x}$) to reach x. For example in the system we saw before,

$$\overset{y}{\bullet} \to \overset{x}{\bullet} \to \overset{z}{\bullet} \circlearrowright$$

the elements x and z have this property, but y doesn't because no element goes to y. (This is a *positive property*. We'll come back to these later.) I claim that accessibility is preserved by the maps in $\mathcal{S}^{\circlearrowright}$. In other words, if we have $X^{\circlearrowright\alpha} \xrightarrow{f} Y^{\circlearrowright\beta}$ in $\mathcal{S}^{\circlearrowright}$ and x is a value of α, then $f(x)$ is a value of β.

To prove this, assume that we have $\bar{x}$ such that $x = \alpha(\bar{x})$ and try to find a $\bar{y}$ such that $f(x) = \beta(\bar{y})$. The natural thing to try is $f(\bar{x}) = \bar{y}$. It is immediate to show that this works, i.e.

$$\beta(\bar{y}) = \beta f \bar{x} = f \alpha \bar{x} = f x$$

OMER: Can you do it the other way, putting $y = f \alpha \bar{x} = \beta f \bar{x}$?

Yes, and your way helps to discover what to take as $\bar{y}$.

This talk about positive properties is to prepare the way to talk about *negative properties*. An example of a negative property of x is *not being a fixed point*, i.e. $\alpha(x) \neq x$. Negative properties tend not to be preserved, but instead they tend to be *reflected*. To say that a map $X^{\circlearrowright\alpha} \xrightarrow{f} Y^{\circlearrowright\beta}$ in $\mathcal{S}^{\circlearrowright}$ reflects a property means that if the value of f at x has the property, then x itself has the property. In the case of not being a fixed point this means that if $f(x)$ is not a fixed point (i.e. $\beta(f(x)) \neq f(x)$), then x also is not a fixed point (i.e., $\alpha(x) \neq x$.) This is obvious since f has been proved to preserve fixed points. Exercise 4 illustrates that negative properties tend not to be preserved, by asking you to find an example that proves that the property of *not* being a fixed point is not always preserved. The hint says that the simplest example will suffice. We must choose $X^{\circlearrowright\alpha}$ having at least one non-fixed point, and $Y^{\circlearrowright\beta}$ having at least one fixed point. The simplest example is

$$X^{\circlearrowright\alpha} = \boxed{\bullet \longrightarrow \bullet\circlearrowright} \quad \text{and} \quad Y^{\circlearrowright\beta} = \boxed{\bullet\circlearrowright}$$

There is only one possible map $X^{\circlearrowright\alpha} \xrightarrow{f} Y^{\circlearrowright\beta}$ in $\mathbf{S}^{\circlearrowright}$, namely

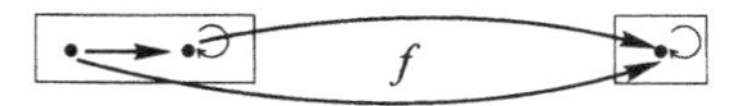

DANILO: If you want to do an example with numbers maybe you can use the identity for the endomap β.

Yes. Take, for example, $X = \mathbb{Z}$, and $\alpha = \textit{adding } 2$ (i.e. $\alpha(n) = 2 + n$) and $f = \textit{parity}$ (i.e. $f(n) = \textit{even}$ or $\textit{odd}$, depending on what n is). Then we can take for Y the set $\{\textit{even}, \textit{odd}\}$ and for β the identity map.

$$\mathbb{Z}^{\circlearrowright 2+(\,)} \xrightarrow{f = \textit{parity}} \{\textit{even}, \textit{odd}\}^{\circlearrowright \mathrm{Id}}$$

Notice that adding 2 doesn't change the parity of a number, which means that f is a map in the category $\mathbf{S}^{\circlearrowright}$, and also that no point is fixed in X, but all points are fixed in Y, so that f takes non-fixed points to fixed points.

Another example is the map

$$\mathbb{Z}^{\circlearrowright 2\times(\,)} \xrightarrow{f = \textit{parity}} \{\textit{even}, \textit{odd}\}^{\circlearrowright \beta}$$

where this β is the map that sends both *even* and *odd* to *even*.

Try Exercise 5 yourself.

SESSION 15

Objectification of properties in dynamical systems

1. Structure-preserving maps from a cycle to another endomap

Let $X^{\circlearrowright\alpha}$ and $Y^{\circlearrowright\beta}$ be the following objects of the category $\mathbf{S}^{\circlearrowright}$:

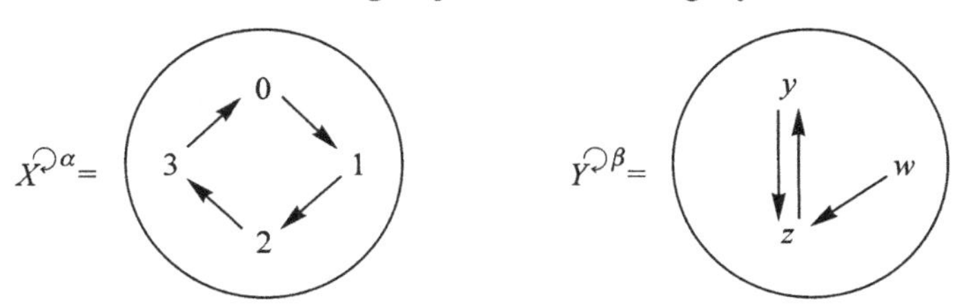

We want to find a map f from $X^{\circlearrowright\alpha}$ to $Y^{\circlearrowright\beta}$ in $\mathbf{S}^{\circlearrowright}$ that sends 0 to y. There are $3^3 = 27$ maps from X to Y that take 0 to y. How many of them are structure-preserving?

For f to be structure-preserving (i.e. $f(\alpha(x)) = \beta(f(x))$ for every x in X) we must have that $f(1)$ is z, because

$$f(1) = f(\alpha(0)) = \beta(f(0)) = \beta(y) = z$$

By the same token $f(2)$ must be $\beta(z)$, which is y, and $f(3) = z$; so f is

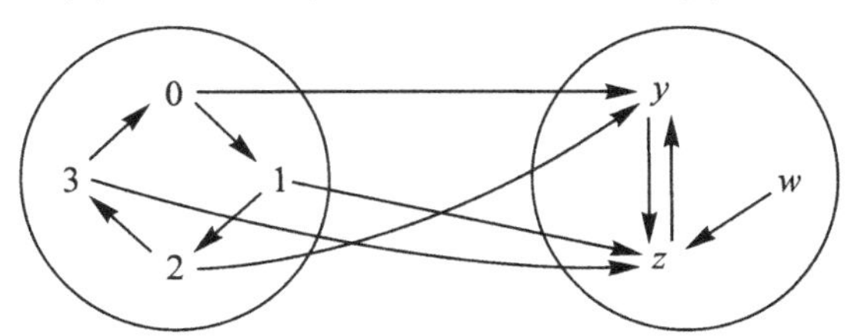

But this is based on the *assumption* that the map f is structure-preserving. We must check that the two maps $f \circ \alpha$ and $\beta \circ f$ are equal, i.e. that they agree at all four elements of their common domain X:

$$\begin{aligned}
&\text{on } 0\text{:}\ f\alpha 0 = f1 = z \ \text{ and } \ \beta f0 = \beta y = z\\
&\text{on } 1\text{:}\ f\alpha 1 = f2 = y \ \text{ and } \ \beta f1 = \beta z = y\\
&\text{on } 2\text{:}\ f\alpha 2 = f3 = z \ \text{ and } \ \beta f2 = \beta y = z\\
&\text{on } 3\text{:}\ f\alpha 3 = f0 = y \ \text{ and } \ \beta f3 = \beta z = y
\end{aligned}$$

We checked the first three elements as we constructed f; only $f\alpha 3 = \beta f 3$ needed checking. Thus the two maps $f \circ \alpha$ and $\beta \circ f$ agree on the four elements of X, showing that f is structure-preserving. In fact, there are exactly *two* structure-preserving maps from $X^{\circlearrowleft\alpha}$ to $Y^{\circlearrowleft\beta}$ one which sends 0 to y, and one which sends 0 to z, but none which sends 0 to w. Do you see why?

These maps also illustrate that structure-preserving maps do not preserve negative properties: every element x of $X^{\circlearrowleft\alpha}$ has the negative property $\alpha^2(x) \neq x$, but the image of x has to be y or z and neither of these has this negative property. On the other hand, the positive property $x = \alpha^4(x)$ is preserved by a structure-preserving map; since 0 has this property, $f(0)$ must also have it.

Note the difference in the type of proof we have for the two facts:

1. *all maps in* $\mathbf{S}^{\circlearrowleft}$ *preserve positive properties*, and
2. *some maps in* $\mathbf{S}^{\circlearrowleft}$ *do not preserve negative properties.*

The proof of (1) is algebraic, while the proof of (2) is by 'counterexample.'

IAN: Is there a general rule to tell how many structure-preserving maps there are between two given sets-with-endomap?

There is no simple general rule, but in some particular cases it is easy to find the number of maps. For example, the number of maps from the $X^{\circlearrowleft\alpha}$ above to any other $Y^{\circlearrowleft\beta}$ is equal to the number of elements in $Y^{\circlearrowleft\beta}$ which have period four. We say that an element y in $Y^{\circlearrowleft\beta}$ has **period** four if $\beta^4(y) = y$. All elements that have period one or two are *included* in this, because if $\beta(y) = y$ or $\beta^2(y) = y$, then also $\beta^4(y) = y$. Now, as we saw before, a map of $\mathbf{S}^{\circlearrowleft}$ with domain $X^{\circlearrowleft\alpha}$ is completely determined by its value at 0, and this can be any element of $Y^{\circlearrowleft\beta}$ of period four. Thus maps from $X^{\circlearrowleft\alpha}$ to any $Y^{\circlearrowleft\beta}$ correspond exactly to elements of period four in $Y^{\circlearrowleft\beta}$. The number of these can also be expressed as a sum of three numbers:

$$\begin{aligned}&(\textit{number of fixed points in } Y^{\circlearrowleft\beta})\\ &+ 2 \times (\textit{number of cycles of length } 2 \textit{ in } Y^{\circlearrowleft\beta})\\ &+ 4 \times (\ \textit{of cycles of length } 4 \textit{ in } Y^{\circlearrowleft\beta})\end{aligned}$$

2. Naming the elements that have a given period by maps

We define 'the cycle of length n,' for any natural number n, as the set of n elements $\{0, 1, 2, \ldots, n-1\}$ with the 'successor' endomap, the 'successor' of $n-1$ being 0.

$C_n =$ 0 → 1 → 2 → 3 → • ... → $n-1$

The object $X^{\circlearrowleft\alpha}$ in our example was C_4, since according to this definition

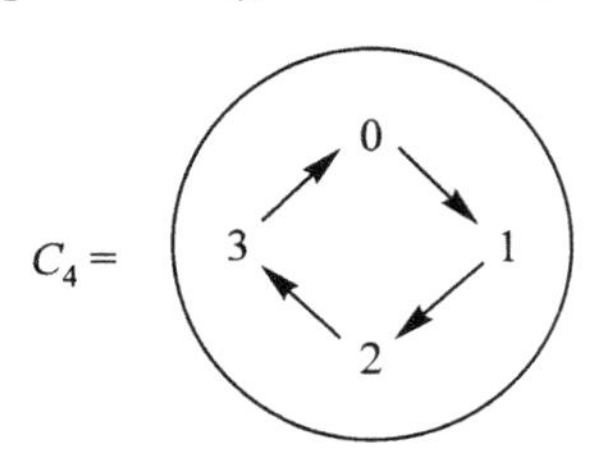

Just as with C_4, maps from C_n to any object $Y^{\circlearrowright\beta}$ correspond precisely to the elements y of $Y^{\circlearrowright\beta}$ having period n. Notice that if an element has period n it also has period equal to every multiple of n. In particular the fixed points have period 1 and therefore they also have period n for every positive integer n.

Thus a fundamental property of the cycle C_n is that the maps from it to any object $Y^{\circlearrowright\beta}$ 'name' exactly the elements of period n in $Y^{\circlearrowright\beta}$. Each map $C_n \longrightarrow Y$ names the element $f(0)$ in Y. This bijective correspondence is expressed symbolically by

$$\frac{C_n \longrightarrow Y^{\circlearrowright\beta}}{\textit{elements } y \textit{ in } Y^{\circlearrowright\beta} \textit{ having period } n}$$

In particular, since the elements of period 1 are precisely the fixed points,

$$\frac{C_1 \longrightarrow Y^{\circlearrowright\beta}}{\textit{fixed points of } B}$$

DANILO: I don't see how there can be any map from C_5 to C_2.

Right, there are none, because C_2 has no element of period five. There is a general pattern worth noticing. If an element has any positive period, it must have a smallest period. In fact, all its periods are multiples of this smallest one! (Of course, every element of C_n has n as its smallest period.)

Exercise 1:
Show that an element which has both period 5 and period 7 must be a fixed point.

3. Naming arbitrary elements

Can we find an object $X^{\circlearrowright\alpha}$ of the category $\boldsymbol{\mathcal{S}}^{\circlearrowright}$ such that the maps from it to any other object name all the elements? Such an $X^{\circlearrowright\alpha}$ must have in it an element with no special positive properties at all, otherwise it could only name elements having the same properties. In particular, it has to be an element without any period at all. An element without a period is one which is not part of a cycle, like the x in

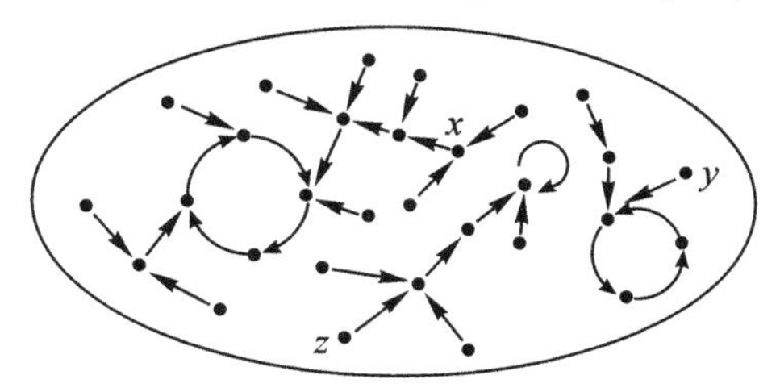

However, the element x here has the property that it 'enters a cycle in three steps.' Is it possible to have an element without any special positive property at all?

DANILO: Using addition?

How is addition an endomap?

DANILO: Take the natural numbers with the endomap which adds one.

That's a good idea. The 'successor map' $\sigma : \mathbb{N} \longrightarrow \mathbb{N}$ defined by $\sigma(n) = n + 1$ looks like this (indicating the entire dynamical system by N):

$$N = \mathbb{N}^{\circlearrowright \sigma} = \boxed{0 \to 1 \to 2 \to 3 \to \ldots}$$

Here the element 0 has no positive property and indeed maps in $\boldsymbol{S}^{\circlearrowright}$ from $\mathbb{N}^{\circlearrowright \sigma}$ to any object $Y^{\circlearrowright \beta}$ give precisely the elements of Y, by evaluating at 0. This can be proved as follows.

From a map $\mathbb{N}^{\circlearrowright \sigma} \xrightarrow{f} Y^{\circlearrowright \beta}$ in $\boldsymbol{S}^{\circlearrowright}$ we get the element $y = f(0)$ of Y. In this way from different elements we get different maps: if we got the same element from f and g, i.e. $f(0) = g(0)$, then since f and g are maps of $\boldsymbol{S}^{\circlearrowright}$, we can deduce that

$$f(1) = f(\sigma(0)) = \beta(f(0)) = \beta(g(0)) = g(\sigma(0)) = g(1)$$

and similarly $f(2) = g(2)$, and in general

$$f(n+1) = f(\sigma(n)) = \beta(f(n)) = \beta(g(n)) = g(\sigma(n)) = g(n+1)$$

so that $f = g$ since they agree at every input. For every element y of Y there is a map $\mathbb{N}^{\circlearrowright \sigma} \xrightarrow{f} Y^{\circlearrowright \beta}$ in $\boldsymbol{S}^{\circlearrowright}$ such that $f(0) = y$, defined by $f(1) = \beta(y), f(2) = \beta^2(y)$, and in general for any natural number n, $f(n) = \beta^n(y)$. This can easily be checked to be a map in $\boldsymbol{S}^{\circlearrowright}$.

An element y of $Y^{\circlearrowright \beta}$ has period four precisely when the corresponding map $\bar{y} : \mathbb{N}^{\circlearrowright \sigma} \longrightarrow Y^{\circlearrowright \beta}$ (with $\bar{y}(0) = y$) factors through the unique map from $\mathbb{N}^{\circlearrowright \sigma}$ to C_4 which sends 0 to 0, like this

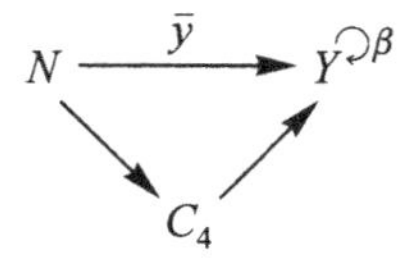

This illustrates that one can express facts about and properties of dynamical systems without resorting to anything outside $\boldsymbol{S}^{\circlearrowright}$; any complicated dynamical system can be 'probed' by maps from simple objects like $\mathbb{N}^{\circlearrowright \sigma}$ and C_n.

Exercise 2:
Find all the maps from $\mathbb{N}^{\circlearrowright \sigma}$ to C_4, the cycle of length 4.

For any $Y^{\circlearrowright\beta}$, we found two processes (maps of sets)

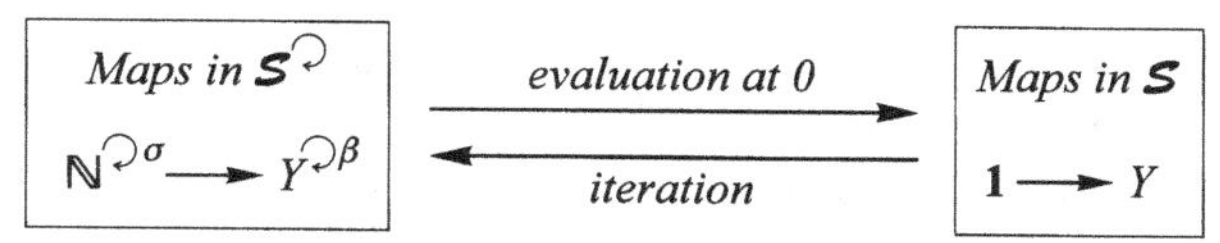

'Iteration' assigns to each y in Y the map f given by $f(n) = \beta^n(y)$.

Exercise 3:
Show that *evaulation at* 0 and *iteration* are inverse (to each other).

Now, having found a way to 'recapture' Y from information entirely within the category $\mathcal{S}^{\circlearrowright}$, we would also like to recapture β. We can do this too! The next two exercises show you how.

Exercise 4:
For any dynamical system $X^{\circlearrowright\alpha}$, show that α is itself a map of dynamical systems $X^{\circlearrowright\alpha} \xrightarrow{\alpha} X^{\circlearrowright\alpha}$.

In particular $\mathbb{N}^{\circlearrowright\sigma} \xrightarrow{\sigma} \mathbb{N}^{\circlearrowright\sigma}$ is a map in $\mathcal{S}^{\circlearrowright}$.

Exercise 5:
Show that if $\mathbb{N}^{\circlearrowright\sigma} \xrightarrow{f} Y^{\circlearrowright\beta}$ corresponds to y, then $\mathbb{N}^{\circlearrowright\sigma} \xrightarrow{f\circ\sigma} Y^{\circlearrowright\beta}$ corresponds to $\beta(y)$.

The results of Exercises 3 through 5 show that we have, for any dynamical system $S = Y^{\circlearrowright\beta}$, the correspondence

$$\frac{\textit{states of } S\ (= \textit{elements } y_0 \textit{ of } Y)}{\textit{maps of dynamical systems } N \xrightarrow{y} S}$$

and that if y_0 in Y corresponds (by $y(n) = \beta^n(y_0)$) to

$$N \xrightarrow{y} S$$

then the 'next state' $\beta(y_0)$ corresponds to $y \circ \sigma$:

$$N \xrightarrow{\sigma} N \xrightarrow{y} S$$

This suggests an alternate scheme of notation in which a single letter, say S or T, stands for an entire dynamical system, and the single letter σ is used for the endomap of every dynamical system, but with a subscript: σ_S for the endomap of the system S, σ_T for T, etc. (Here σ is thought of as the universal 'act of pressing the button' – or 'next state', in the dynamical system view – and the subscript tells us to which system the act is being applied.) In this notation, the observation above becomes $\sigma_S(y) = y \circ \sigma$; the act of pressing the button becomes the act of precomposing with $N \xrightarrow{\sigma} N$.

In 1872 Felix Klein proposed that the way to study an object is to investigate all its automorphisms, which he called *symmetries*. Indeed, investigating symmetries proved to be very useful, in crystallography and elsewhere; but 'probing' an object by means of maps from a few standard objects has proved to be even more useful. In our dynamical systems, this utility comes from the fact that while $Y^{\circlearrowright\beta}$ may have very few symmetries (perhaps only one, the identity map) it will always have enough maps from $\mathbb{N}^{\circlearrowright\sigma}$ into it to describe it completely, as the exercises have shown.

4. The philosophical role of N

In Session 6 we emphasized the notion that in studying a large objective category $\mathcal{X} = \mathcal{S}$, the category of all abstract sets and maps, a bare minimum $\mathcal{C}$ which is adequate is the category with eight maps whose two objects are a one-point set $\mathbf{1}$ and a two-point set $\mathbf{2}$; this is because:

1. the maps $\mathbf{1} \longrightarrow X$ are the points of X;
2. the maps $\mathbf{2} \longrightarrow X$ are the pairs of points of X;
3. the maps $X \longrightarrow \mathbf{2}$ are sufficient to express all the *yes/no* properties of points of X;
4. precomposing with a map $\mathbf{2} \longrightarrow \mathbf{2}$ exchanges the roles of two points in a pair;
5. following by a map $\mathbf{2} \longrightarrow \mathbf{2}$ effects negating a property; and
6. composing $\mathbf{1} \longrightarrow X \longrightarrow \mathbf{2}$ records in $\mathcal{C}$ whether a particular point has a particular property.

These are sufficient basic ingredients to analyze any map $X \longrightarrow Y$ in $\mathcal{X}$. If we add a three-point set to $\mathcal{C}$ (getting a category with only 56 maps, most of which can be expressed as composites of a wisely chosen few), then our strengthened 'subject' will be adequate in an even stronger sense, at least for arbitrarily large finite sets X. Then *and* and *or* and other crucial logical operations on properties become internal to $\mathcal{C}$.

This method turns out to apply to all our deeper examples of objective categories $\mathcal{X}$ as well – for example, to $\mathcal{X} = \mathcal{S}^{\circlearrowright}$ the category of discrete dynamical systems. After some initial investigation of some of the simpler objects of $\mathcal{X}$, we can make a wise determination of an appropriate small subcategory $\mathcal{C}$ to recognize and keep as our subjective instrument for the further study of the more complex objects in $\mathcal{X}$.

To show that a proposed $\mathcal{C}$ is adequate, we must, of course, prove appropriate theorems. In the case $\mathcal{X} = \mathcal{S}^{\circlearrowright}$, we have seen that any subcategory $\mathcal{C}$ which includes the special object

$$N = \mathbb{N}^{\circlearrowright\sigma} = \boxed{0 \to 1 \to 2 \to 3 \to \dots}$$

will be adequate to discuss the states of any $S = Y^{\circlearrowright\beta}$ because:

1. the maps $N \xrightarrow{y} S$ 'are' the states of S, and
2. precomposing with $N \xrightarrow{\sigma} N$ effects the 'next state' operation.

Questions about a state such as 'Does it return to itself after seven units of time?' can also be *objectified within the subjective* if we include in $\mathcal{C}$ objects such as C_7. Other questions such as 'Does the state become a rest state after three units of time?' can also be objectified within the subjective if we include in $\mathcal{C}$ systems such as

$$\boxed{\bullet \to \bullet \to \bullet \to \bullet\circlearrowright}$$

Maps $X \longrightarrow B$ with B in $\mathcal{C}$ may express very important properties of states. For example, if we consider the two-state system with one fixed point

$$B = \boxed{\begin{array}{c} \bullet\ \textit{male} \\ \downarrow \\ \circlearrowright\bullet\ \textit{female} \end{array}}$$

and if $X^{\circlearrowright m}$ represents the matrilineal aspect of a society (i.e. X is the set of people, present and past, in a society and $m(x) =$ *the mother of* x), then:

Exercise 6:
Show that the gender map $X^{\circlearrowright m} \xrightarrow{g} B$ is a map in the category $\mathcal{S}^{\circlearrowright}$.

Including this object B in $\mathcal{C}$ will permit beginning the *objectification in the subjective* of gender as property.

The inclusion in $\mathcal{C}$ of the dynamical system

$$\Omega = \boxed{\circlearrowright\overset{0}{\bullet} \leftarrow \overset{1}{\bullet} \leftarrow \overset{2}{\bullet} \leftarrow \overset{3}{\bullet} \leftarrow \cdots \overset{\infty}{\bullet}\circlearrowleft}$$

with two rest states (0 and ∞) and an infinite number of 'finite' states, each of which eventually becomes the rest state 0, permits recording in $\mathcal{C}$ any stable property of states in any discrete dynamical system, in a sense which we will discuss more systematically later, and thus plays a role similar to the role of the inclusion of **2** in $\mathcal{S}$, but is more powerful.

The fundamental role of Ω is to describe *stable* properties of states of $X^{\circlearrowright\alpha}$: those properties which are not lost by applying α. The question: 'How many steps of the dynamical endomap from the given state x are required to make the property *become* true?', is answered by a number, a state of Ω! Thus a stable property is a map $X \longrightarrow \Omega$ in $\mathcal{S}^{\circlearrowright}$; only if the property *never* becomes true of x does this map take the value 'false' ($= \infty$) at x. (See Session 33 for further discussion of Ω-objects.)

In general such a 'subjective contained in objective' interpretation of an inclusion of categories, $\mathcal{C}$ contained in $\mathcal{X}$, induces (at least) a four-fold division of kinds of maps:

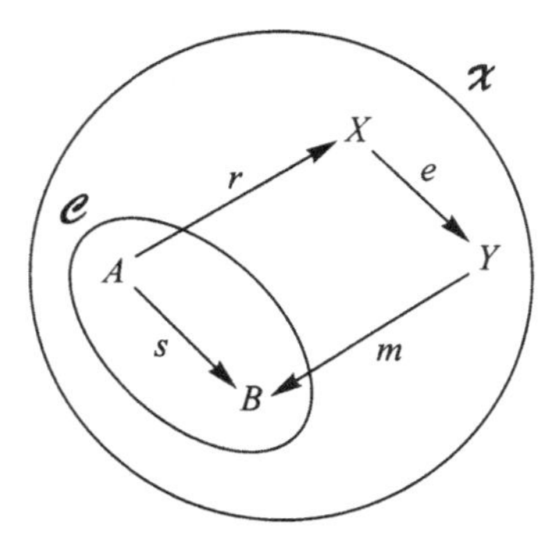

In the case of dynamical systems ($\mathcal{X} = \mathcal{S}^{\circlearrowright}$), a rough description of some of the many possible uses of this division can be expressed in words as:

equivariant map e of dynamical systems;
modeling m or simulation of a natural system in a theoretical system;
interpretation or simulation s of a theory in a computer;
realization r of a design for a machine.

Making these rough descriptions more precise may require a category of systems with a deeper structure than just that given by a single endomap, so at this stage these words are only suggestive.

5. Presentations of dynamical systems

Is there a simple rule to determine the number of $\mathcal{S}^{\circlearrowright}$-maps from $X^{\circlearrowright\alpha}$ to $Y^{\circlearrowright\beta}$? This question deserves more attention than we gave it earlier. The answer depends on how $X^{\circlearrowright\alpha}$ and $Y^{\circlearrowright\beta}$ are 'presented' to us; but there is a systematic way to describe a finite system which is very useful. Here is an example:

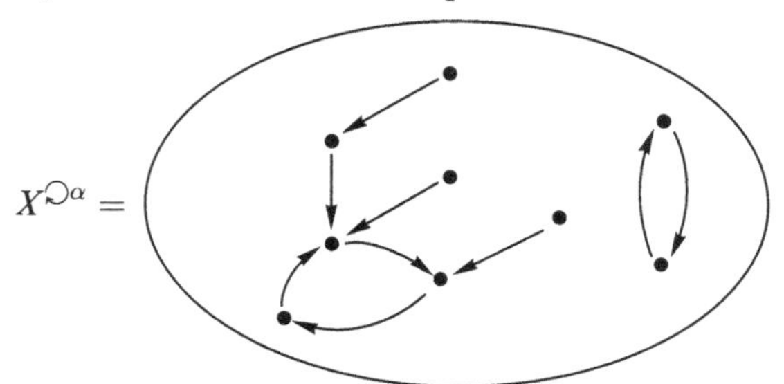

1. Choose names for *some* of the elements; if a cycle has hair, label only the ends, but if it has no hair, label one of the elements of the cycle.

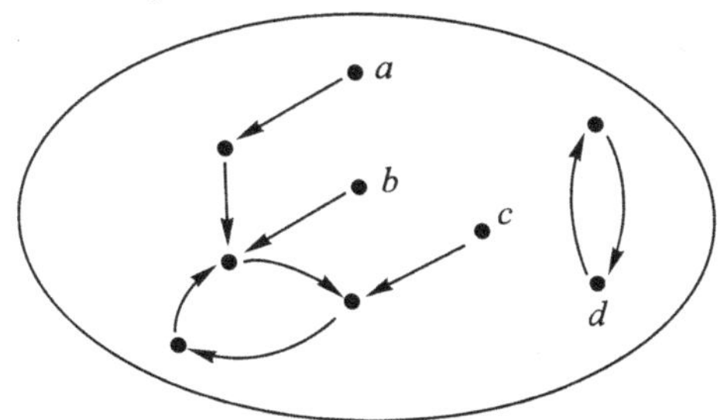

The labelled elements are called 'generators' for X.

2. Choose an order in which to list the generators; here a, b, c, d seems a reasonable order.
3. Start from the first element in the list, and apply α until you would get a repetition by going further: $a, \alpha a, \alpha^2 a, \alpha^3 a, \alpha^4 a$ are distinct, but we stop here because

   (i) $\alpha^5 a = \alpha^2 a$

4. Now pass to the next element in the list a, b, c, d and continue as before: $a, \alpha a, \alpha^2 a, \alpha^3 a, \alpha^4 a, b$ are distinct, but we stop here because

   (ii) $\alpha b = \alpha^2 a$

5. Repeat step (4) until the list a, b, c, d is exhausted.

If you do this correctly, you will get this list of labels:

$$\textbf{(L)} \quad a, \alpha a, \alpha^2 a, \alpha^3 a, \alpha^4 a, b, c, d, \alpha d$$

and you will have found these equations:

$$\begin{aligned} \textbf{(R)} \quad &\text{(i)} & \alpha^5 a &= \alpha^2 a \\ &\text{(ii)} & \alpha b &= \alpha^2 a \\ &\text{(iii)} & \alpha c &= \alpha^3 a \\ &\text{(iv)} & \alpha^2 d &= d \end{aligned}$$

These equations are called 'relations' among the generators. From the way we constructed it, the list (**L**) labels each element of X exactly once:

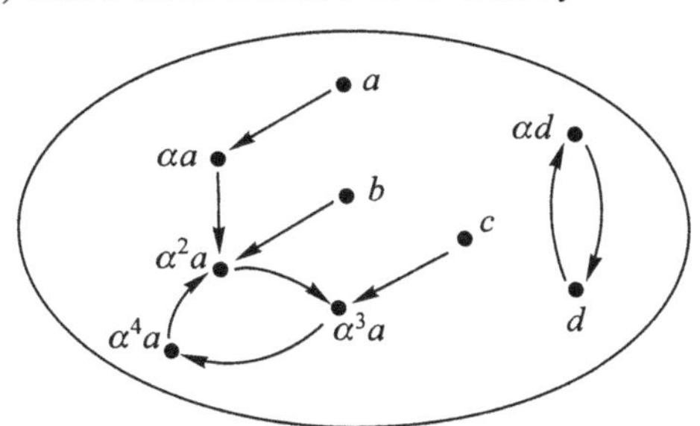

In finding the equations, it is helpful to do this labelling as you go along.

Now, of course, any map $X^{\circlearrowright\alpha} \xrightarrow{f} Y^{\circlearrowright\beta}$ sends a, b, c, and d to elements $f(a) = \bar{a}$, $f(b) = \bar{b}$, etc. in Y satisfying the 'same relations':

$$
\begin{array}{lll}
(\bar{\mathbf{R}}) & \text{(i)} & \beta^5\bar{a} = \beta^2\bar{a} \\
& \text{(ii)} & \beta\bar{b} = \beta^2\bar{a} \\
& \text{(iii)} & \beta\bar{c} = \beta^3\bar{a} \\
& \text{(iv)} & \beta^2\bar{d} = \bar{d}
\end{array}
$$

The surprise is that this procedure can be reversed: given *any* elements $\bar{a}, \bar{b}, \bar{c}, \bar{d}$, in Y satisfying the relations $(\bar{\mathbf{R}})$, there is exactly one f with $f(a) = \bar{a}$, $f(b) = \bar{b}$, $f(c) = \bar{c}$, and $f(d) = \bar{d}$. Symbolically:

$$
\frac{\mathcal{S}^{\circlearrowright}\text{– maps} \quad X^{\circlearrowright\alpha} \xrightarrow{f} Y^{\circlearrowright}}{\text{lists } \bar{a}, \bar{b}, \bar{c}, \bar{d} \text{ in } Y \text{ satisfying } (\bar{\mathbf{R}})}
$$

A family of labels (like a, b, c, d) for elements of X, together with a family of equations (like $(\mathbf{R})$) which these satisfy, is called a **presentation** of $X^{\circlearrowright\alpha}$ if it has the surprising 'universal property' above: that maps from $X^{\circlearrowright\alpha}$ to each $Y^{\circlearrowright\beta}$ correspond exactly to families in Y satisfying the 'same' equations. The method we described gives a 'minimal' presentation of $X^{\circlearrowright\alpha}$.

Does all this really help to find all the maps $X^{\circlearrowright\alpha} \xrightarrow{f} Y^{\circlearrowright\beta}$, and, in particular, to count how many maps there are? I find that it does. Suppose $X^{\circlearrowright\alpha}$ is the system we pictured earlier, and

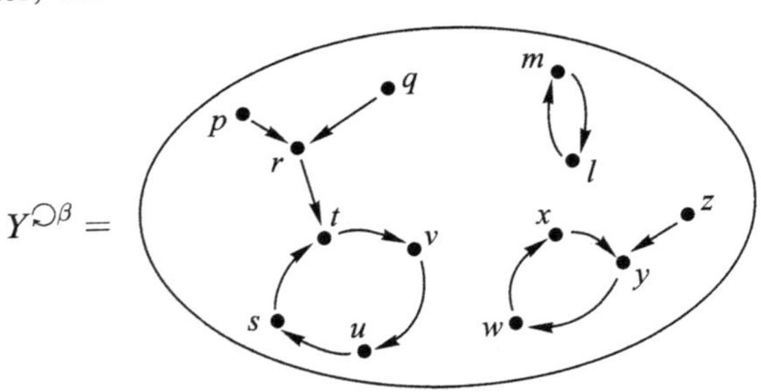

One systematic way to find all the maps $X^{\circlearrowright\alpha} \xrightarrow{f} Y^{\circlearrowright\beta}$ is this:

1. Find all possible choices for $f(a)$; i.e. elements $\bar{a}$ satisfying the relation
$$\beta^5(\bar{a}) = \beta^2(\bar{a})$$
You will find that w, x, y, and z satisfy this equation, but $l, m, p, q, \ldots v$ do not.
2. For each of the choices in (1), look for the elements $\bar{b}$ which satisfy equation (ii). For instance, if we choose $\bar{a} = w$, we see that
$$\beta^2\bar{a} = \beta^2 w = y$$
so the equation (ii) becomes $\beta\bar{b} = y$, which means $\bar{b}$ must be x or z.

3. For each choice of $\bar{a}$ and $\bar{b}$, find all the choices of $\bar{c}$ satisfying (iii).
4. Then go on to $\bar{d}$.

Exercise 7:
Find all the maps from the $X^{\circlearrowright\alpha}$ above to the $Y^{\circlearrowright\beta}$ above. (Unless I made a mistake, there are 14 of them.)

Here is a word of advice: To follow blindly an 'algorithm,' a systematic procedure like this, is always a bit tedious and often inefficient. If you keep your wits about you while you are doing it, though, you often discover interesting things. For instance, in our example the choices for $\bar{d}$ are completely independent of the choices you made for $\bar{a}$, $\bar{b}$, and $\bar{c}$; but the choices you made for $\bar{b}$ and $\bar{c}$ depend on the earlier choices. Is this related to the obvious feature of the picture of $X^{\circlearrowright\alpha}$, which seems to show a 'sum' of two simpler systems? Also, $Y^{\circlearrowright\beta}$ is such a sum of three parts. Is this helpful?

Exercise 8:
Draw some simple dynamical systems and find presentations for them. (It is more interesting if you start from a dynamical system that arises from a real problem!)

Exercise 9:
Our procedure treated $X^{\circlearrowright\alpha}$ and $Y^{\circlearrowright\beta}$ very differently. Suppose that in addition to a presentation of $X^{\circlearrowright\alpha}$ you had a presentation of $Y^{\circlearrowright\beta}$. Try to find a method to calculate the solutions of the equations ($\bar{\mathbf{R}}$) without having to draw the picture of $Y^{\circlearrowright\beta}$, but just working with a presentation. One can even program a computer to find all the maps f, starting from presentations of $X^{\circlearrowright\alpha}$ and $Y^{\circlearrowright\beta}$.

Even infinite dynamical systems may have finite presentations. For example, $\mathbb{N}^{\circlearrowright\sigma}$ is presented by one generator, 0, and *no* equations!

Exercise 10:
Find a presentation for this system, which continues to the right forever:

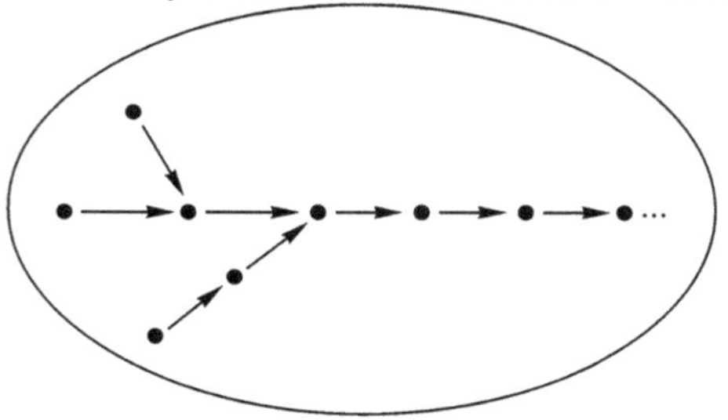

Exercise 11:
Think about presentations of graphs. If you don't get any good ideas, think about them again after Session 22.

Exercise 12:
A non-autonomous dynamical system S is one in which the 'rule of evolution' $\mathbb{N} \times S \xrightarrow{r} S$ itself depends on time. These can be studied by reducing to the ordinary, or autonomous, system on the state space $X = \mathbb{N} \times S$ with dynamics given by $\rho(n, s) = \langle n+1, r(n, s) \rangle$. Show that for any r there is exactly one sequence u in S for which $u(n+1) = r(n, u(n))$ and for which $u(0) = s_0$ is a given starting point. (Hint: Reduce this to the universal property of $N = (\mathbb{N}, \sigma)$ in $\mathbf{S}^{\circlearrowright}$.)

SESSION 16

Idempotents, involutions, and graphs

1. Solving exercises on idempotents and involutions

In Article III are some exercises about automorphisms, involutions, and idempotents. Exercise 4 asks whether the endomap α of the integer numbers, defined by assigning to each integer its negative: $\alpha(x) = -x$, is an involution or an idempotent, and what its fixed points are. What is an involution?

OMER: An endomap that composed with itself gives the identity.

Right. This means that for each element x its image is mapped back to x, hence it goes in a cycle of length 2 unless it is a fixed point. The internal diagram of an involution consists of some cycles of length 2 and some fixed points:

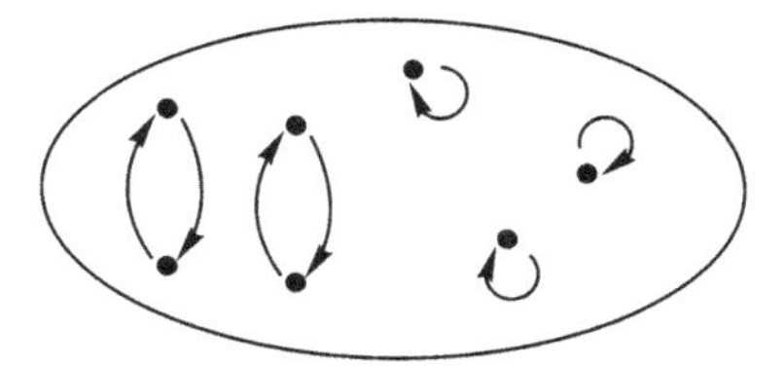

On the other hand, an idempotent endomap is one that applied twice (i.e. composed with itself) has the same effect as applied once. This means that the image of any element is a fixed point and therefore any element, if not already a fixed point, reaches a fixed point in one step. The internal diagram of an idempotent endomap consists of some fixed points that may have some hairs attached, as in the picture:

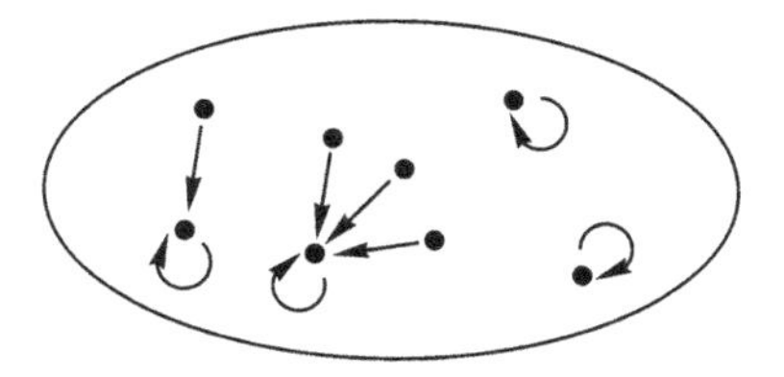

This picture represents *one* endomap which is idempotent and not an involution, even though it has some common features with the involution pictured above, namely some fixed points.

The exercises illustrate these types of maps with some examples with sets of numbers, in particular, the involution of the set $\mathbb{Z}$ of integers mentioned at the beginning, $\alpha(x) = -x$. The proof that this is an involution is based on the fact that the opposite of the opposite is the same number, i.e. $-(-x) = x$, so that

$$\alpha^2(x) = \alpha(\alpha(x)) = -(-x) = x$$

and therefore $\alpha^2 = 1_{\mathbb{Z}}$. The internal diagram of this involution is

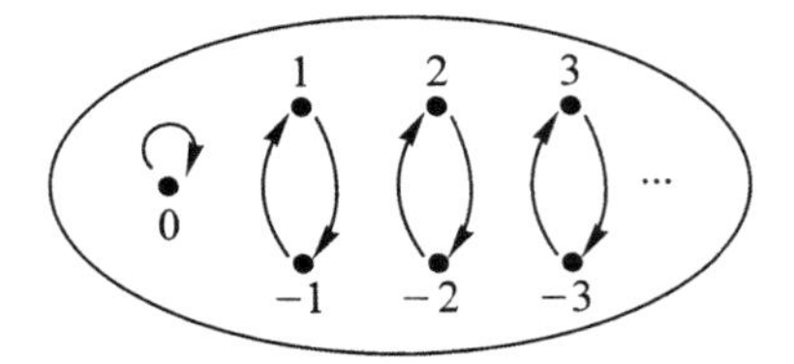

which indeed has only cycles of length 2 and fixed points (just one fixed point in this case).

Exercise 5 asks the same questions about the 'absolute value' map on the same set. The map changes the sign of the negative integers but leaves the other numbers unchanged. What would this map be?

CHAD: Idempotent.

Right. All the images are non-negative, and therefore are left unchanged. The internal diagram of this map is

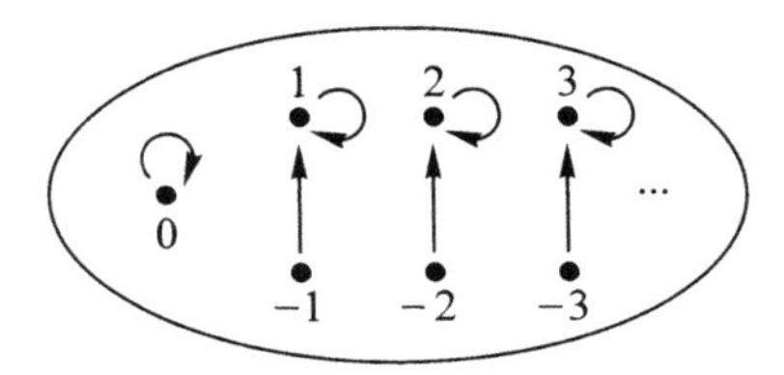

one bare fixed point and an infinite number of fixed points with one 'hair' each.

Exercise 6 is about the endomap of the integers which 'adds 3,' i.e. map α given by $\alpha(x) = x + 3$. Is this an involution or an idempotent?

CHAD: Neither.

Right. If you add 3 twice to a number you get the same as adding 6; the result is neither the number you started with (thus it is not an involution) nor the same as adding 3 only once (thus it is not idempotent). Is this map an automorphism?

OMER: Yes, because it has an inverse.

What is the inverse?

OMER: The map 'adding -3.'

That's right. If $\alpha(x) = x + 3$ then the inverse is given by $\alpha^{-1}(x) = x - 3$. What about Exercise 7 where we have the map $\alpha(x) = 5x$? This one is not an automorphism of $\mathbb{Z}$ because the inverse map would in particular satisfy

$$2 = \alpha(\alpha^{-1}(2)) = 5\alpha^{-1}(2)$$

but there is no integer x with $2 = 5x$. But integers are part of the rationals and this map 'extends' to the set $\mathbb{Q}$ of rational numbers:

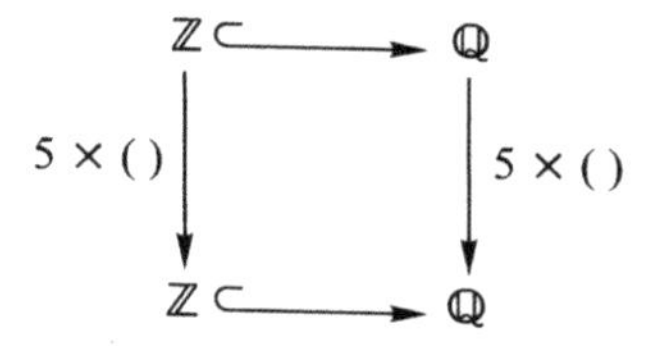

The extended map has an inverse:

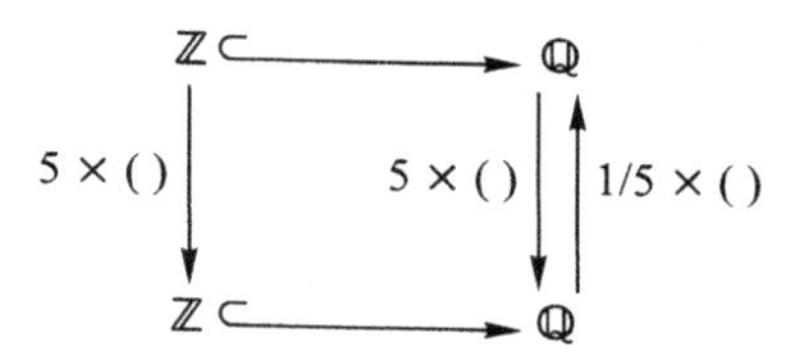

2. Solving exercises on maps of graphs

Exercise 11 in Article III is about irreflexive graphs. Recall that an irreflexive graph is a pair of maps with the same domain and the same codomain, like

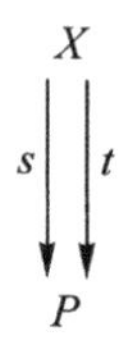

The domain can be interpreted as the set of arrows of the graph and the codomain as the set of dots, while the maps are interpreted as 'source' and 'target,' i.e. $s(x)$ is the dot that is the source of the arrow x, while $t(x)$ is the target dot of the same arrow. Interpreting things this way every such pair of parallel maps can be pictured as a graph such as

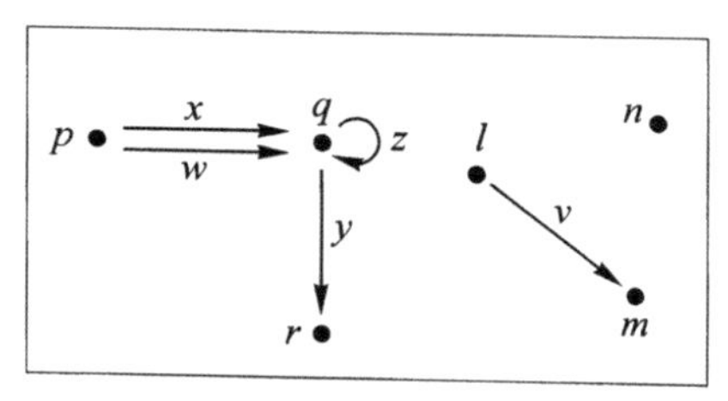

while given such a graph we can always reconstruct the sets and the maps. For example, for the graph pictured above the sets would be $X = \{x, y, z, v, w\}$ and $P = \{p, q, r, l, m, n\}$, while the maps are given by the table

X	source	target
x	p	q
y	q	r
z	q	q
v	l	m
w	p	q

By looking at this table we can answer all the questions about the graph. For example: is there a loop? All we have to do is to look for an arrow whose source and target are the same. We find that the arrow z is a loop.

Now we want to talk about maps of graphs; i.e. a way to picture one graph inside another. A map in the category of irreflexive graphs $\mathbf{S}^{\downarrow\downarrow}$

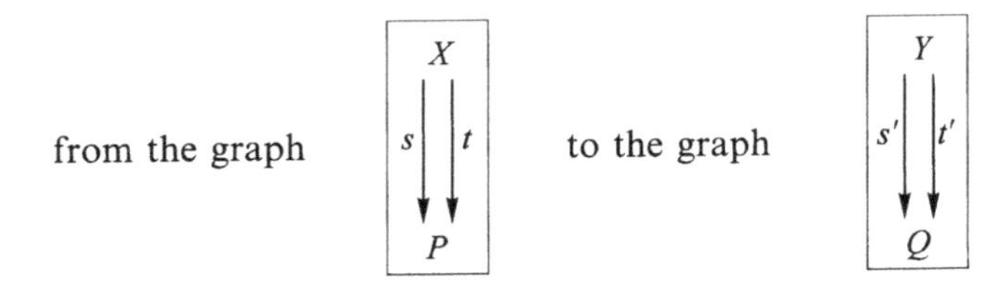

is a pair of set maps, one from arrows (X) to arrows (Y) (to be denoted with the subscript 'A' for 'arrows') and the other from dots (P) to dots (Q) (to be denoted with the subscript 'D' for 'dots'). Therefore

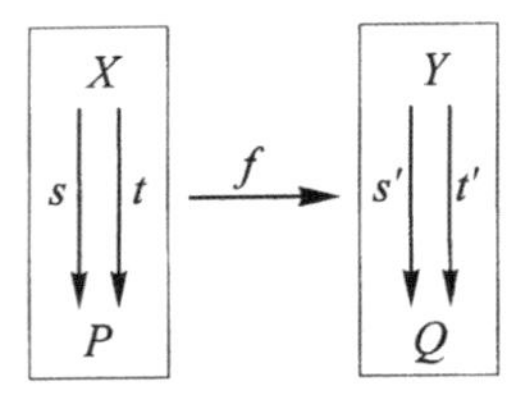

in $\mathbf{S}^{\downarrow\downarrow}$ means

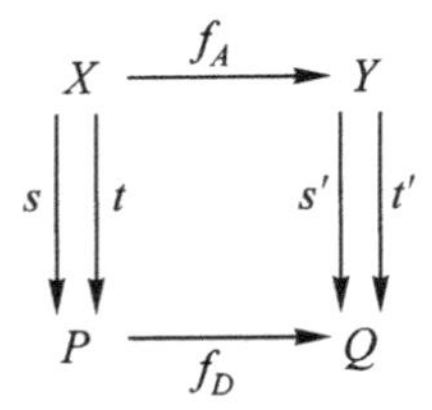

two maps, but not arbitrary. They must satisfy two conditions, namely, to preserve source and to preserve target. These conditions are two equations which are very easily remembered by just looking at the above diagram. They are:

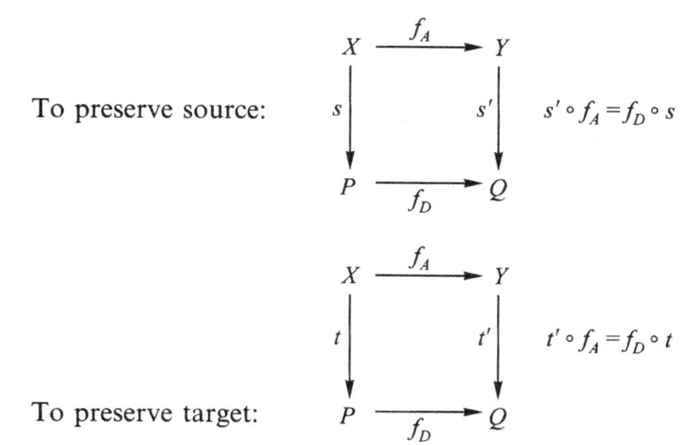

As usual we should show that if we have two such maps one after the other like this:

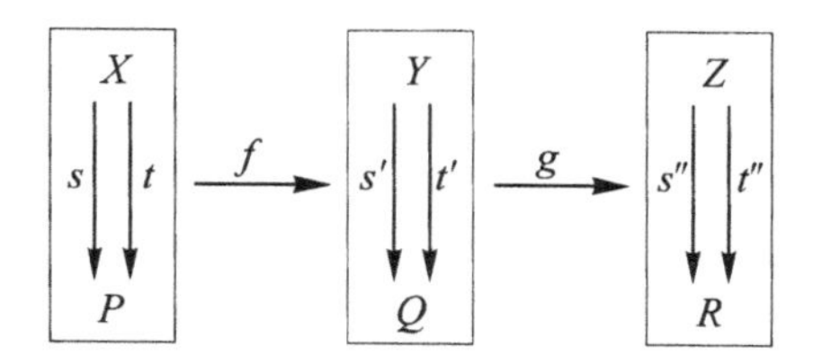

they can be composed. Both the definition of composition and the proof that the composite is again a map of graphs are easy if you draw the diagrams this way:

$$\begin{array}{ccccc} X & \xrightarrow{f_A} & Y & \xrightarrow{g_A} & Z \\ s \downarrow\downarrow t & & s' \downarrow\downarrow t' & & s'' \downarrow\downarrow t'' \\ P & \xrightarrow[f_D]{} & Q & \xrightarrow[g_D]{} & R \end{array}$$

The equations we know are true are:

$$s' \circ f_A = f_D \circ s, \quad t' \circ f_A = f_D \circ t, \quad s'' \circ g_A = g_D \circ s', \quad t'' \circ g_A = g_D \circ t'$$

The only reasonable maps to take from X to Z and from P to R are the composites

$$X \xrightarrow{g_A \circ f_A} Z \quad \text{and} \quad P \xrightarrow{g_D \circ f_D} R$$

This is the same as defining $(g \circ f)_A = g_A \circ f_A$ and $(g \circ f)_D = g_D \circ f_D$, and the equations that we have to prove are

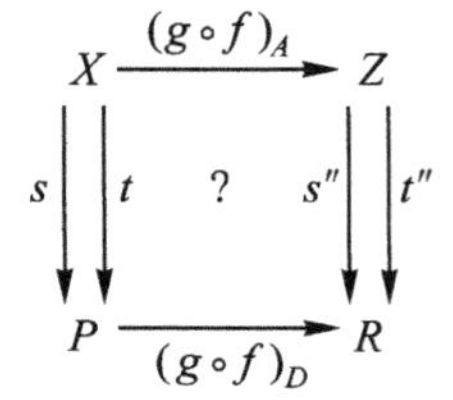

$$\begin{array}{ccc} X & \xrightarrow{(g\circ f)_A} & Z \\ s \downarrow\downarrow t & ? & s'' \downarrow\downarrow t'' \\ P & \xrightarrow[(g\circ f)_D]{} & R \end{array}$$

$$s'' \circ (g \circ f)_A \stackrel{?}{=} (g \circ f)_D \circ s \quad \text{and} \quad t'' \circ (g \circ f)_A \stackrel{?}{=} (g \circ f)_D \circ t,$$

or, using the above definitions,

$$s'' \circ (g_A \circ f_A) \stackrel{?}{=} (g_D \circ f_D) \circ s \quad \text{and} \quad t'' \circ (g_A \circ f_A) \stackrel{?}{=} (g_D \circ f_D) \circ t.$$

These are to be proved using the known equations. The proof is very easy by just following the arrows in the diagrams. For example, the proof that the sources are preserved goes like this: $s'' \circ (g_A \circ f_A)$ is the composite

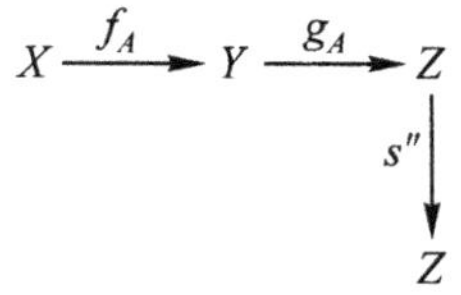

$$\begin{array}{ccccc} X & \xrightarrow{f_A} & Y & \xrightarrow{g_A} & Z \\ & & & & \downarrow s'' \\ & & & & Z \end{array}$$

which by the known equations is equal to

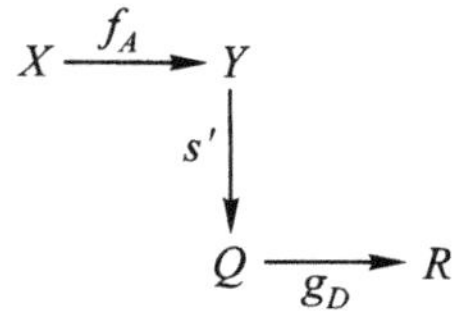

$$\begin{array}{ccccc} X & \xrightarrow{f_A} & Y & & \\ & & \downarrow s' & & \\ & & Q & \xrightarrow[g_D]{} & R \end{array}$$

and this in turn is equal to

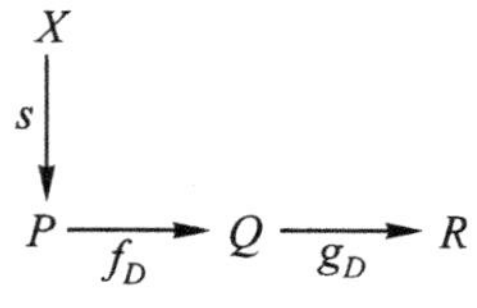

$$\begin{array}{ccccc} X & & & & \\ \downarrow s & & & & \\ P & \xrightarrow[f_D]{} & Q & \xrightarrow[g_D]{} & R \end{array}$$

Therefore we have proved $s'' \circ (g_A \circ f_A) = (g_D \circ f_D) \circ s$. The other is proved in much the same way. To show that we have a category, we must also decide what the identity maps are, and check the identity and associative laws. All these are easy, but you should do them.

Exercises 15 and 16 are a little different. There, instead of irreflexive graphs we deal with reflexive graphs, which are a richer structure. Recall that a reflexive graph is the same as an irreflexive graph, but with the additional structure given by a map $i : P \longrightarrow X$ that assigns to each dot a special or 'preferred' arrow that has that dot as source and as target, and therefore it is a loop on that dot. In other words, the map i is required to satisfy the equations,

$$s \circ i = 1_P \quad \text{and} \quad t \circ i = 1_P$$

and therefore i is a section for s as well as for t.

Thus, the structure of reflexive graph involves two sets and three structural maps. What would be involved in a map of reflexive graphs such as

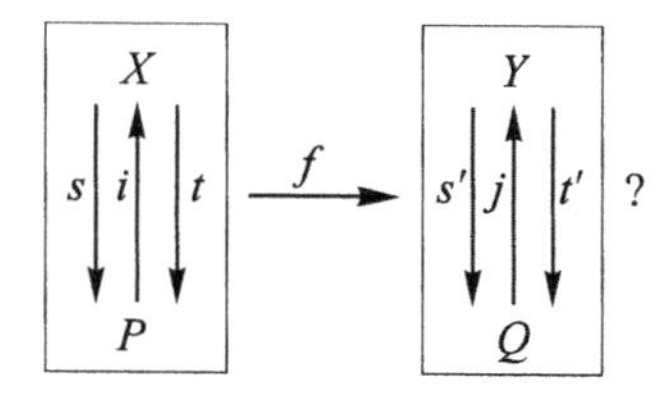

Again, this involves two maps of sets

$$X \xrightarrow{f_A} Y$$

$$P \xrightarrow[f_D]{} Q$$

but now they have to satisfy three equations, which express that f preserves sources, preserves targets, and preserves the preferred loop at each dot. From the diagram

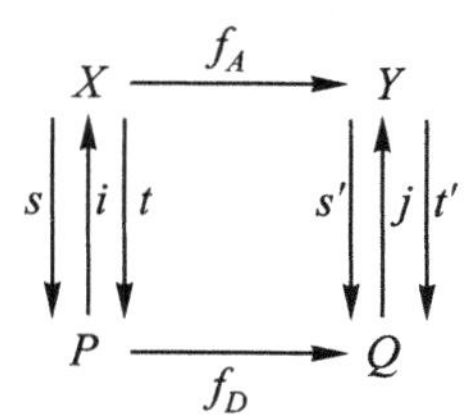

we easily read the three equations, which are the same as for irreflexive graphs, but include one additional equation: $f_A \circ i = j \circ f_D$, or

$$\begin{array}{ccc} X & \xrightarrow{f_A} & Y \\ {\scriptstyle i}\uparrow & & \uparrow{\scriptstyle j} \\ P & \xrightarrow[f_D]{} & Q \end{array}$$

Exercise 15 is about the idempotent maps we get by composing *source* and *target* with their common section i. These are called: $e_0 = i \circ s$ and $e_1 = i \circ t$. The exercise is to prove that not only are e_0 and e_1 idempotent endomaps of the set of arrows X but that furthermore they satisfy the equations

$$e_0 \circ e_1 = e_1 \quad \text{and} \quad e_1 \circ e_0 = e_0$$

Adding to these the equations that say that they are idempotent, we can summarize the four equations by saying that by composing any two of the endomaps e_0, e_1 in any order, the result is always the right-hand one. Or with symbols:

$$e_i \circ e_j = e_j$$

where the subindices can be 0 or 1.

Exercise 16 is to prove that to specify a map of reflexive graphs it is sufficient to give the map at the arrows level (i.e. to give f_A) because this automatically determines the map at the dots level (f_D). How does this happen? The answer is that for reflexive graphs each dot 'is' (in a certain sense) a special type of arrow. Thus, to evaluate the map at the level of dots all we need is to identify each dot with the preferred arrow at that dot and evaluate there the map at the level of arrows. Formally an answer to this exercise is as follows.

Suppose we have a map of reflexive graphs

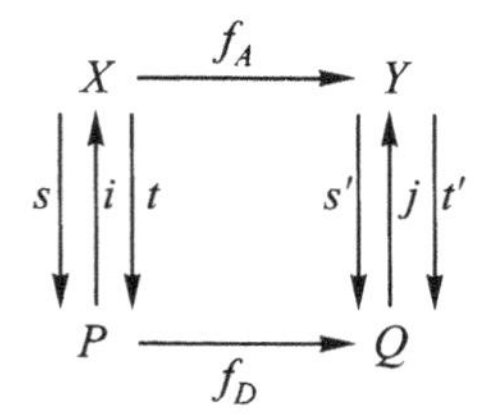

Then in particular $s' \circ j = 1_Q$ and $j \circ f_D = f_A \circ i$, therefore we have

$$f_D = 1_Q \circ f_D = s' \circ j \circ f_D = s' \circ f_A \circ i$$

This shows that at the level of dots f can be evaluated by just knowing f at the level of arrows and the structural maps s' and i. There is nothing special about s' that t' doesn't have, and a similar reasoning shows that we can also evaluate f_D as $t' \circ f_A \circ i$.

One can go on in the same spirit to show that within the category of reflexive graphs, a point of a graph G (i.e. a map $\mathbf{1} \longrightarrow G$ from the terminal object) corresponds to a *preferred* loop (not an arbitrary loop as in the case of irreflexive graphs), or equivalently to just a dot of G. Thus the distinction between dots and points disappears, and indeed when working in the category of reflexive graphs, a slightly different internal picture is often used: only the non-preferred loops are drawn *as* loops, the preferred loop at a point being considered as implicit in the point. A crucial feature of the category of reflexive graphs is that a map $G \longrightarrow H$ can make an arrow in G 'degenerate' into a point of H.

Exercise 1:
For a given object G in a category $\mathcal{C}$, the category $\mathcal{C}/G$ has, as objects, objects of $\mathcal{C}$ equipped with a given $\mathcal{C}$-sorting $X \xrightarrow{s} G$ and, as maps, commutative triangles in $\mathcal{C}$

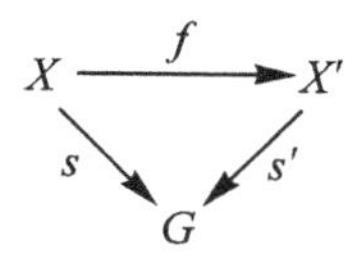

For example, in Session 12, Exercise 3, a category modeling kinship relations was considered as $\mathcal{C}/G$ where an object of $\mathcal{C} = \mathbf{S}^{\circlearrowright\cdot\circlearrowleft}$ is thought of as a set of people equipped with *father* and *mother* endomaps and G is the object of genders. On the other hand, in Exercise 17 of Article III, another description was given in terms of *two* sets and four structural maps. Explain in what sense these two descriptions give the 'same' category.

SESSION 17

Some uses of graphs

1. Paths

What sort of problem might suggest using the category of irreflexive graphs, $\mathbf{S}^{\downdownarrows}$? Remember that an object of this category looks like this:

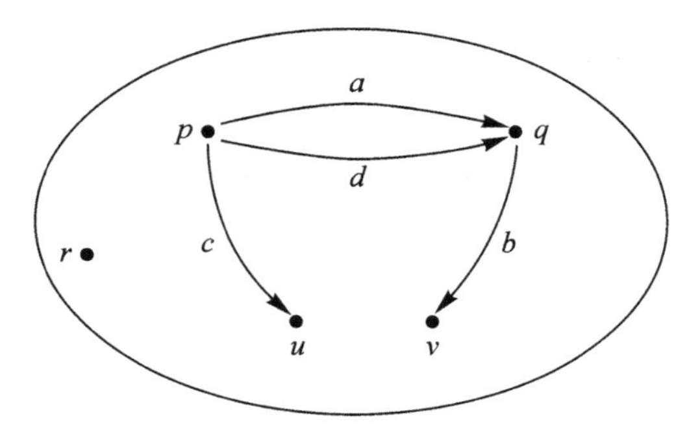

Why is this kind of picture useful? The dots may stand for physical locations and the arrows may represent roads joining them, so that this picture is a schematic road map and may be useful to plan a trip. (In practice, a two-way road is really two one-way roads, separated by a patch of grass or at least a line of paint.) Dots may represent states of a physical system, and arrows the various simple operations you can perform to bring it from one state to another. Even games can be represented by graphs. For example, a game I played when I was a child involves a board with holes placed like this

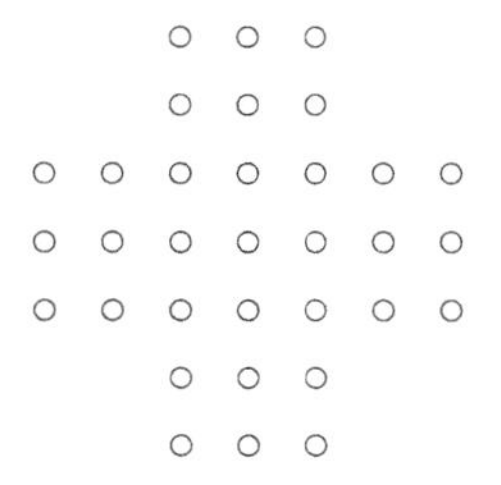

and marbles that can rest in these holes. The game starts with every hole occupied except the central one, and the goal is to remove all the marbles except one, by means of allowed moves only. (In the expert version, this last marble should be in the central hole.) An allowed move makes a marble jump over another situated in one of the four adjacent holes, and removes from the board the marble that was jumped over. Thus, one of the four allowed initial moves is:

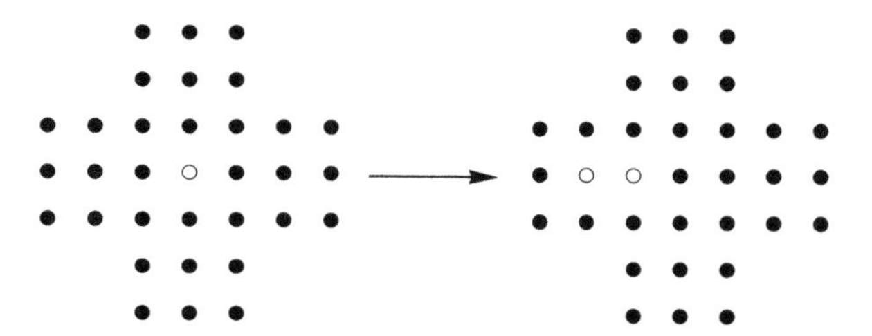

What does this game have to do with graphs? Well, there is a graph associated with this game, whose dots are all the 2^{33} possible distributions of marbles on the board, with an arrow from one distribution to another indicating that the second can be obtained from the first by means of a legal move. Note that there is at most one arrow connecting any two given dots and that this graph does not have a loop or cycle since a legal move produces a state having one less marble. So, the graph associated with this game looks something like this:

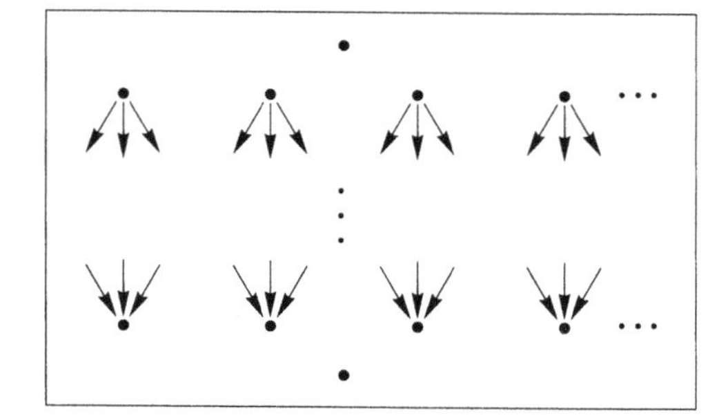

The object of the game is to find a path in this graph from the beginning configuration b to the desired end configuration e:

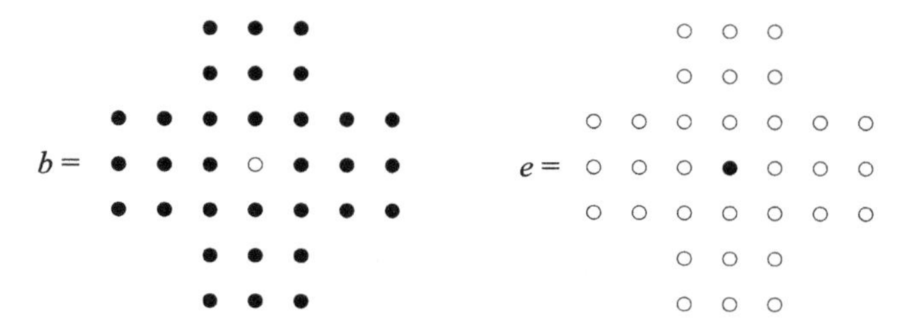

Some of the reasonable questions that one can ask about a graph are: given two nodes b and e, is there a path from b to e? How many? What is the shortest? For example, in the graph given at the beginning, there are two shortest paths from p to v. (In fact they are the only two paths from p to v.)

But in the graph

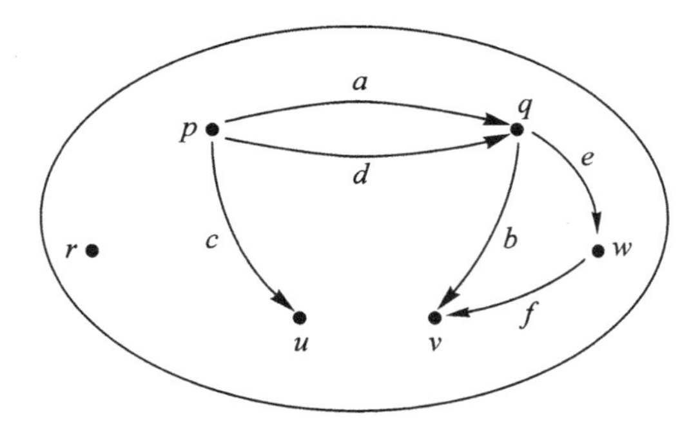

there are also longer paths from p to v, and in the graph below, there are arbitrarily long paths from p to v (since the loop g can be repeated)

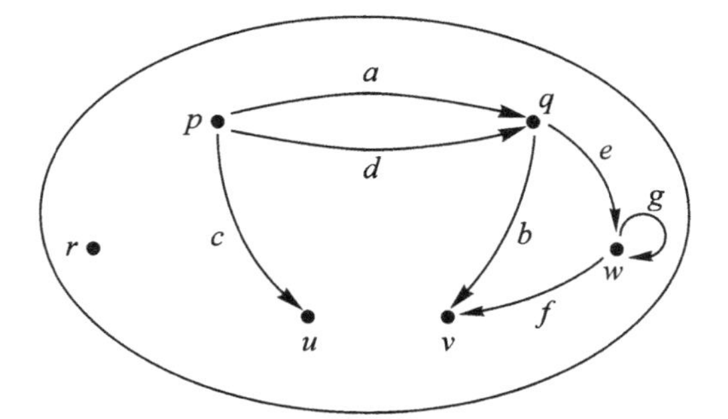

Our categorical framework is well adapted to describe questions such as these, and the resulting clarity is often the key to finding the answers. As an example consider the idea of an arrow or path of length one between two nodes of a graph G. This concept is contained in the category as a graph morphism from the graph $\boxed{\bullet \rightarrow \bullet}$ to G, and in the same way, a two-step path is just a map of graphs

$$\boxed{\bullet \rightarrow \bullet \rightarrow \bullet} \longrightarrow G$$

An example of a two-step path is indicated with the darker arrows d and b:

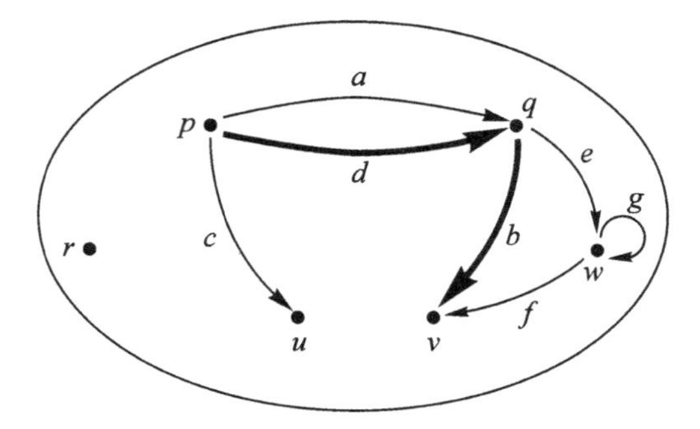

Notice, though, that a path is more than just which arrows are used; we want to count the order in which they are used, too. Just darkening arrows won't give this information if there are 'cycles' in the graph, so the description of a path as a map of graphs is the right one. (Compare this with Galileo's idea that a motion of a particle in space is not simply its track, but is a map from an interval of time into space.)

DANILO: Can we say that a two-step path is a composite of two one-step paths?

Yes, and it is a good idea to make a category from the given graph. An object in this category is just a dot of the graph, and a map from one dot to another is a path, of whatever length. You must be careful to include paths of length zero to serve as identity maps. It is not difficult to check all the axioms for a category, and this is called the *free category* on the given graph.

ALYSIA: How do you use all this to solve problems?

DANILO: You can use graphs to represent chemical reactions.

Yes. The first step is to *formulate* the problem clearly, and for this it is very helpful to have a common setting, that of categories, in which most problems and their solutions can be expressed. One of the many advantages of such a common setting is that when two problems, one familiar and the other not, are formulated in the same way, we see more precisely their common features so that experience with one guides you toward the solution of the other. In this way we build up a small family of concepts and methods which can be used to solve further problems.

One further point: solving particular problems is not the only, or even the principal, goal of science. *Understanding* things, and having clear ideas about them, is a goal as well. Think of Newton's discovery of a single general principle governing the fall of an apple and the motions of the planets for example. The search for clear understanding of motion is what led us to the possibility of space travel.

OMER: It seems to me that categories are for science what a compass is for a navigator.

Yes. Of course, the first time you explore new territory a compass doesn't seem an adequate guide, but at least it helps you draw a map so that the next time you can find your way around more easily.

Exercise 1:
Danilo noticed that from a graph G we can build a category $\mathcal{F}(G)$, the *free category on the graph G*. An object is a dot of G, and a map is a path in G. For which of the following graphs does Danilo's category have a terminal object?

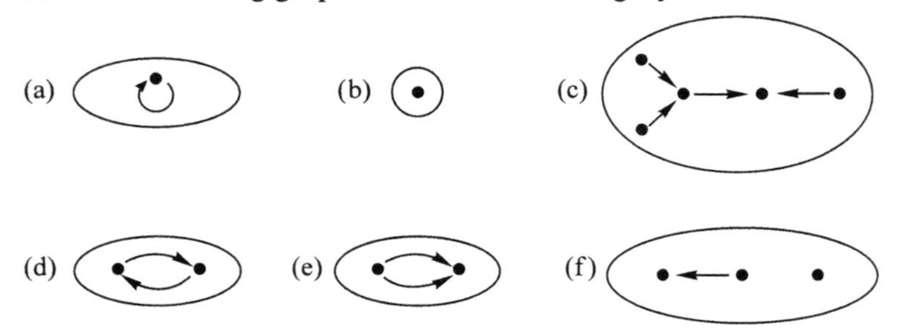

2. Graphs as diagram shapes

The resemblance between graphs and our external diagrams of objects and maps in a category suggests one of the principal uses of graphs. If G is a graph, for example

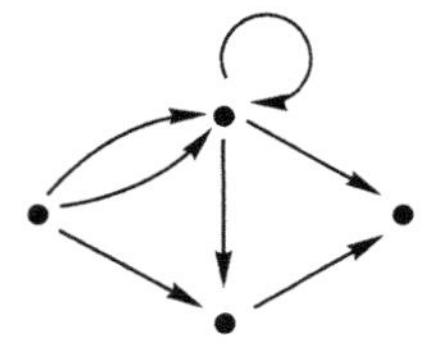

then in any category $\mathcal{C}$ we can have diagrams of shape G in $\mathcal{C}$. Such a diagram assigns to each dot in G an actual object of $\mathcal{C}$ and to each arrow in G an actual map in $\mathcal{C}$ with the appropriate domain and codomain. It could be called a 'figure of shape G in $\mathcal{C}$.' It might only use two *different* objects, like this:

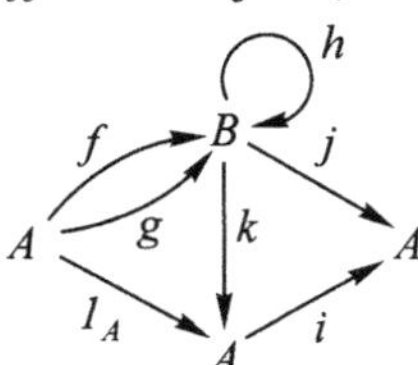

In our 'figure' analogy, we could call this a *singular* diagram, because several dots are sent to the same object A. For now, it may be easiest to think of $\mathcal{C}$ as the category of sets, but you can see that we can have diagrams of shape G in any category.

Now if we have a path in G, for instance the path darkened below

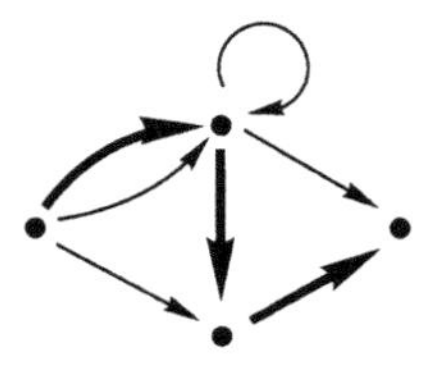

our diagram in $\mathscr{C}$ allows us to 'interpret' this as an actual map in $\mathscr{C}$, the map ikf. This works even for paths of length zero: the path of length zero from the leftmost dot in G to itself is interpreted as the identity map on A.

3. Commuting diagrams

Definition: *We say that a diagram of shape G in* $\mathscr{C}$ **commutes** *if for each pair p, q of dots in G, all paths in G from p to q are interpreted as the same map in* $\mathscr{C}$.

Here are some examples of graphs G, and for each graph, what it means to say a diagram of shape G commutes.

Example 1

A diagram of shape (graph) in $\mathscr{C}$ looks like (diagram)

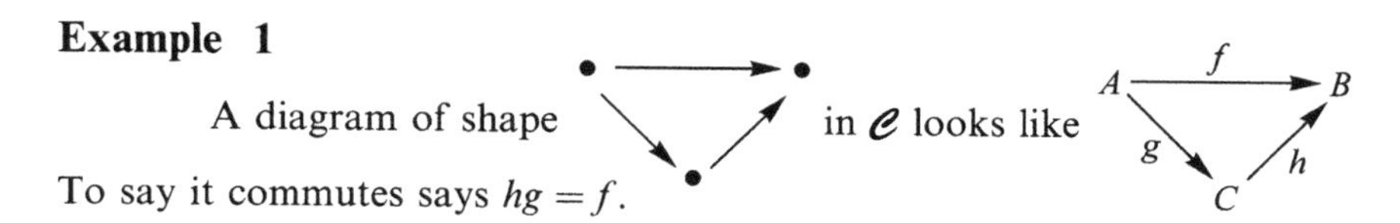

To say it commutes says $hg = f$.

Example 2

A diagram of shape $\boxed{\bullet\circlearrowright}$ is a dynamical system in $\mathscr{C}$, i.e. $A^{\circlearrowright f}$, an object together with an endomap. In the graph $\boxed{\bullet\circlearrowright}$ there are infinitely many paths from the dot to itself, one of each of the lengths 0, 1, 2, ... These are interpreted as the maps $1_A, f, ff, fff, \ldots$. We can abbreviate these as $f^0, f^1, f^2, f^3, \ldots$. Notice that we only use exponents on endomaps; if $A \xrightarrow{f} B$, then f^2, f^3, etc. would be nonsense, and it would not be clear *which* identity map f^0 ought to stand for. (This makes it all the more surprising that we do sensibly use the symbol f^{-1} for the inverse of *any* map that has one.) For our diagram to commute, all these endomaps must be the same. It seems that we need infinitely many conditions:

$$1_A = f, \quad f = f^2, \quad f^2 = f^3, \quad \text{etc.}$$

But we don't really need them all: the equation $1_A = f$ implies all the others!

Example 3

Since in the arrow-graph $\boxed{\bullet \longrightarrow \bullet}$ there is at most one path from any dot to any other, *every* diagram of this shape commutes – even if the diagram happens to be $A \xrightarrow{f} A$. If you compare this with Example 2, you will see that you have to look at the shape-graph, and not just at the $\mathscr{C}$-maps used, to tell whether a diagram commutes.

Example 4
The shape

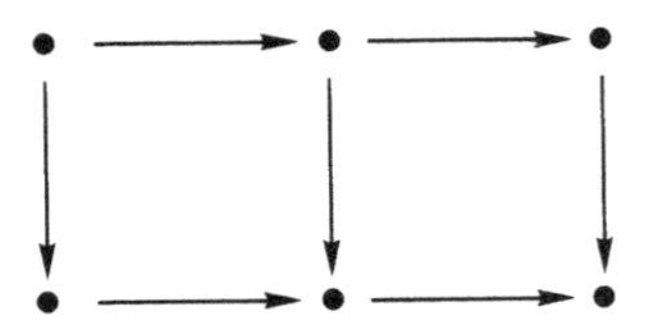

gives us diagrams

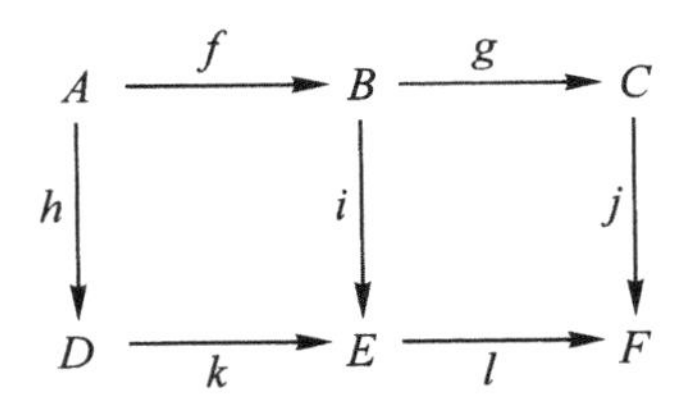

What equations do we need to make such a diagram commute? We have to look at those pairs of dots between which there is more than one path:

1. upper left to lower middle needs $kh = if$;
2. upper middle to lower right needs $li = jg$; and
3. upper left to lower right needs that the three maps jgf, lif, lkh must be equal, but these can be proved equal from (1) and (2). (How?)

As these examples may suggest, for graphs which have cycles it can be a fairly difficult problem to find a shortest list of equations which will imply that a diagram of that shape commutes, while for graphs without cycles it is easier. The cases which arise most often are fortunately not difficult, so we won't need to describe the general theory. In each instance, just check that the equations we prove imply whatever additional equations we use.

Exercise 2:
Show that a diagram of shape $\bullet \rightleftarrows \bullet$ commutes if and only if the maps assigned to the two arrows are inverse.

Exercise 3:
In the diagram

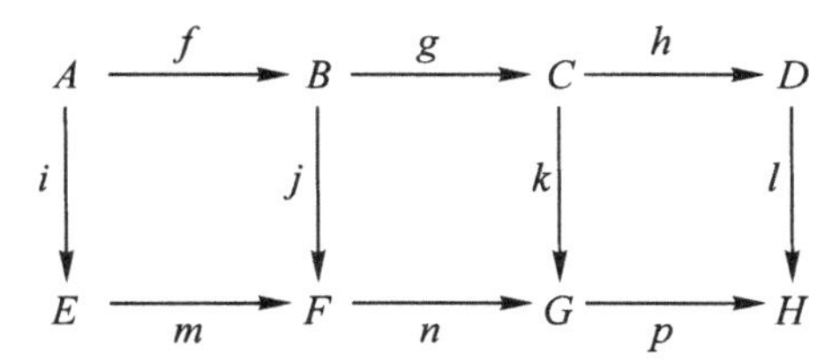

the three equations (1) $jf = mi$, (2) $kg = nj$, (3) $lh = pk$ actually force the diagram to commute; but you are just asked to prove that

$$pnmi = lhgf$$

Exercise 4:
For each of these diagrams, find a shortest list of equations that will make it commute.

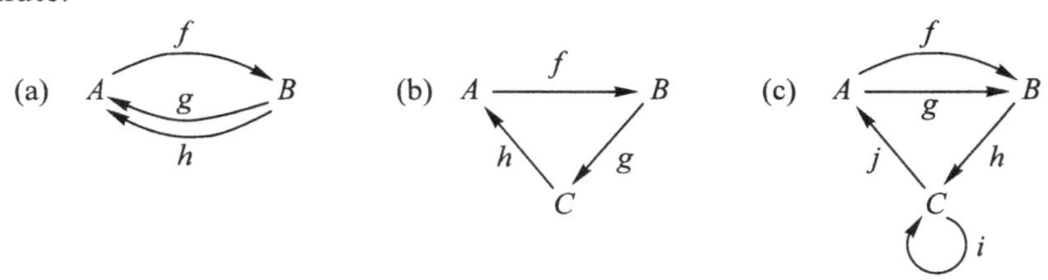

After you have found the answers try to explain clearly how you know, from the equations you chose, that *all* possible paths give equal composites.

4. Is a diagram a map?

If G is a graph, a diagram of shape G in a category $\mathcal{C}$ associates to each dot of G an object in $\mathcal{C}$ and to each arrow of G a map in $\mathcal{C}$; and it respects the structure of G. This suggests that a diagram of shape G in $\mathcal{C}$ is a 'map of graphs' from G to $\mathcal{C}$; but that doesn't make sense! $\mathcal{C}$ is a *category*, not a graph. Still, associated to any category $\mathcal{C}$ there is a big graph, whose dots are the objects in $\mathcal{C}$, arrows are the maps in $\mathcal{C}$, source is domain, and target is codomain. Let's call this big graph $\mathcal{U}(\mathcal{C})$. This graph forgets how to compose maps in $\mathcal{C}$, and only records what the objects and maps are, and what are the domain and codomain of each map. So the diagram *is* a map. A diagram of shape G in $\mathcal{C}$ is a map of graphs, but from G to $\mathcal{U}(\mathcal{C})$, which in fact extends uniquely to a functor from the free category $\mathcal{F}(G)$ to $\mathcal{C}$. The operations $\mathcal{U}$ and $\mathcal{F}$ allow an efficient treatment of the basic relationship between graphs and categories.

Test 2

1. Suppose

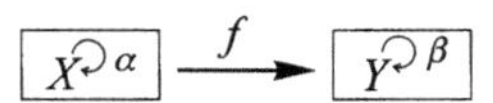

is a map in $\mathcal{S}^{\circlearrowleft}$. Show that if α has a fixed point, then β must also have a fixed point.

2. Find all maps of (irreflexive) graphs from

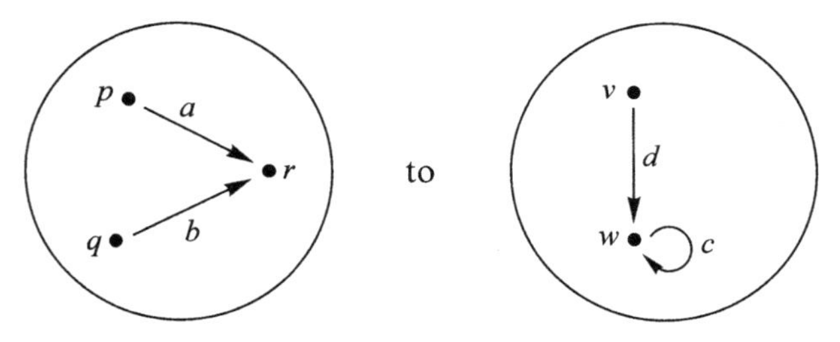

(There are not more than a half-dozen of them.)

3. Find an example of a set X and an endomap $X \xrightarrow{\alpha} X$ with $\alpha^2 = \alpha^3$ but $\alpha \neq \alpha^2$.

SESSION 18

Review of Test 2

The history of science demonstrates that precision in the fundamental ideas develops slowly, and is then crystallized – at least in mathematics – in precise definitions. These then play an important role in developing the subject further, so that in studying, mastery of the definitions is an essential step. Tests illustrate this: to get started, we must know the precise definitions, and if we know the definitions, a simple calculation will often bring us to a solution.

Now Danilo will show us his solution to the first problem, Katie hers to the third, and Omer his to the second.

DANILO: (1) Suppose $X^{\circlearrowleft\alpha} \xrightarrow{f} Y^{\circlearrowleft\beta}$ and that α has a fixed point, then show that β must also have a fixed point.

Answer: Suppose x is a fixed point of α and let y be $f(x)$:

$$\alpha x = x$$
$$y = fx$$

I'll show that $\beta y = y$.

$$\begin{aligned}
\beta y &= \beta(fx) && \text{by choice of } y\\
&= (\beta f)x\\
&= (f\alpha)x && \text{because } f \text{ is a map in } \boldsymbol{\mathcal{S}}^{\circlearrowleft}\\
&= f(\alpha x)\\
&= fx\\
&= y
\end{aligned}$$

KATIE: (3) $X \longrightarrow X$

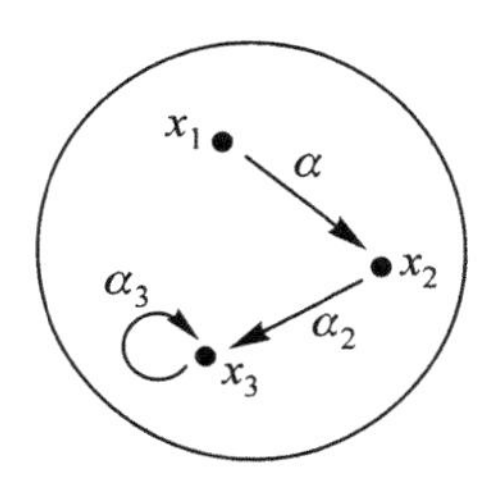

Show $\alpha \neq \alpha^2$:

$$\alpha^2 x_1 = \alpha(\alpha x_1) = \alpha(x_2) = x_3 \neq x_2 = \alpha x_1$$

so $\alpha \neq \alpha^2$.

Show $\alpha^2 = \alpha^3$:

$$\begin{aligned}
\alpha^2 x_1 &= \alpha(\alpha(x_1)) = \alpha(x_2) = x_3 =\\
&= \alpha(x_3) = \alpha(\alpha(x_2)) = \alpha(\alpha(\alpha(x_1))) =\\
&= \alpha^3(x_1).
\end{aligned}$$

Katie drew the internal diagram and gave each arrow a different label, but it is rather the whole diagram that is the endomap α. When we draw the internal diagram of a map $f : A \longrightarrow B$,

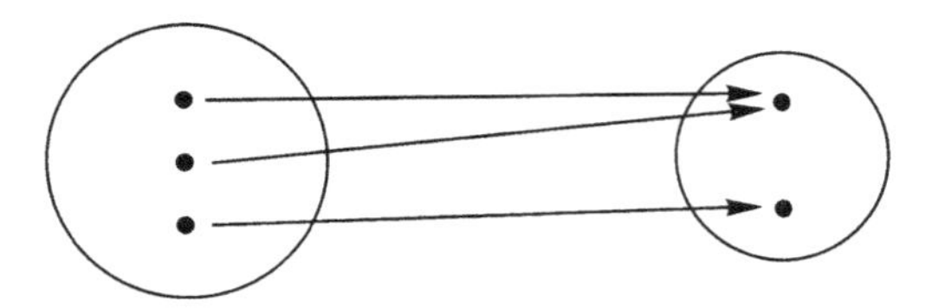

we do not label each arrow differently; it is the whole diagram that is f. Katie's endomap could be drawn:

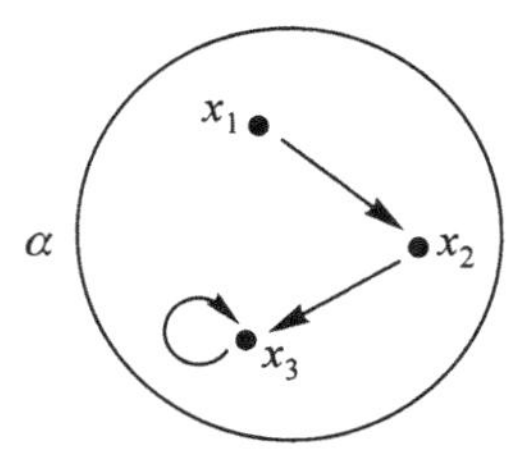

Now she sees that α and α^2 do different things to x_1. This proves that $\alpha \neq \alpha^2$. Next, we wish to prove that $\alpha^2 = \alpha^3$. How does one prove that two maps are equal?

CHAD: Check for every input that they give the same output.

Right. In this case we have three inputs, and we have to check three things:

$$\alpha^2(x_1) = \alpha^3(x_1), \quad \alpha^2(x_2) = \alpha^3(x_2), \quad \text{and} \quad \alpha^2(x_3) = \alpha^3(x_3)$$

Katie has shown the first, but didn't do the other two. They are as easy as the first, but they have to be done. First check: $\alpha^2(x_2) = x_3$, and then it follows that $\alpha^3(x_2) = \alpha(\alpha^2(x_2)) = \alpha(x_3) = x_3$. The last equation is even easier since $x_3 = \alpha(x_3)$, so that $\alpha^2(x_3) = x_3$ and also $\alpha^3(x_3) = x_3$.

CHAD: She also missed checking the other two in the first part.

No! In the first part she is doing the opposite of checking that two maps are equal, she is proving that two maps are not equal by giving a counterexample. It is similar to this: If I say that everybody in this room is a male, one example (of a female) suffices to prove me wrong.

It is remarkable that all the people who answered this question gave the same example. Is this by chance? No, there is a definite thought process that leads to this answer. We want a point, call it x_1, with $\alpha^3(x_1) = \alpha^2(x_1)$. Decide not to make x_1 special in any other way, and we gradually build our example:

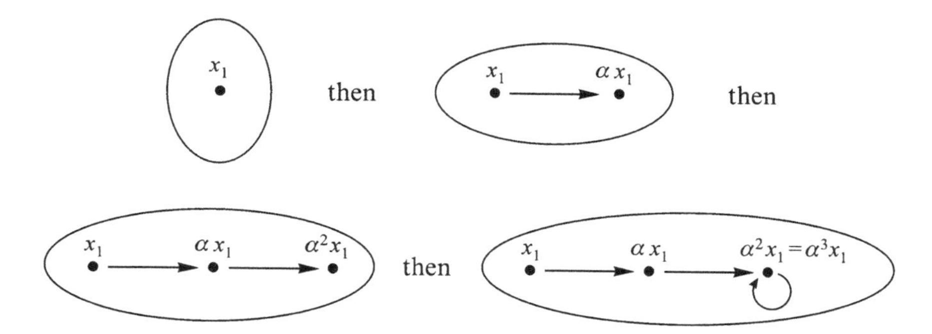

We could say that this is the *generic* dynamical system with a point x_1 satisfying $\alpha^3(x_1) = \alpha^2(x_1)$. (If you studied *presentations* in Session 15, you will see that this dynamical system is presented by the single generator x_1, together with the single relation $\alpha^3(x_1) = \alpha^2(x_1)$.) This idea already generates the example. Thinking of it in this way also simplifies some of the rest of the calculation. Since every point x is of the form $x = \alpha^r(x_1)$ for some natural number r, we can prove

$$\alpha^3 x = \alpha^3 \alpha^r x_1 = \alpha^r \alpha^3 x_1 = \alpha^r \alpha^2 x_1 = \alpha^2 \alpha^r x_1 = \alpha^2 x$$

Thus we do not have to check $\alpha^3 x = \alpha^2 x$ for each of the points x by a separate calculation. If the problem were to produce an α with $\alpha^{100} = \alpha^{200}$ but $\alpha^{77} \neq \alpha^{99}$, you would really appreciate the saving!

Let's now go to Problem 2, which probably was the hardest. Omer will show us his own elegant scheme to picture a map of graphs. Let's remember that an (irreflexive) graph is two sets and two maps arranged as in this diagram::

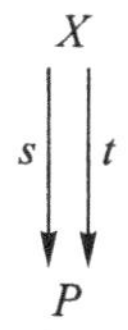

A map of (irreflexive) graphs is a pair of maps, one between the arrows sets, $f_A : X \longrightarrow Y$, and the other between the dots sets $f_D : P \longrightarrow Q$, satisfying two equations: 'respecting sources,' $f_D s = s' f_A$, and 'respecting targets,' $f_D t = t' f_A$.

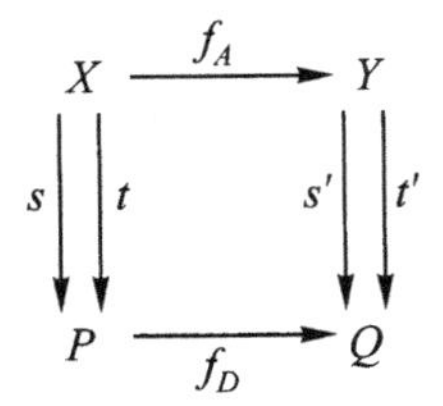

In the test we specified the two graphs by these *internal* diagrams:

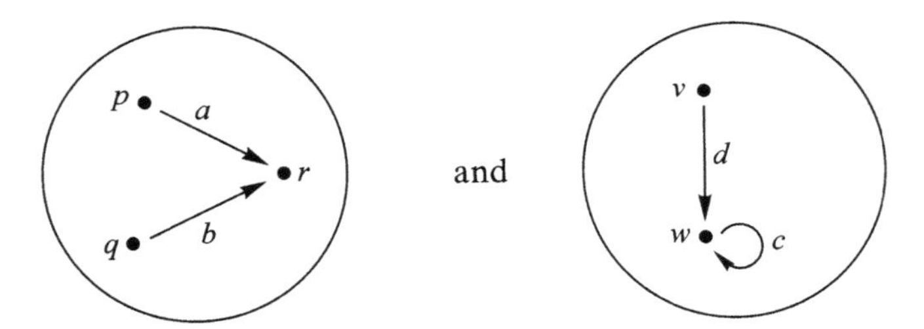

The question is: How many maps of graphs are there from the first graph to the second one? To give a graph map we need two maps, f_A and f_D, satisfying the equations given above. Omer found a way to incorporate all this information in one picture. We need two maps defined on the first graph, one acting on the dots and the other acting on the arrows.

OMER: (2)

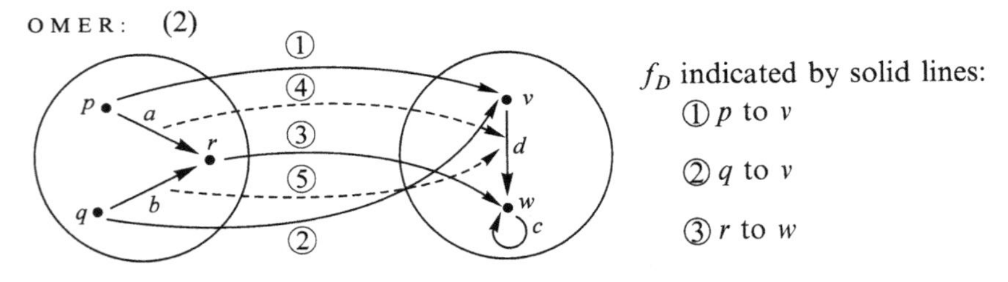

dots to dots, arrows to arrows

f_D indicated by solid lines:

① p to v

② q to v

③ r to w

f_A indicated by dotted lines:

④ a to d

⑤ b to d

His picture shows arrows going to arrows and dots going to dots, and we need only check that these maps satisfy the properties of a morphism of graphs, i.e. $f_D s = s' f_A$ and $f_D t = t' f_A$. How many things do we have to check to verify $f_D s = s' f_A$? The two maps $f_D s$ and $s' f_A$ have only two inputs, since their domain is the set with elements a, b. Therefore we have to verify two things

$$f_D s(a) = s' f_A(a) \quad \text{and} \quad f_D s(b) = s' f_A(b)$$

These are very easy to check in Omer's picture. Then check that the other two composites are equal

$$f_D t(a) = t' f_A(a) \quad \text{and} \quad f_D t(b) = t' f_A(b)$$

is also immediate, so Omer has given a genuine map of graphs.

There are more maps between these graphs. To discover them easily, notice that in his example (as opposed to one in which the domain graph has a dot which is neither the source nor the target or any arrow), as soon as you say where the arrows go, the images of the dots are forced by the conditions that sources and targets be preserved. For example, suppose that we want to send both arrows, a, b to the arrow c. Using Omer's type of diagram,

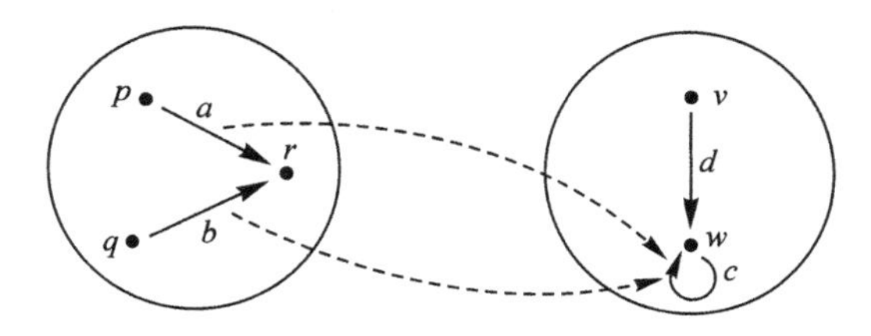

Then this forces

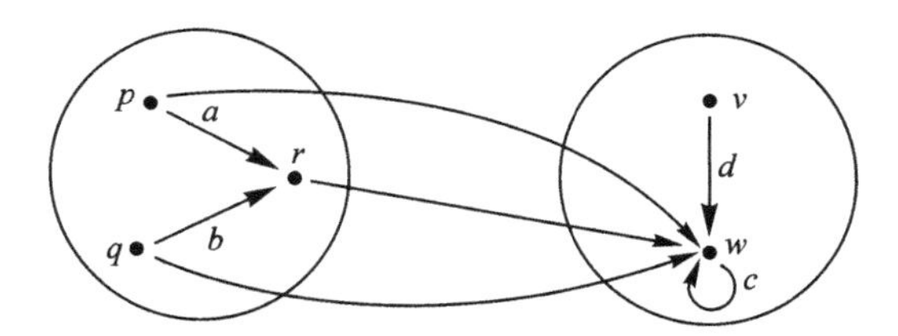

because p, being the source of a, must map to the source of c. Since r is the target of a, it must go to the target of c, and so on. Note that the conditions required for a map of graphs are checked as we go along.

Another possibility is to send a to c and b to d, which forces the map of graphs

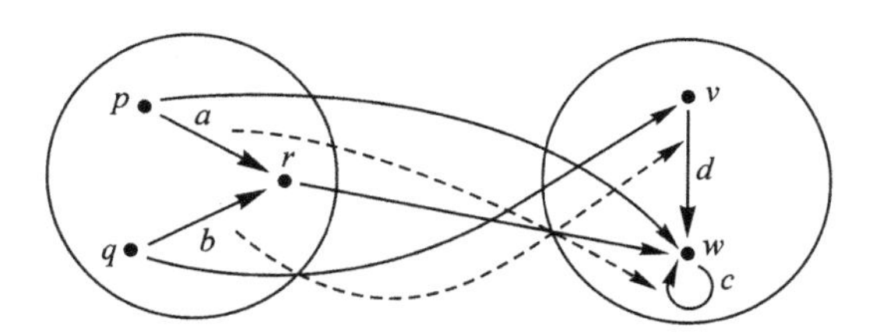

which you should check really is a map of graphs, because sometimes we are not so lucky as to be able to map any arrow the way we want.

For example, if the arrow c in the codomain had a different dot as target, there would be no map of graphs taking a to c and b to d.

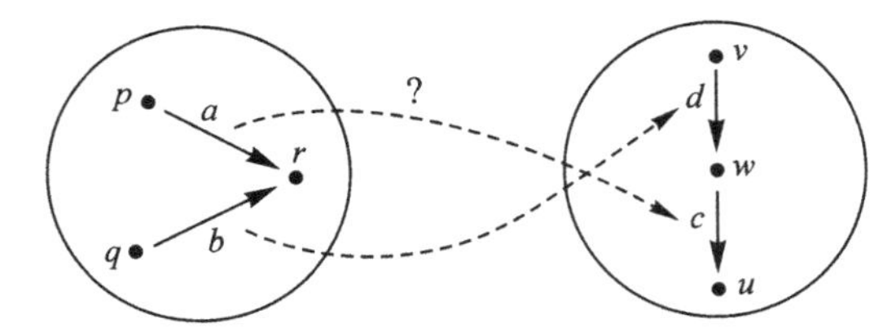

Since r is the target of both a and b, the image of r must be the target of both c and d, which is impossible.

Find the fourth map of graphs between the graphs of Test 2 yourself. There is a method which enables you to discover the four maps, and that there are no others, without having to try many fruitless possibilities. The idea behind it is that the first graph is the generic graph having two arrows with the same target. Somewhat more precisely, it is *presented* by the list of two generator-arrows a, b, together with the one relation $t(a) = t(b)$, so that we have the invertible correspondence

$$\frac{\textit{maps from our graph to any graph } G}{\textit{pairs } \bar{a}, \bar{b} \textit{ of arrows in } G \textit{ with the same target}}$$

Intuitively, you can think of a map of graphs as a way to lay the arrows and dots of the domain graph physically on top of the arrows and dots of the codomain graph, without tearing the domain graph apart.

PART IV

Elementary universal mapping properties

We find there is a single definition of multiplication of objects, and a single definition of addition of objects, in all categories. The relations between addition and multiplication are found to be surprisingly different in various categories.

ARTICLE IV

Universal mapping properties

Terminal and initial objects
Product and sum of a pair of objects

1. Terminal objects

In the category $\mathcal{S}$ of abstract sets, any one-point object **1** has exactly one map from each object X to **1**; in a great many other categories $\mathcal{C}$ of interest, there are also special objects having the same property relative to all objects of $\mathcal{C}$, even though these special objects may be intuitively much more complicated than 'single element.'

Definition: *An object S in a category $\mathcal{C}$ is said to be a* **terminal object** *of $\mathcal{C}$ if for each object X of $\mathcal{C}$ there is exactly one $\mathcal{C}$-map $X \longrightarrow S$.*

This definition is often called a 'universal' property, since it describes the nature of a particular object S in terms of its relation to 'all' objects X of the category $\mathcal{C}$. Moreover, the nature of the relation of S to other objects X is described in terms of maps in the category, more precisely, saying that 'there is exactly one' map satisfying given conditions; *terminal object* is the simplest universal mapping property, because the given conditions here are merely the domain/codomain condition stated in '$X \longrightarrow S$,' but in other universal mapping properties there will be further conditions.

Proposition: (Uniqueness of Terminal Objects) *If S_1, S_2 are both terminal objects in the category $\mathcal{C}$, then there is exactly one $\mathcal{C}$-map $S_1 \longrightarrow S_2$, and that map is a $\mathcal{C}$-isomorphism.*

Proof: Since S_2 is terminal, there is for each object X in $\mathcal{C}$ exactly one $\mathcal{C}$-map $X \longrightarrow S_2$. In particular, for $X = S_1$, there is exactly one $S_1 \longrightarrow S_2$. Since S_1 is also terminal in $\mathcal{C}$, there is for each Y in $\mathcal{C}$ exactly one $\mathcal{C}$-map $Y \longrightarrow S_1$; for example, taking $Y = S_2$ we have exactly one $\mathcal{C}$-map $S_2 \longrightarrow S_1$. To complete the proof we show that the latter map $S_2 \longrightarrow S_1$ is a two-sided inverse for the former map $S_1 \longrightarrow S_2$, i.e. that the composites

$$S_1 \longrightarrow S_2 \longrightarrow S_1$$
$$S_2 \longrightarrow S_1 \longrightarrow S_2$$

are respectively 1_{S_1} and 1_{S_2}, the identity maps of these objects. Observe that any terminal object S has the property that the only map $S \longrightarrow S$ is 1_S; for applying the definition a third time with $X = S$ itself, we get that there is exactly one map $S \longrightarrow S$. Since both S_1 and S_2 are terminal in $\mathcal{C}$, we can apply this observation to the case $S = S_1$ and then to $S = S_2$ to conclude that the composite $S_1 \longrightarrow S_2 \longrightarrow S_1$ is 1_{S_1} and that $S_2 \longrightarrow S_1 \longrightarrow S_2$ is 1_{S_2}. Thus the map $S_1 \longrightarrow S_2$ has as inverse the map $S_2 \longrightarrow S_1$, and hence is an isomorphism as claimed.

The proposition says a bit more than that any two terminal objects in the same category are isomorphic. Usually, when two objects are isomorphic, there are many isomorphisms establishing that fact, but for terminal objects there is only one. Any two terminal objects have in common all properties that can be expressed by maps in their category, so we often imagine that one terminal object has been chosen and called '**1**.' The detailed proof above will be the outline or basis for similar proofs for more involved universal properties.

Even terminal objects themselves are not completely trivial; while counting maps $X \longrightarrow \mathbf{1}$ whose *codomain* is terminal may be considered trivial (since the answer is always 'exactly one'), counting maps $\mathbf{1} \longrightarrow X$ whose *domain* is terminal gives particular information about the codomain object X.

Definition: *If* **1** *is a terminal object of a category* $\mathcal{C}$ *and if* X *is any object of* $\mathcal{C}$, *then a* $\mathcal{C}$*-map* $\mathbf{1} \longrightarrow X$ *is called a* **point** *of* X.

Exercise 1:
1 has one point. If $X \xrightarrow{f} Y$ and x is a point of X then fx is a point of Y.

Exercise 2:
In the category $\mathcal{S}$ of abstract sets, each point of X 'points to' exactly one element of X and every element of X is the value of exactly one point of X. (Here X is any given abstract set.)

Exercise 3:
In the category $\mathcal{S}^{\circlearrowright}$ of discrete dynamical systems, a point of an object 'is' just a fixed point of the endomap (i.e. an 'equilibrium state' of the dynamical system); thus most states do not correspond to any $\mathcal{S}^{\circlearrowright}$-map from the terminal object.

Exercise 4:
In the category $\mathcal{S}^{\downdownarrows}$ of (irreflexive) graphs, a 'point' of graph X 'is' just any loop in X. Hint: Determine what a terminal object looks like, using the definition of 'map in $\mathcal{S}^{\downdownarrows}$'.

Exercise 5:
The terminal object **1** in $\mathcal{S}$ has the further property of 'separating arbitrary maps'. If $X \underset{g}{\overset{f}{\rightrightarrows}} Y$ and if for each point x of X we have $fx = gx$, then $f = g$. This further property is NOT true of the terminal object in either $\mathcal{S}^{\circlearrowright}$ or $\mathcal{S}^{\downdownarrows}$. Give counterexample in each.

2. Separating

Although the terminal graph does not separate arbitrary graph maps, there are a few (non-terminal) graphs that do. Consider the two graphs whose internal pictures are as indicated

$A = \boxed{\bullet \longrightarrow \bullet}$ the generic arrow

$D = \boxed{\bullet}$ the naked dot

Then for any graph X, each arrow in X is indicated by exactly one $\mathcal{S}^{\downarrow\downarrow}$-map $A \longrightarrow X$ and each dot in X is indicated by exactly one $\mathcal{S}^{\downarrow\downarrow}$-map $D \longrightarrow X$. It follows that:

Let X, Y be any two graphs and $X \overset{f}{\underset{g}{\rightrightarrows}} Y$ any two graph maps. If $fx = gx$ for all $A \xrightarrow{x} X$ with domain A and also $fx = gx$ for all $D \xrightarrow{x} X$ with domain D, then $f = g$.

In most of our examples of categories there will be a few objects sufficient to separate maps as A, D do for graphs and as $\mathbf{1}$ does for sets; i.e. if $X \overset{f}{\underset{g}{\rightrightarrows}} Y$ with $f \neq g$, there will exist some $B \xrightarrow{x} X$ with $fx \neq gx$ and with B among the chosen few – we say that x *separates* f from g. In most categories the terminal object alone is not sufficient to separate in this sense.

Exercise 6
Show that in the category $\mathcal{S}^{\circlearrowleft}$ of discrete dynamical systems, there is a single object N such that the $\mathcal{S}^{\circlearrowleft}$-maps from N are sufficient to separate the maps $X \rightrightarrows Y$ of arbitrary objects. Hint: The object N must have an infinite number of states and may be taken to be the basic object of 'arithmetic.'

3. Initial object

Many general definitions of kinds of objects or maps in a category can be 'dualized' by reversing all arrows and compositions in the definition, in particular interchanging domains and codomains. For example, the concept 'dual' to that of terminal object is the following:

Definition: S is an **initial** *object of $\mathcal{C}$ if for every object X of $\mathcal{C}$ there is exactly one $\mathcal{C}$-map $S \longrightarrow X$.*

Exercise 7:
If S_1, S_2 are both initial in $\mathcal{C}$ then the (unique) map $S_1 \longrightarrow S_2$ is an isomorphism.

Exercise 8:
In each of $\mathcal{S}$, $\mathcal{S}^{\downdownarrows}$, and $\mathcal{S}^{\circlearrowright}$, if $\mathbf{0}$ is an initial object and $X \xrightarrow{f} \mathbf{0}$ is a map, then both

(a) for any $X \xrightarrow{g} \mathbf{0}$, $g = f$; and
(b) X itself is initial.

Exercise 9:
Define the category $\mathbf{1}/\mathcal{S}$ of *pointed sets*: an *object* is a map $\mathbf{1} \xrightarrow{x_0} X$ in $\mathcal{S}$, and a *map* from $\mathbf{1} \xrightarrow{x_0} X$ to $\mathbf{1} \xrightarrow{y_0} Y$ is a map $X \xrightarrow{f} Y$ in $\mathcal{S}$ for which

$$fx_0 = y_0 \qquad \begin{array}{ccc} & \mathbf{1} & \\ {}^{x_0}\swarrow & & \searrow^{y_0} \\ X & \xrightarrow[f]{} & Y \end{array}$$

Show that in $\mathbf{1}/\mathcal{S}$ any terminal object is also initial and that part (b) of the previous exercise is false.

Exercise 10:
Let $\mathbf{2}$ be a fixed 2-point set. Define the category $\mathbf{2}/\mathcal{S}$ of *bipointed sets* to have as objects the $\mathcal{S}$-maps $\mathbf{2} \xrightarrow{\bar{x}} X$ and as maps the $\mathcal{S}$-maps satisfying $f\bar{x} = \bar{y}$

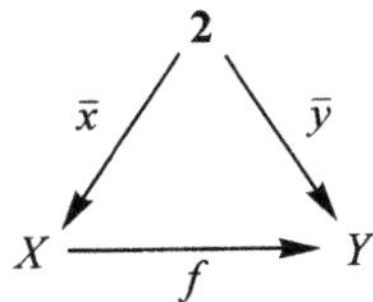

Show that in $\mathbf{2}/\mathcal{S}$ 'the' initial object is the identity map $\mathbf{2} \longrightarrow \mathbf{2}$ and that part (a) of Exercise 8 is false, i.e. an object can have more than one map to the initial object.

Exercise 11:
Show that in the category $\mathcal{S}$, if an object X is not an initial object, then X has at least one point (map from a terminal object). Show that the same statement is false in both the categories $\mathcal{S}^{\circlearrowright}$ and $\mathcal{S}^{\downdownarrows}$.

4. Products

Now we discuss an important universal mapping property which may be considered to be the objective content of the word 'and', as in Galileo's observation that a motion in space is equivalent to a motion on the horizontal plane *and* a motion on the vertical line.

Suppose B_1 and B_2 are given objects in a category $\mathcal{C}$ and that $P \xrightarrow{p_1} B_1$, $P \xrightarrow{p_2} B_2$ are given $\mathcal{C}$-maps. Then, of course, any $\mathcal{C}$-map $X \xrightarrow{f} P$ gives rise by composition to a new pair of $\mathcal{C}$-maps $X \xrightarrow{p_1 f} B_1$, $X \xrightarrow{p_2 f} B_2$. By careful choice of P, p_1, p_2 we might achieve the 'converse':

Definition:
An object P together with a pair of maps $P \xrightarrow{p_1} B_1$, $P \xrightarrow{p_2} B_2$ is called a **product** *of B_1 and B_2 if for each object X and each pair of maps $X \xrightarrow{f_1} B_1$, $X \xrightarrow{f_2} B_2$, there is exactly one map $X \xrightarrow{f} P$ for which both $f_1 = p_1 f$ and $f_2 = p_2 f$.*

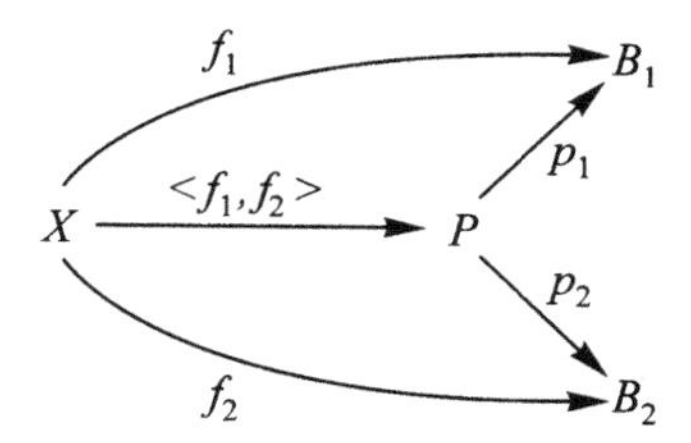

This map f, since it is uniquely determined by f_1 and f_2, can be denoted by $\langle f_1, f_2 \rangle$. The maps p_1 and p_2 are called **projection** maps for the product.

Exercise 12:
If P, p_1, p_2 and also Q, q_1, q_2 are both products of the same pair of objects B_1, B_2 in a given category, then the unique map

$$P \xrightarrow{f} Q$$

for which $p_1 = q_1 f$ and $p_2 = q_2 f$ is an *isomorphism*.

Since this exercise shows that different choices of product for B_1 and B_2 are isomorphic, we often imagine that we have chosen a specific product and denote it by $B_1 \times B_2$, p_1, p_2.

Exercise 13:
In a category $\mathcal{C}$ with products and a terminal object, each point of $B_1 \times B_2$ is uniquely of the form $\langle b_1, b_2 \rangle$, where b_i is a point of $B_i (i = 1, 2)$; and any pair of points of B_1, B_2 are the projections of exactly one point of $B_1 \times B_2$.

If T is an object corresponding to 'time,' so that a map $T \longrightarrow X$ may be called a 'motion in X,' and if P is equipped with projections to B_1, B_2 making it a product, then a motion in P corresponds uniquely to a pair of motions in the factors, and conversely. We show this briefly by

$$\frac{T \longrightarrow B_1 \times B_2}{T \longrightarrow B_1, \quad T \longrightarrow B_2}$$

where it is understood that the correspondence between the single maps above the line and the pairs of maps below the line is mediated via composition with given projection maps.

Recall that 'points' (maps from terminal objects) give important information, thus denying the apparent triviality of terminal objects. Similarly, maps whose domain is a product

$$B_1 \times B_2 \xrightarrow{f} C$$

express important information that cannot be expressed in terms of the factors separately, because the determination of the values of f involves an *interaction* of the elements of the factors. Two special cases are particularly important:

Definitions: *A* **binary operation** *on an object A is a map*

$$A \times A \longrightarrow A$$

An **action** *of an object A on an object X is a map*

$$A \times X \longrightarrow X$$

For example, if $\mathbb{N} = \{0, 1, 2, \ldots\}$ is the set of natural numbers considered as an object of $\boldsymbol{\mathcal{S}}$, then addition is a binary operation on $\mathbb{N}$

$$\mathbb{N} \times \mathbb{N} \xrightarrow{\alpha} \mathbb{N}$$

where $\alpha\langle x, y\rangle = x + y$ for each $\langle x, y\rangle$ in $\mathbb{N} \times \mathbb{N}$. Multiplication

$$\mathbb{N} \times \mathbb{N} \xrightarrow{\mu} \mathbb{N}$$

is another binary operation on $\mathbb{N}$. An action $A \times X \xrightarrow{\alpha} X$ of A on X may be considered as an 'A-parameterized family of endomaps of X,' because for each $\mathbf{1} \xrightarrow{a} A$, α gives rise to an endomap of X

$$X \xrightarrow{\langle \bar{a}, 1_X\rangle} A \times X \xrightarrow{\alpha} X$$

where $\bar{a}$ is the 'constant map' $X \longrightarrow \mathbf{1} \xrightarrow{a} A$. For example, an action of $\mathbf{1}$ on X 'is' just a given endomap of X, since '$\mathbf{1} \times X = X$.'

Indeed, our example $\boldsymbol{\mathcal{S}}^{\circlearrowright}$ of a category can be generalized to $\boldsymbol{\mathcal{S}}^A$ for any given set A, as follows: An *object* of $\boldsymbol{\mathcal{S}}^A$ is a set X together with any action $A \times X \xrightarrow{\xi} X$ of A on X. A *map* from X, ξ to Y, η is any $\boldsymbol{\mathcal{S}}$-map $X \xrightarrow{f} Y$ which respects the actions of A in the sense that

$$f(\xi(a, x)) = \eta(a, f(x)) \quad \text{for all } a, x$$

This can be expressed in another way if we define $1_A \times f$ to be the map

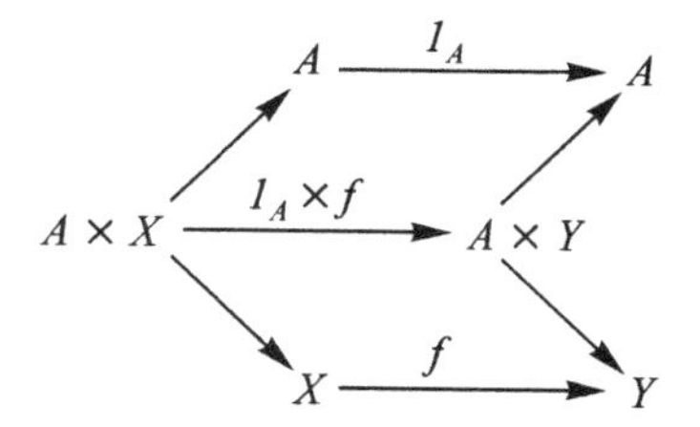

whose A-projection from $A \times Y$ is the A-projection from $A \times X$, but whose Y-projection is f following the X-projection, as indicated. The condition that f respect the given A-actions can be restated as follows:

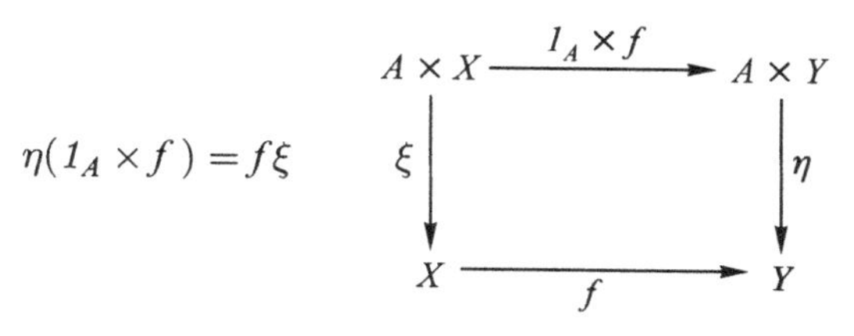

> **Exercise 14:**
> Define composition of maps in $\mathcal{S}^A$ and show that it is a category.

If A is already equipped with a preferred binary operation $A \times A \xrightarrow{\alpha} A$ and point $\mathbf{1} \xrightarrow{a_0} A$, then we may restrict the notion of 'action of A on X' to those actions which are 'compatible with α and a_0,' in the sense that under the action α corresponds to composition of endomaps of X and a_0 acts as 1_X, i.e.

$$\xi(\alpha(a,b),x) = \xi(a,\xi(b,x)) \qquad \text{for all } a,b,x$$
$$\xi(a_0,x) = x \qquad \text{for all } x$$

> **Exercise 15:**
> Express these equations as equations between maps
>
> $$A \times A \times X \rightrightarrows X,\ X \rightrightarrows X$$
>
> constructed by using ξ and the universal mapping property of products.

These equations are frequently considered when the given binary operation on A is associative and the given point is neutral for it; in other words, when A, α, a_0 together constitute a monoid (see Session 13). In that case the actions satisfying these compatibility equations constitute a subcategory of $\mathcal{S}^A$ called the category of all actions of the monoid on sets.

5. Commutative, associative, and identity laws for multiplication of objects

For multiplication of *numbers*, you may have seen that the basic laws

$$a \times b = b \times a \quad \text{(commutative law)}$$
$$1 \times a = a \quad \text{and} \quad a \times 1 = a \quad \text{(identity laws)}$$
$$a \times (b \times c) = (a \times b) \times c \quad \text{(associative law)}$$

can be shown (with some effort) to imply more complicated laws, such as

$$(z \times ((1 \times x) \times x)) \times (p \times q) = p \times (((q \times x) \times z) \times x)$$

That is, in a product of several factors:

grouping does not matter;
order does not matter; and
trivial factors (factors which are 1) can be omitted.

The product is completely determined by what the non-trivial factors are, provided that we take account of repetitions.

For multiplication of *objects*, in any category $\mathcal{C}$ having products of pairs of objects and a terminal object, the laws mentioned above are also valid (after replacing 'equals' by 'is isomorphic to'). To see this, it is not necessary to prove the simpler laws first and then deduce the more complicated laws. We can *directly* define the product of any family of objects by a universal mapping property, without having to list the objects in an order, nor having to multiply them two at a time. It turns out, as you will see below, that the proof of the uniqueness theorem for products of pairs of objects works equally well for any family of objects.

First we need some notation for 'families.' Let I (for 'indices') be a set, and for each i in I, let C_i be an object of $\mathcal{C}$. (Repetitions are allowed: for distinct indices i and j we allow C_i and C_j to be the same object. Also, the set I of indices is allowed to be empty!) Together these data constitute an (indexed) *family* of objects of $\mathcal{C}$.

Definition: *A* **product** *of this indexed family is an object P together with maps $P \xrightarrow{p_i} C_i$ (one for each i), having the universal property:*

Given any object X and any maps $X \xrightarrow{f_i} C_i$ (one for each i), there is exactly one map $X \xrightarrow{f} P$ such that all the triangles below commute, i.e. such that $p_i f = f_i$ for each i in I.

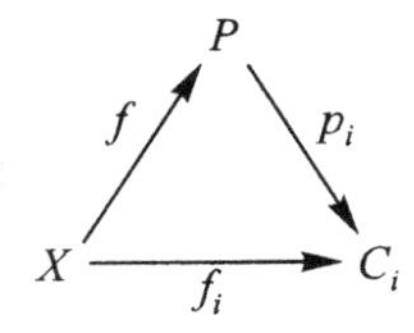

The discussion of products of pairs of objects can be copied almost verbatim for products of families.

Theorem: (Uniqueness of products) *If the maps $P \xrightarrow{p_i} C_i$ and $Q \xrightarrow{q_i} C_i$ make both P and Q products of this family, then (because Q is a product) there is exactly one map $P \xrightarrow{f} Q$ for which $q_i f = p_i$ for each i in I. Moreover, the map f is an isomorphism.*

Notation: We can assume that we have chosen a particular product for the family; we denote it by $\prod_I C_i$ and call the projection maps p_i.

The commutative, associative, and identity laws (and their more complicated consequences) all follow from the uniqueness theorem, together with the use of 'partial' products: To multiply a family of objects you can group it into subfamilies, and calculate the product of the products of these subfamilies. The exercise below asks you to carry out explicitly the proof of the special case of a family of three objects, indexed by

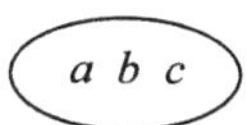

grouped into two subfamilies as

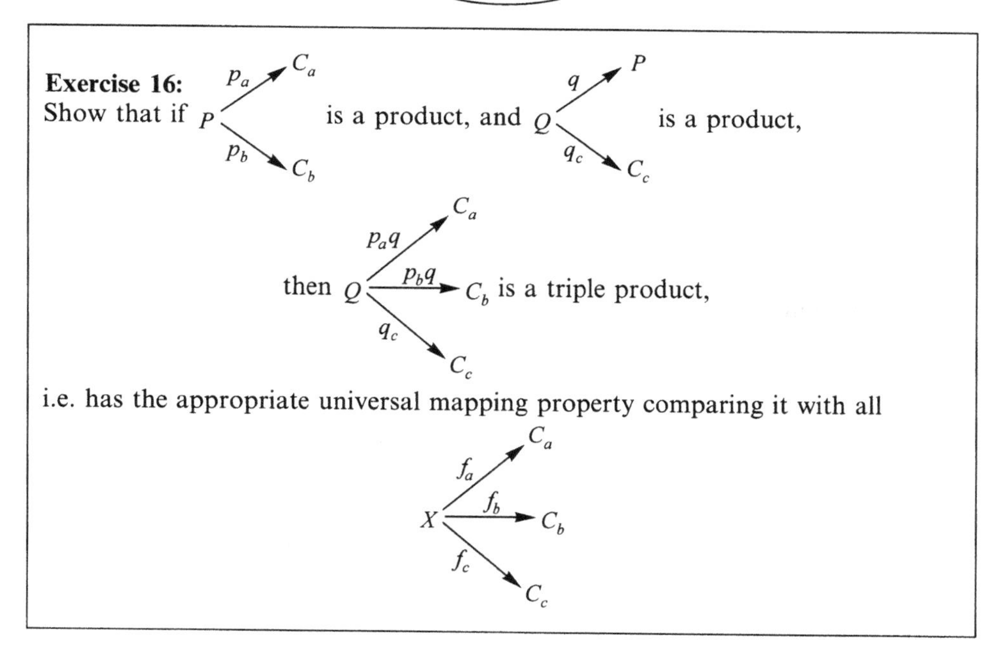

This exercise shows that the iterated product $(C_a \times C_b) \times C_c$ is a triple product of this family; in particular, if $\mathcal{C}$ has products of pairs, it also has triple products. A similar proof shows that $C_a \times (C_b \times C_c)$ is also a triple product. The uniqueness theorem then implies that these two are isomorphic, which is the associative law. Of course, the uniqueness theorem does more; it gives a specific isomorphism compatible with the projection maps.

6. Sums

Dualizing the notion of product-projections, we get:

Definition: *A pair $B_1 \xrightarrow{j_1} S$, $B_2 \xrightarrow{j_2} S$, of maps in a category makes S a* **sum** *of B_1 and B_2 if for each object Y and each pair $B_1 \xrightarrow{g_1} Y$, $B_2 \xrightarrow{g_1} Y$, there is exactly one map $S \xrightarrow{g} Y$ for which both $g_1 = gj_1$ and $g_2 = gj_2$.*

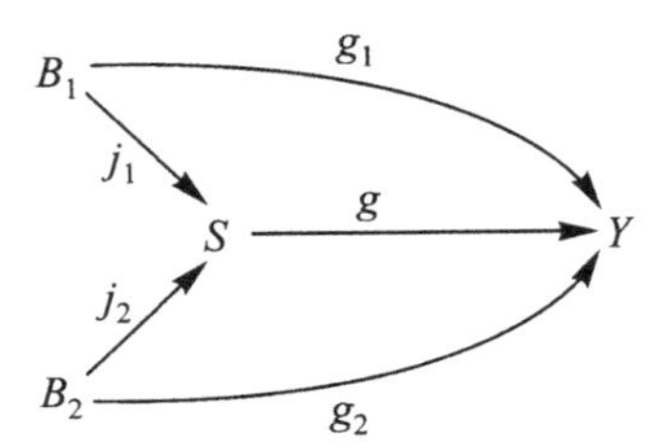

Note: The maps j_1, j_2 are called the **injection** maps for the sum. As with products, we often choose a special sum of B_1 and B_2 and denote it by $B_1 + B_2, j_1, j_2$.

Exercise 17:
In $\mathbf{S}$, $\mathbf{S}^{\downarrow\downarrow}$, and $\mathbf{S}^{\circlearrowleft}$, sums have the property that any point of $B_1 + B_2$ comes via injection from a point of exactly one of B_1, B_2.

Exercise 18:
In $\mathbf{S}$, there are many maps $X \longrightarrow \mathbf{1} + \mathbf{1}$ (if $X \neq \mathbf{0}, \mathbf{1}$) which do not factor through either injection. (Give examples.)

Exercise 19:
Show that in a category with sums of pairs of objects, the 'iterated sums'

$$(A + B) + C \quad \text{and} \quad A + (B + C)$$

are isomorphic.

7. Distributive laws

We have seen that the algebraic laws for multiplication of objects (commutative, associative, and identity laws) are valid in any category which has products; and likewise addition of objects satisfies the corresponding laws. Surprisingly, the usual laws relating addition to multiplication, called distributive laws

$$(a \times b) + (a \times c) = a \times (b + c)$$

and

$$0 = a \times 0$$

are false in many categories!

There is at least a map comparing the two sides of the expected equations. In any category having both sums (and initial objects) and products, there are standard maps

$$(A \times B) + (A \times C) \longrightarrow A \times (B + C)$$
$$\mathbf{0} \longrightarrow A \times \mathbf{0}$$

constructed using only the implied injections and projections and universal mapping properties.

Definition: *A category is said to satisfy the* **distributive law** *if the standard maps above are always isomorphisms in the category.*

For example $\mathbf{S}$, $\mathbf{S}^{\circlearrowright}$, and $\mathbf{S}^{\downdownarrows}$ all satisfy the distributive law, as is not difficult to see. A proof using exponentiation will be discussed in Part V.

Exercise 20:
The category $\mathbf{1}/\mathbf{S}$ of pointed sets does not satisfy the distributive law. Hint: First determine the nature of sums within the category $\mathbf{1}/\mathbf{S}$.

Exercise 21:
If A, D denote the generic arrow and the naked dot in $\mathbf{S}^{\downdownarrows}$, show that

$$A \times A = A + D + D$$

Hint: Besides counting the arrows and dots of an arbitrary graph X (such as $X = A \times A$) via maps $A \longrightarrow X$, $D \longrightarrow X$, the actual internal structure of X can be calculated by composing these maps with the two maps $D \overset{s}{\underset{t}{\rightrightarrows}} A$.

8. Guide

Universal properties have been seen to be the source of both multiplication and addition of objects; the more extended discussion of these constructions in Sessions 19–28 will illustrate ways in which they are used, and will show how to calculate them. Relations, such as the distributive law, *between* addition and multiplication are deeper; discussion of these begins at the end of Session 25. Following Session 28 are some sample tests. In Session 29, we study a further property of products which will be raised to a higher level in Part V.

SESSION 19

Terminal objects

Now we will discuss ONE, the unit or identity for multiplication. You have met several different things called 'one.' First, the number 1, the unit for multiplication of numbers, satisfying:

$$\text{for each number } x, \quad 1 \times x = x$$

Second, the identity map of a set A, the map $1_A : A \longrightarrow A$ defined by

$$1_A(x) = x \qquad \text{for each member } x \text{ of } A$$

which satisfies the identity laws:

$$\text{for each map } f \text{ with codomain } A, \quad 1_A \circ f = f, \text{ and}$$
$$\text{for each map } g \text{ with domain } A, \quad g \circ 1_A = g$$

Third, and this is the starting point for our topic, you have met 'singleton sets,' sets with exactly one member.

Our goal is to understand everything in terms of maps and their composition, so we should ask ourselves: what special property do singleton sets have? We want the answer to involve maps. Any ideas?

OMER: There is only one map to a singleton set.

Good. A singleton set such as $\{Alysia\}$ has the property that *for each set X*, there is exactly one map from X to $\{Alysia\}$.

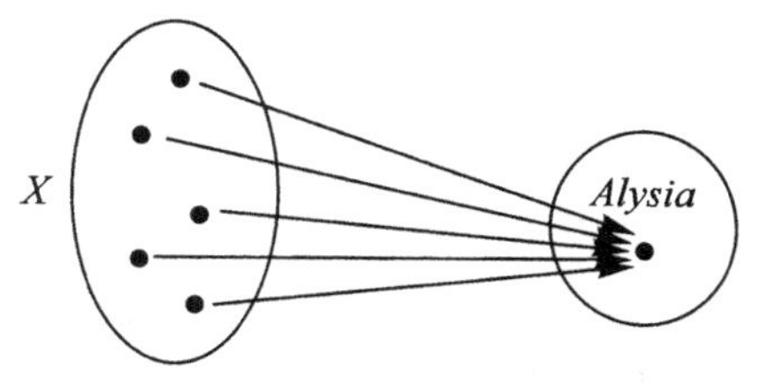

Remember that this works even if the domain X is empty; then the internal diagram of the map is

with no arrows, since X has no members.

We have succeeded in finding a special property of singleton sets, a property which is expressed entirely in terms of maps, without mentioning members. Why do we want to describe the singleton sets entirely in terms of maps? The reason is that in other categories, say dynamical systems or graphs, it is not so clear what a 'member' should be, but properties expressed in terms of maps and composition (such as Omer's property) still make sense in any category. Therefore we define:

Definition: *In any category $\mathcal{C}$, an object T is a* **terminal object** *if and only if it has the property:*

for each object X in $\mathcal{C}$ there is exactly one map from X to T.

The 'X' in the definition is a pronoun. We could have said, 'T is a terminal object if and only if for each object in $\mathcal{C}$ there is exactly one map from that object to T'; but to ensure that the phrase 'that object' is unambiguous, we give it a name 'X' when it is first mentioned. It doesn't make any difference what letter we use. To say 'For each object Y in $\mathcal{C}$ there is exactly one map from Y to T,' would say exactly the same thing about T.

Let's look for examples of terminal objects in other categories. Is there any terminal object in $\mathcal{S}^{\circlearrowright}$?

OMER: A set with one member.

That's a good idea. But a set by itself is not an object of $\mathcal{S}^{\circlearrowright}$; we must specify an endomap of the set. What endomap should we choose?

ALYSIA: The member goes to itself?

Exactly. In fact this is the only endomap our singleton set has. So we try

$$T = \boxed{\bullet\circlearrowright}$$

Is this really a terminal object in $\mathcal{S}^{\circlearrowright}$? That's a lot to ask of T. It must satisfy: for each dynamical system $X^{\circlearrowright\alpha}$ in $\mathcal{S}^{\circlearrowright}$ (no matter how complicated), there is exactly one map $X^{\circlearrowright\alpha} \longrightarrow T$. What is a map $X^{\circlearrowright\alpha} \longrightarrow Y^{\circlearrowright\beta}$, in $\mathcal{S}^{\circlearrowright}$?

OMER: A map of sets such that $f \circ \alpha = \beta \circ f$.

Right. How many maps from the set X to a singleton set $\mathbf{1}$ are there, independently of whether they satisfy their extra condition? Yes, precisely one. Does it satisfy the condition $f \circ \alpha = \beta \circ f$?

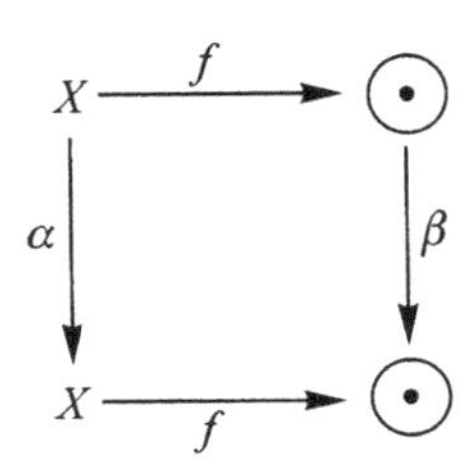

Yes, because both $f \circ \alpha$ and $\beta \circ f$ are maps (of sets) from X to $\mathbf{1}$, and there is only one such map. That finishes it: we have shown that this dynamical system

$$\boxed{\bullet\circlearrowright}$$

is a terminal 'set-with-endomap'; i.e. a terminal object in the category $\mathcal{S}^{\circlearrowright}$.

Let's now go to the category of irreflexive graphs. An object is a pair of sets X, P and a pair of maps from X to P. Thus a picture of one object is

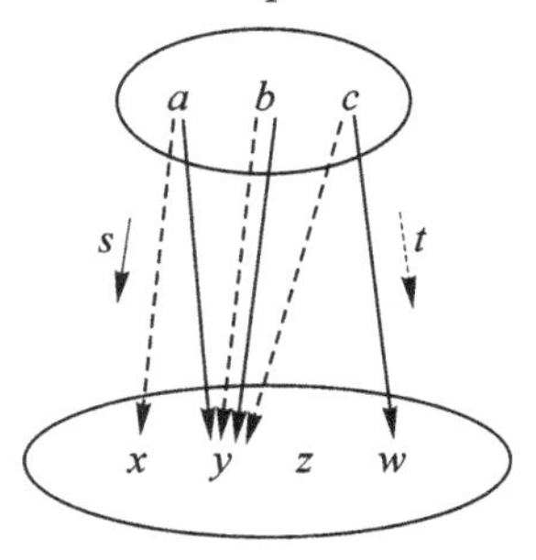

where we draw the map s with solid arrows and the map t with dotted arrows. But people in computer science, in electrical engineering or in traffic control who use graphs all the time do not draw them like that. They draw them like this:

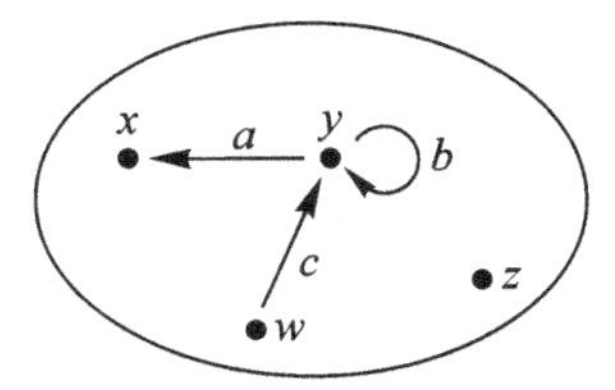

This is why the elements of X are called arrows: they are pictured as arrows of the graph, while the elements of P are drawn as dots, and the structural maps are suggestively called 'source' and 'target.'

In the second picture one can see right away that to go from w to x we first have to go from w to y and then from y to x. Or we can go to y, then take one or more round trips back to y and then go to x. On the other hand, in order to study maps between graphs the first picture may help. What is a map

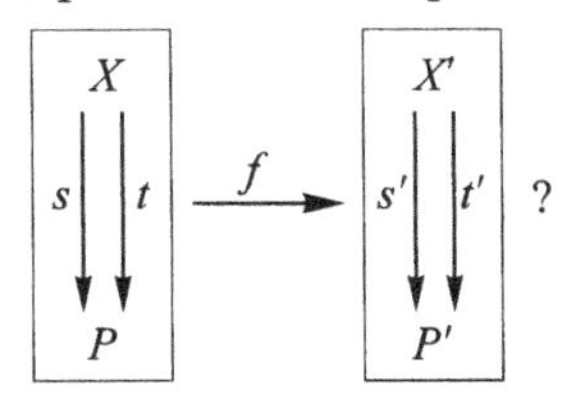

OMER: Send X to X' and P to P'.

That's right. A map in this category is two maps of sets, $f_A : X \longrightarrow X'$ and $f_D : P \longrightarrow P'$, but not any two maps. They must satisfy the equations

$$f_D \circ s = s' \circ f_A \quad \text{and} \quad f_D \circ t = t' \circ f_A$$

one equation for each type of structural map that the objects involve. We must decide what to put in the box to the right

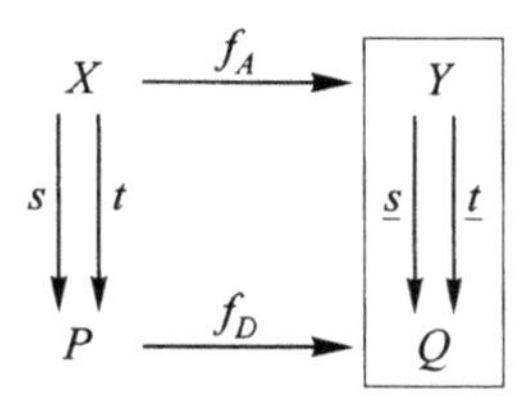

so that there is precisely one map of graphs from any graph to the one in that box. Does anybody have any ideas?

CHAD: Y and Q with the same elements as X and P?

No. Y and Q will be fixed once and for all; they can't depend on what X and P are.

CHAD: Put Y and Q with one member each.

That's a good idea. Let $Y = \boxed{a}$ and $Q = \boxed{p}$. What should the maps $\underline{s}$ and $\underline{t}$ be?

DANILO: There is only one possibility.

Yes, there is only one map from $\boxed{a}$ to $\boxed{p}$. Is the graph $\boxed{p \bullet\!\circlearrowleft a}$, with only one arrow and only one dot, a terminal object? We must check that from any graph there is precisely one graph map to this one. But no matter what X is, there is only one possible choice for the map $f_A : X \to \boxed{a}$, and there is only one possible choice for the map $f_D : P \to \boxed{p}$. The question is whether these maps satisfy the equations that say that these maps respect the source and the target. Now, the first equation ($f_D \circ s \stackrel{?}{=} \underline{s} \circ f_A$) involves two maps that go from X to $\boxed{p}$:

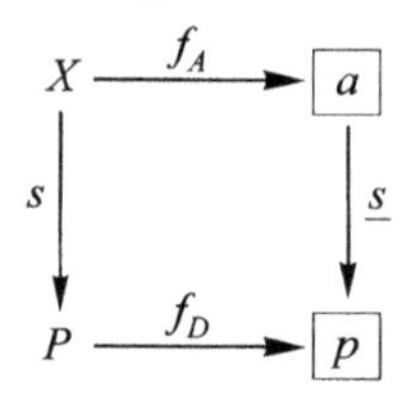

Are these two maps equal?

CHAD: They have to be, because there is only one map from X to $\boxed{p}$.

That's right! And the other equation (the one with t instead of s) is true for the same reason. So this graph $\boxed{\bullet\!\circlearrowleft}$ is the terminal object in the category of graphs.

One might have thought that the terminal object is just one dot without arrows. That is, in our two ways of picturing graphs:

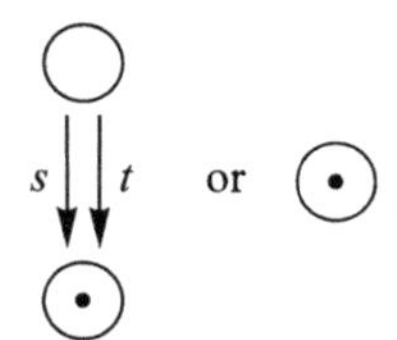

but this doesn't work.

DANILO: In the case of one dot without arrows there won't be any maps to it.

That's right. As long as the domain graph has an arrow, it won't have any maps to this graph because f_A maps to an empty set. This proves that this graph with one dot and no arrow does not work as terminal graph. There is another proof based on the following general theorem:

Theorem: *Suppose that $\mathcal{C}$ is any category and that both T_1 and T_2 are terminal objects in $\mathcal{C}$. Then T_1 and T_2 are isomorphic; i.e. there are maps $f : T_1 \longrightarrow T_2$, $g : T_2 \longrightarrow T_1$ such that $g \circ f$ is the identity of T_1 and $f \circ g$ is the identity of T_2.*

Let's try to work out the proof.

> **Proof:** To show that T_1 and T_2 are isomorphic we need first of all a map $T_1 \longrightarrow T_2$.

How can we get such a map?

DANILO: There is only one map from one terminal object to another.

Does Danilo's remark use the fact that T_1 is terminal, or the fact that T_2 is terminal?

OMER: T_2.

Right. So, the proof continues like this:

> Because T_2 is terminal there is a map $f : T_1 \longrightarrow T_2$. We need a map $g : T_2 \longrightarrow T_1$. Again there is one because T_1 is terminal. But this does not prove yet that T_1 is isomorphic to T_2. These two maps have to be proved to be inverse to each other.

Is it true that the composite $gf : T_1 \longrightarrow T_1$ is equal to 1_{T_1}?

KATIE: Yes, because there is only one map from T_1 to T_1.

Right.

> Because T_1 is terminal, there is only one map from T_1 to T_1. Therefore, $g \circ f = 1_{T_1}$.

I leave as an exercise the proof that the other composite is equal to the corresponding identity map 1_{T_2}; then the proof of the theorem will be complete.

Notice that in this proof we use separately the two aspects of the defining property of a terminal object T, namely that for each object X

1. there is at least one map $X \longrightarrow T$, and
2. there are not two different maps $X \longrightarrow T$.

Statement (1) is used to get maps $T_1 \rightleftarrows T_2$ and (2) is used to prove that they are inverses of each other.

SESSION 20

Points of an object

Everything that can be said about sets can be expressed in terms of maps and their composition. As we have stressed before, that includes everything about 'elements' of sets. We are going to extend this point of view to categories other than that of abstract sets using what we call 'figures.' Recall that to specify an element of the set

for example the element *Emilio*, we use the following map e from a terminal set $\mathbf{1}$ to the given set:

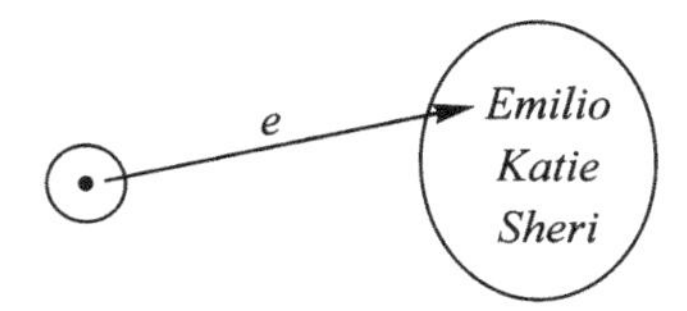

Everything we may want to say about *Emilio* as an element of this set, we can express in terms of this map. For example, to evaluate the gender map g

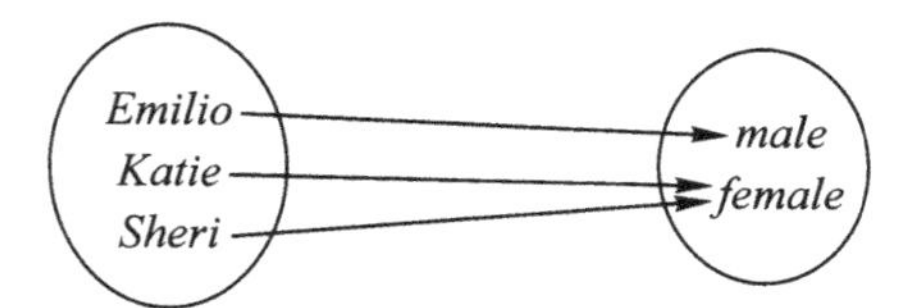

at the element *Emilio*, we simply compose the maps, to obtain

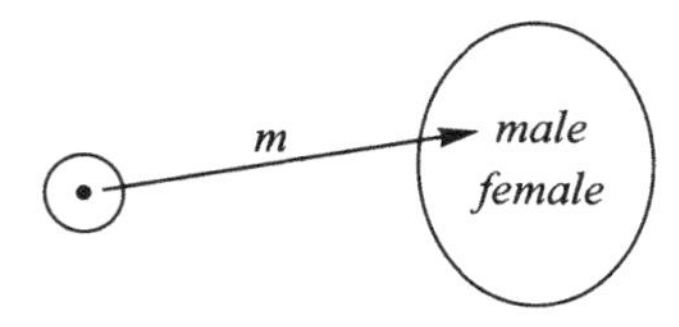

which shows that

$$g \circ e = m$$

We can call the map e by the same name as the corresponding element and say that the element *Emilio* of the set {*Emilio*, *Katie*, *Sheri*} is just the map

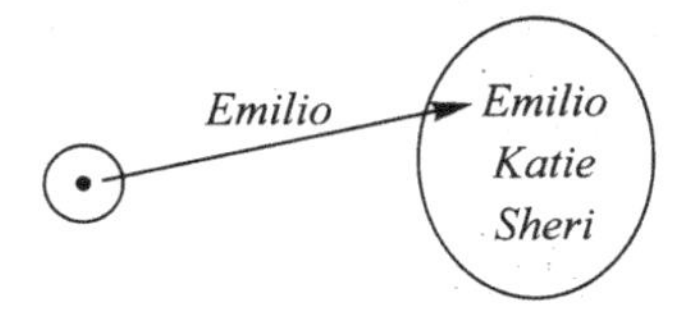

so that we can write

$$g \circ Emilio = male$$

Thus 'evaluation is composition.' We do not need to remember two separate rules, the associative law and a rule for composition of maps. They are the same thing!

ALYSIA: In the set of the first example, is there also a map for the element '*Katie*,' and for every element of the set?

Yes. Each element is a map from the terminal object, so that in the equation $(g \circ f) \circ x = g \circ (f \circ x)$ the map x can be any element.

OMER: Is it always the last map that represents an element?

That case arises most often, but you can compose maps in any order as long as the domain and codomain match. For example, you can compose the following maps

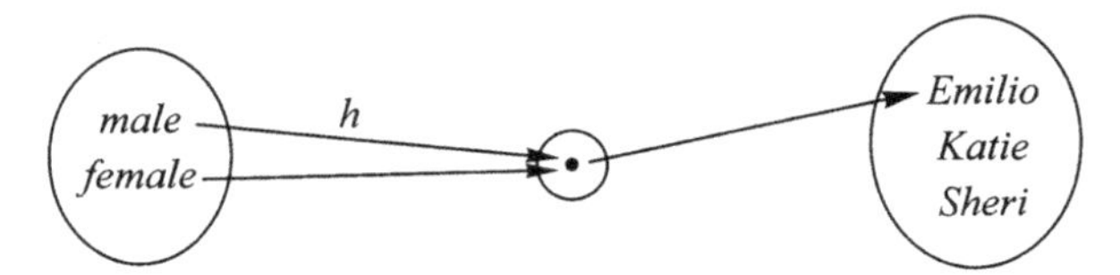

and get a constant map.

OMER: What's the one-member set exactly?

It is any *terminal* set. You can think of it as the set $\boxed{Omer}$, when you are the one who is referring to an element of a set X, the element you are talking about is a map $\boxed{Omer} \longrightarrow X$, which is 'you pointing to the element.'

OMER: But the one-element set also has one element; if every element is a map, what is the map behind the element of the one-element set?

This is a very good question. The answer is: the identity map of the terminal set. We *start* with the idea of terminal set, which does not need the idea of element, but only the idea of map. The basic theorem that makes all this work is:

In any category $\mathcal{C}$, *any two terminal objects are uniquely isomorphic*,

which we proved in Session 19. In the category of sets this result seems obvious, because terminal sets are sets with only one member. But in other categories they are not so simple and the result is not so obvious. For example, in the category $\mathbf{S}^{\circlearrowleft}$ of endomaps of sets, or dynamical systems, the terminal endomap was the endomap $T = \boxed{p\bullet\circlearrowleft}$ (i.e. $\boxed{p\bullet} \xrightarrow{1} \boxed{\bullet p}$), while in the category of irreflexive graphs are terminal graph was the graph with one dot and one arrow, $T = \boxed{p\bullet\circlearrowleft a}$.

CHAD: What is T?

T is any terminal object, i.e. an object in the category such that for each object X in the same category there is exactly one map in the category from X to T.

CHAD: So T is the other object?

Well, I wouldn't say it that way. The object we are describing is T, but we describe it by saying how it relates to every object in our 'universe,' our category. The definition of terminal object uses a 'universal property.' Here's an example: to say that Chad is 'universally admired' means:

For every person X in the world, X admires Chad.

FATIMA: If you want to translate terminal object into arithmetic it would have to be the number 1.

This is a very good point to which we will come back later because it is a remarkable theorem that the terminal object behaves like the number 1 for multiplication. So you must promise that you will raise this point again when we talk about multiplication.

If in a particular category we have determined what the terminal object is, then we can determine what the points of any object are. Suppose that T is a terminal object in the category $\mathcal{C}$; then any $\mathcal{C}$-map from T to another object X of this category is called a **point** of X.

Definition: *A* **point** *of X is a map $T \longrightarrow X$ where T is terminal.*

This means that in the category of sets, the points of a set are precisely the elements of that set, since we have found that the elements of a set X are the maps from a terminal (singleton) set to X. Our next task is to find out what the points of the objects of other categories are. The first example is the category $\mathbf{S}^{\circlearrowleft}$ of endomaps of sets. One may guess that the points of an endomap are the elements of the underlying set, but this is not right. For example, let's look at the following endomap:

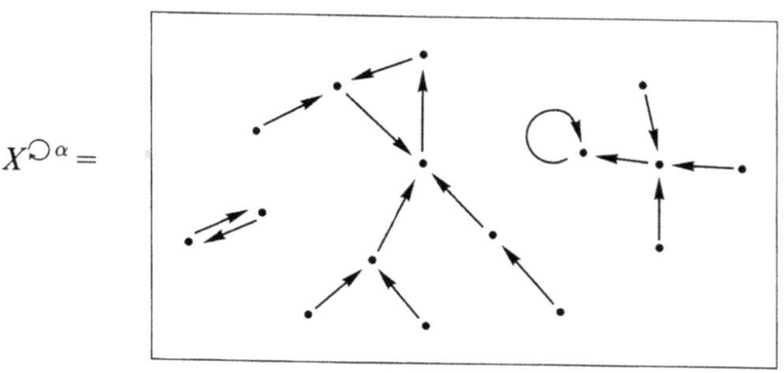

Can anybody find a $\mathbf{S}^{\circlearrowright}$-map from the terminal object $T = \boxed{\bullet\circlearrowright}$ to that $X^{\circlearrowright\alpha}$?

FATIMA: Map this element to the one with the loop in X.

That's right! That is the *only* map $T \longrightarrow X^{\circlearrowright\alpha}$ in this category. This object $X^{\circlearrowright\alpha}$, as complicated as it looks, has only one point, and the point is the map

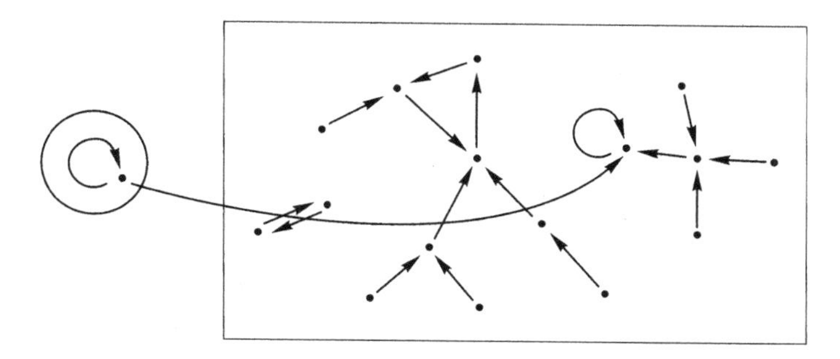

This should be obvious since we have seen that in this category every map takes a fixed point to a fixed point. The conclusion is that in this category 'point' means what we have been called 'fixed points.'

DANILO: But if this has one point, how do we describe the other dots?

Good question! Yes, it seems unfortunate that the maps from the terminal object only describe the fixed points, not as in sets where they give all the elements. However, in this category we have other objects which give the other dots. Remember the set of natural numbers with the successor endomap, $\mathbb{N}^{\circlearrowright(\,)+1}$, maps from which will give all the dots, as we saw in Session 15.

DANILO: That would only tell you the number of dots.

Right. Another object has to be used to find the number of 2-cycles, still another to describe the 3-cycles, and so on.

OMER: You can map back to $\mathbb{N}^{\circlearrowright(\,)+1}$.

Yes, you can ask about the maps from any object to any object.

OMER: But the loop can't be mapped anywhere in $\mathbb{N}^{\circlearrowright(\,)+1}$.

That's right. And that proves that there are *no* points in $\mathbb{N}^{\circlearrowright(\,)+1}$.

Let's make a little table to collect our information about terminal objects.

Category	Terminal object	'Points of X' means...
$\mathcal{C}$	T	map $T \longrightarrow X$
$\mathcal{S}$	[•]	element of X
$\mathcal{S}^{\circlearrowright}$ endomaps of sets	[•⟲]	fixed point or equilibrium state
$\mathcal{S}^{\downdownarrows}$ irreflexive graphs	a ⇉ b or [p•⟲a]	?

Now let's see an example from the category of irreflexive graphs. Consider the graph

$G=$ [*Ian* *Chad* ; $s=$ *usual seat* ↓ ↓ $t=$ *today's seat* ; (set of chairs in the classroom)]

The internal diagrams of these two maps are the following:

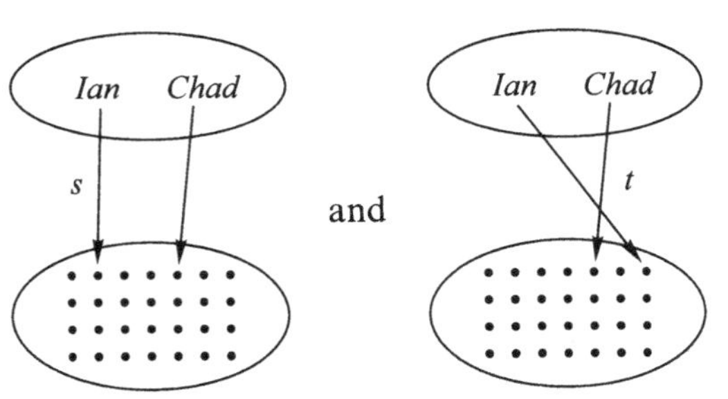

which can be drawn in the same picture either as

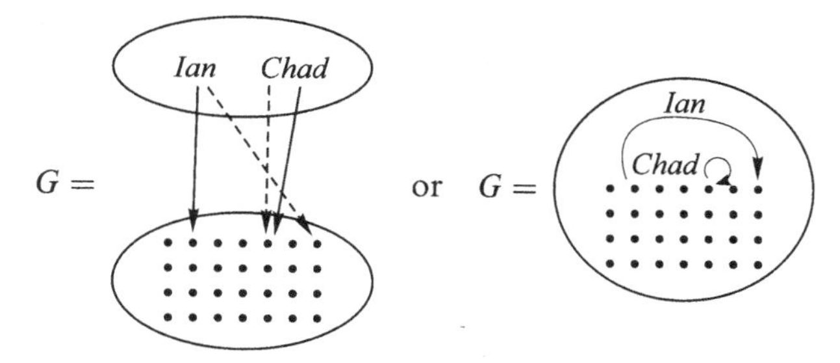

In the last picture one can clearly see that Chad kept the same seat, while Ian moved to the right of his usual seat, but both pictures have the same information in them.

How many maps are there from the terminal graph to this graph G, or in our new terminology, how many points does this graph G have?

OMER: There is only one, isn't there?

Let's see. We must remember what the terminal object of this category is. It must be two sets and two maps, so we must decide what to put in the picture

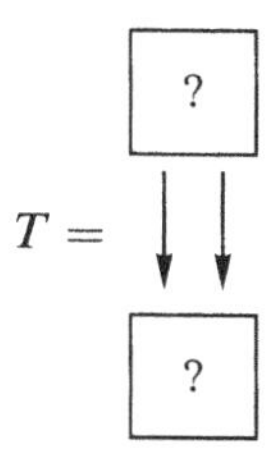

so that is the terminal graph.

CHAD: Put one member in each set.

And the maps? What maps should we put there, Mike?

MIKE: Both are the one which sends the element on top to the element in the bottom set.

That's right, moreover Chad is right: the graph T is terminal, and Omer is also right: G has only one point.

SESSION 21

Products in categories

We want to make precise the notion of product of objects in a category. For this it will be helpful to remember the idea of Galileo that in order to study the motion of an object in space it is sufficient to study instead two simpler motions, the motion of its shadow on a horizontal plane and the motion of its level on a vertical line. The possibility of recombining these two motions to reconstruct the original motion in space is the basis for the notion of product. We are now in a position to make all these ideas precise.

By multiplying the disk times the segment we get their product, the cylinder. The basic ingredients that reveal the cylinder as that product are the two maps 'shadow' and 'level':

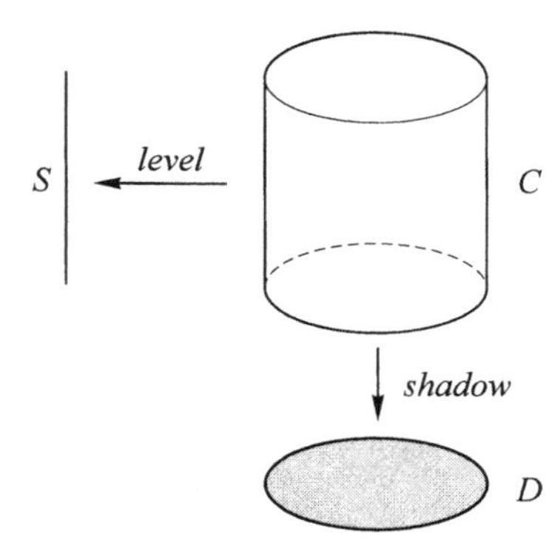

When we multiply two objects we get not only a third object, but also two maps whose domain is the product, one map to each of the two given objects. This suggests that the definition of product in a category should start this way:

A *product* of A and B is

1. an object P, and
2. a pair of maps, $P \xrightarrow{p_1} A$, $P \xrightarrow{p_2} B$

But that cannot be the end of the matter. We need to formulate the principle that a motion in P is uniquely determined by motions in A and in B, and we need to do it in a way applicable to any category. The idea is to replace the interval of time by *each* object in the category. Here is the official definition.

Definition: *Suppose that A and B are objects in a category $\mathcal{C}$. A* **product** *of A and B (in $\mathcal{C}$) is*

1. *an object P in $\mathcal{C}$, and*
2. *a pair of maps, $P \xrightarrow{p_1} A$, $P \xrightarrow{p_2} B$, in $\mathcal{C}$ satisfying:*

for every object T and every pair of maps $T \xrightarrow{q_1} A$, $T \xrightarrow{q_2} B$, there is exactly one map $T \xrightarrow{q} P$ for which $q_1 = p_1 \circ q$ and $q_2 = p_2 \circ q$.

Pictorially:

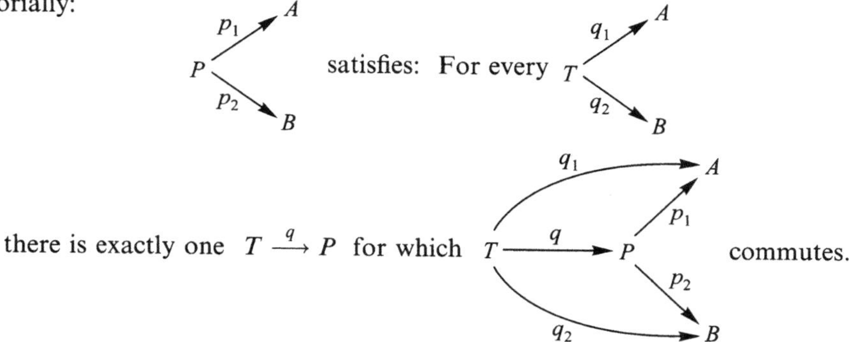

Let's illustrate this with our example of a solid cylinder C as the product of a segment S and a disk D.

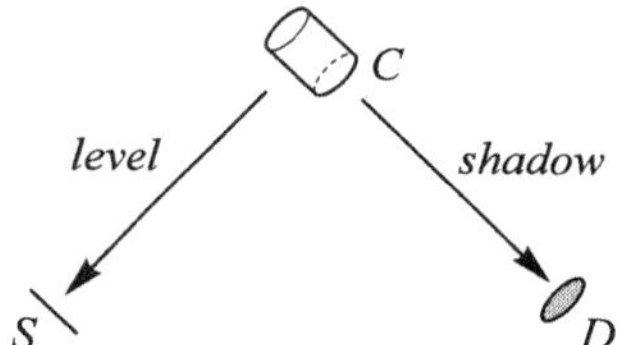

The special property that this pair of maps satisfies is that for *every* object T (in particular for T an interval of time) and for *every* pair of maps

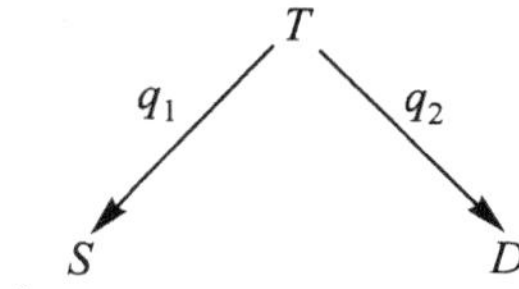

there is exactly one map $T \xrightarrow{q} C$ for which the diagram below commutes:

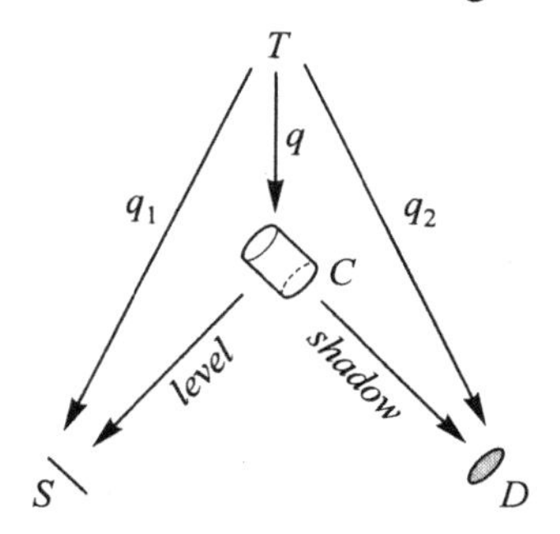

The only way we have enlarged upon Galileo's idea is that we have decided that the principle he applied to an interval of time should apply to *every* object in our category.

Note on terminology: As you know, when you combine numbers by addition (for example $2 + 3 + 7 = 12$) each number (the 2, the 3, and the 7) is called a *summand* and the result (the 12) is called the *sum*. But when you combine numbers by multiplication (as in $2 \times 3 \times 7 = 42$) each number is a *factor* and the result is the *product*. We keep this terminology, so that the objects that are being multiplied are called factors, and the resulting object is called their product.

The definition of multiplication of objects seems long at first, but once you understand it, you see that it is very natural. Just remember that a product is not only an object, but an object with two maps.

In the category of abstract sets and arbitrary maps, you already have a clear picture of the product P of two sets A and B and the two projection maps:

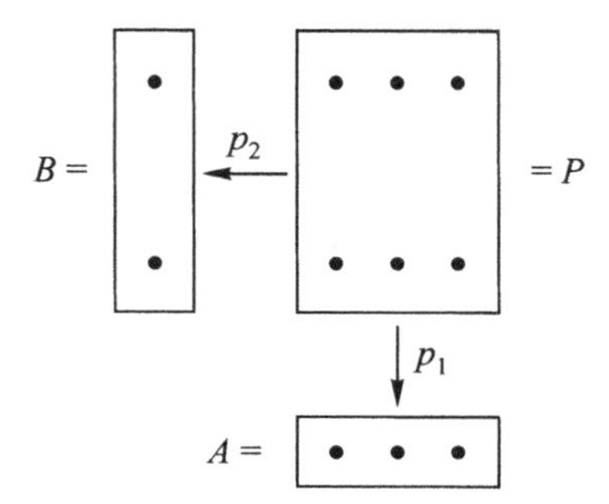

Here we have organized the two projection maps as 'sortings,' projecting the dots either vertically or horizontally.

Does this really have the universal property demanded by our definition of product? Given a set T and a pair of maps $T \xrightarrow{q_1} A$, $T \xrightarrow{q_2} B$, what is the one and only map $T \xrightarrow{q} P$ for which $q_1 = p_1 \circ q$ and $q_2 = p_2 \circ q$? Think it through yourself, until you are convinced that given q_1 and q_2 there is exactly one q satisfying these two equations.

When you were young, you may have been told that the basic idea of multiplication is that it is iterated addition: 3×4 means $4 + 4 + 4$, or perhaps you were told $3 + 3 + 3 + 3$. Such an account of multiplication does not get to the heart of the matter. That account depends on the special feature of the category of finite sets that every object is a sum of ones (and on the distributive law!). The definition of product we have given applies to any category, while still giving the usual result for finite sets; that is, we have the relation between multiplication of objects and of numbers:

$$\#(A \underset{\uparrow}{\times} B) \qquad = \qquad \#A \underset{\uparrow}{\times} \#B$$

multiplication of objects   multiplication of numbers

Surprisingly, not only does such a 6-element set P with the indicated two maps satisfy the definition of product of the A and B above, but (essentially) nothing else does! The following *uniqueness* theorem is true in any category, so that you can apply it also to graphs, dynamical systems, etc.

Theorem: *Suppose that* $A \xleftarrow{p_1} P \xrightarrow{p_2} B$ *and* $A \xleftarrow{q_1} Q \xrightarrow{q_2} B$*, are two products of* A *and* B*. Because* $A \xleftarrow{p_1} P \xrightarrow{p_2} B$ *is a product, viewing* Q *as a 'test object' gives a map* $Q \longrightarrow P$*; because* $A \xleftarrow{q_1} Q \xrightarrow{q_2} B$ *is a product, we also get a map* $P \longrightarrow Q$*. These two maps are necessarily inverse to each other, and therefore the two objects* P, Q *are isomorphic.*

I leave the proof of this for later, but I have stated the theorem so as to suggest most of the proof. One consequence of this theorem is that if I choose a product of two objects and you choose another product of the same objects, we actually get a preferred isomorphism from my object to yours. For that reason, we will usually use the phrase '*the* product of A and B', just as we use '*the* terminal object', when there is one. (In some categories, some pairs of objects do not have a product.)

Let's look at another example. Consider the category $\mathcal{S}^{\circlearrowright}$ of sets-with-endomap, where the maps from $X^{\circlearrowright\alpha}$ to $Y^{\circlearrowright\beta}$ are the set-maps $X \xrightarrow{f} Y$ such that $f \circ \alpha = \beta \circ f$ (so that there are usually fewer $\mathcal{S}^{\circlearrowright}$-maps from $X^{\circlearrowright\alpha}$ to $Y^{\circlearrowright\beta}$ than set-maps from X to Y).

Take the example of the set *Days* of all the days that have been and that ever will be – we imagine this as an infinite set – and consider also the set *Days of the week* = {*Sun, Mon, Tue, Wed, Thu, Fri, Sat*}. These two sets have obvious endomaps that could be called in both cases 'tomorrow,' and can be pictured like this

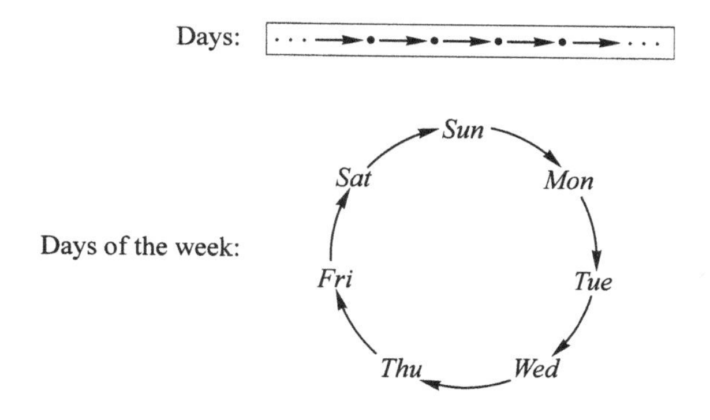

Furthermore we have an obvious map *Days* $\longrightarrow$ *Days of the week*, assigning to each day the corresponding day of the week. This map may be more clearly pictured in a 'sorting' picture in which we place all the days in an infinite helix above a circle like this:

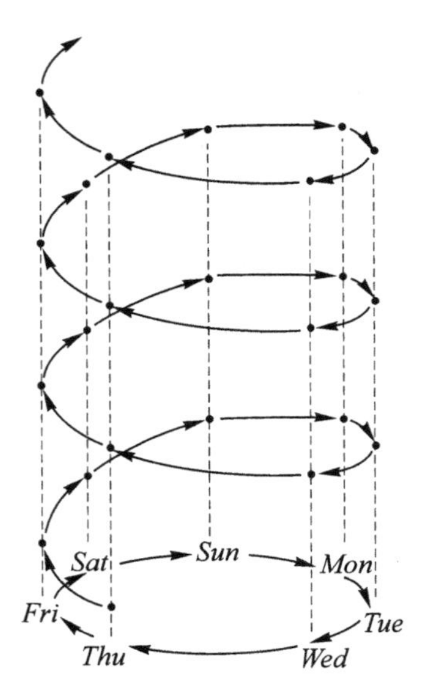

Check that this is really a map in the category $\mathcal{S}^{\circlearrowright}$.

Now I take another example. Imagine a factory where people work in shifts like this:

Night shift:	Midnight to 8am,
Day shift:	8am to 4pm,
Evening shift:	4pm to midnight.

Then we can think of two more sets-with-endomap. One is the set of hours of the day with the obvious endomap of 'next hour':

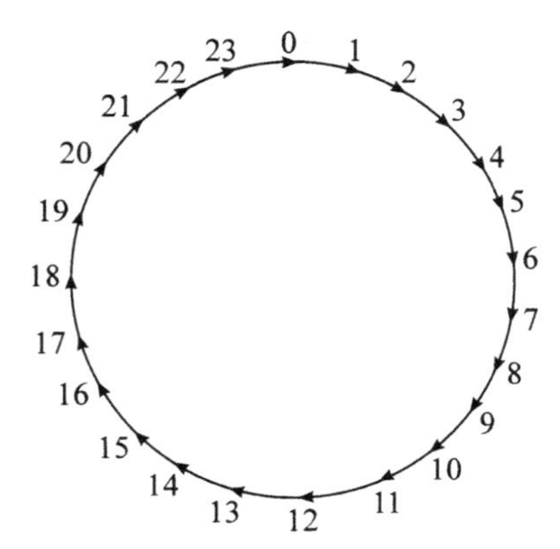

This can be called the 'day clock.' The other object involves the set of eight 'working hours' in a shift, which we can label {0, 1, 2, 3, 4, 5, 6, 7}. Here also there is an obvious endomap which we can picture as

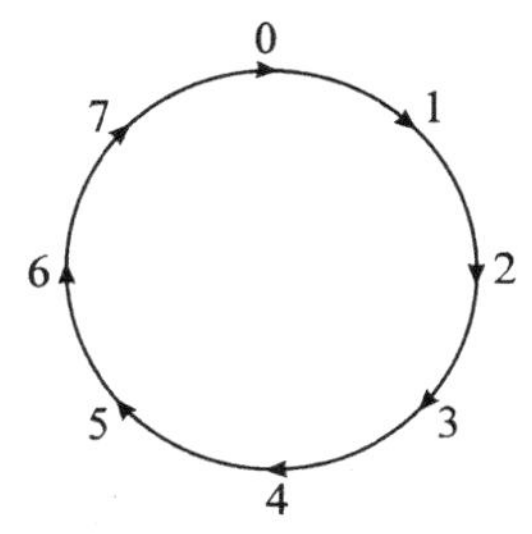

and which can be called the 'shift clock.' Furthermore, we have again a map from one set to the other, which assigns to each hour of the day, the hour in the present shift. This map is more difficult to picture, but it should be obvious that it is also a map in $\boldsymbol{\mathcal{S}}^{\circlearrowleft}$. Part of its internal diagram is

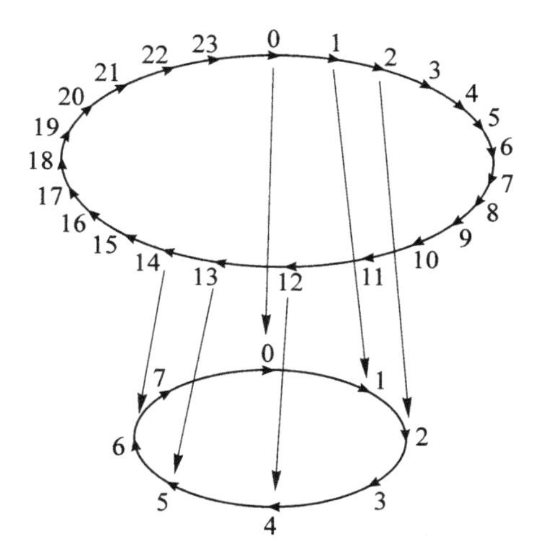

> **Exercise 1:**
> Is there a map in $\boldsymbol{\mathcal{S}}^{\circlearrowleft}$ from the 'day clock' to some $X^{\circlearrowleft\alpha}$ which together with the map above makes the 'day clock' into the product of $X^{\circlearrowleft\alpha}$ and the 'shift clock'?

One of the points of this exercise is that if you ignore the additional structure, you can see that the set of hours in the day is the product of the set of hours in a shift and the set of shifts. This is accomplished by the obvious projection maps:

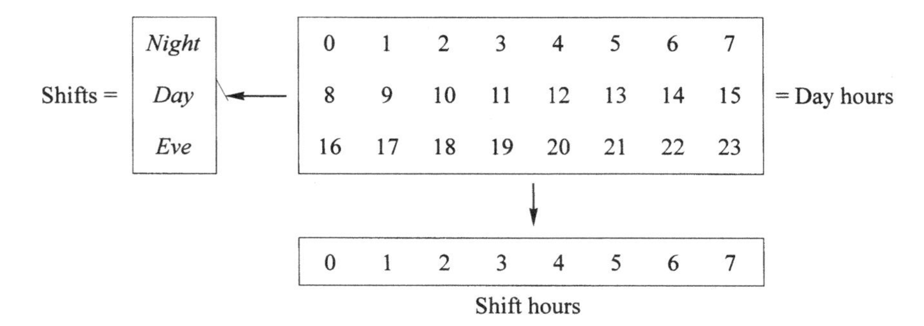

An object $X^{\circlearrowleft\alpha}$ in a solution to Exercise 1 could not have X as the set of shifts with the projection map above, because no endomap on this set admits that projection map from the *Day clock* as structure-preserving. Because 0 goes to *Night* and 1 to *Night* and 0 to 1, we must have *Night* goes to *Night*. Also, 7 goes to *Night*, but 8 to *Day* and 7 to 8, so that we must have *Night* goes to *Day*, contradicting *Night* goes to *Night*. This shows that we must look elsewhere, if we hope to find an object $X^{\circlearrowleft\alpha}$ in $\mathbf{S}^{\circlearrowleft}$ and a map *Day clock* $\longrightarrow X^{\circlearrowleft\alpha}$ *in* $\mathbf{S}^{\circlearrowleft}$ such that the diagram below is a product in the category $\mathbf{S}^{\circlearrowleft}$.

$$X^{\circlearrowleft\alpha} \longleftarrow \textit{Day clock}$$
$$\downarrow$$
$$\textit{Shift clock}$$

I won't tell you now whether there is such a product diagram, but we will investigate the products in $\mathbf{S}^{\circlearrowleft}$ to help you find the answer. What do the products look like in this category? Suppose that $A^{\circlearrowleft\alpha}$ and $B^{\circlearrowleft\beta}$ are two objects in $\mathbf{S}^{\circlearrowleft}$. According to the definition a product of these two objects is another object $P^{\circlearrowleft\gamma}$ and two $\mathbf{S}^{\circlearrowleft}$-maps, $A^{\circlearrowleft\alpha} \xleftarrow{p_1} P^{\circlearrowleft\gamma} \xrightarrow{p_2} B^{\circlearrowleft\beta}$ (this implies $p_1\gamma = \alpha p_1$ and $p_2\gamma = \beta p_2$) such that for any other object $T^{\circlearrowleft\tau}$ and maps $A^{\circlearrowleft\alpha} \xleftarrow{q_1} T^{\circlearrowleft\tau} \xrightarrow{q_2} B^{\circlearrowleft\beta}$ in $\mathbf{S}^{\circlearrowleft}$, there is exactly one map $T^{\circlearrowleft\tau} \xrightarrow{q} P^{\circlearrowleft\gamma}$ that fits in the diagram

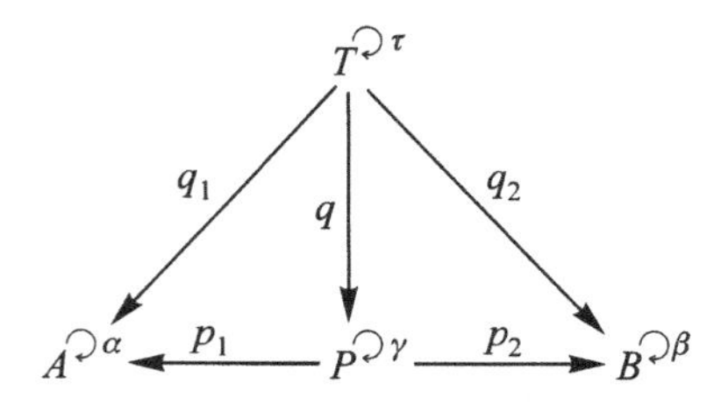

i.e. such that $p_1 q = q_1$ and $p_2 q = q_2$.

That seems rather long, but it turns out to be precisely what we need in order to calculate what $P^{\circlearrowleft\gamma}$ must be. You'll remember that the elements of P correspond precisely to the $\mathbf{S}^{\circlearrowleft}$-maps $\mathbb{N}^{\circlearrowleft()+1} \longrightarrow P^{\circlearrowleft\gamma}$, which tells us that they are the pairs of $\mathbf{S}^{\circlearrowleft}$-maps

$$A^{\circlearrowright\alpha} \longleftarrow \mathbb{N}^{\circlearrowright()+1} \longrightarrow B^{\circlearrowright\beta}$$

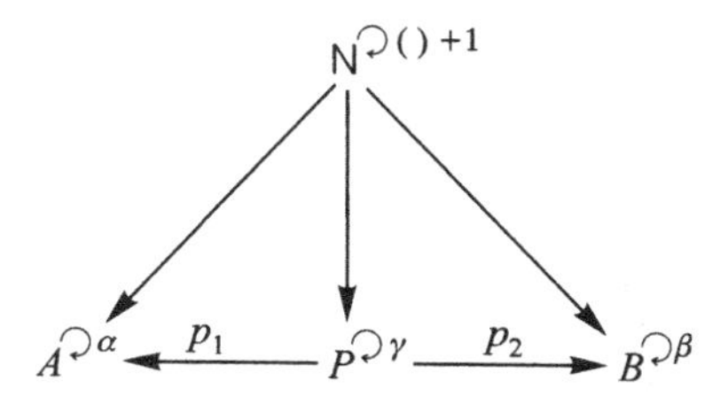

which in turn correspond to the pairs of elements (a, b) where a is from A and b is from B. Therefore, the set P must be the product (in the category of sets) of A and B. We now need to determine what the endomap γ of P must be, but this is also not hard. The solution suggests itself, since for a pair (a, b) we can apply α to a and β to b, so that we can write

$$\gamma(a, b) = (\alpha(a), \beta(b))$$

In fact, this endomap works very well because it makes the usual 'set projections' $A \xleftarrow{p_1} P \xrightarrow{p_2} B$ preserve the structure of the endomap, so that we have all the ingredients of a product in $\mathcal{S}^{\circlearrowright}$ (i.e. $A^{\circlearrowright\alpha} \xleftarrow{p_1} P^{\circlearrowright\gamma} \xrightarrow{p_2} B^{\circlearrowright\beta}$); and it is not hard to prove that indeed this is a product. The idea behind this product is that for each pair of arrows in the endomaps α and β we get an arrow in the 'product' endomap γ. We can picture this in the following way, where we have drawn only part of each internal diagram:

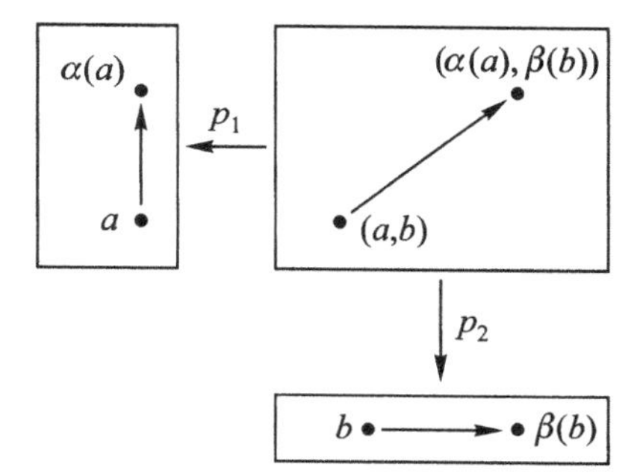

To gain some practice in understanding products in $\mathcal{S}^{\circlearrowright}$ it is good to work this out in an example. Let's take the endomaps

$A^{\circlearrowright\alpha} =$ [Omer ⇄ Chad] and $B^{\circlearrowright\beta} =$ [Alysia → Katie → Fatima → Alysia]

Then their product is

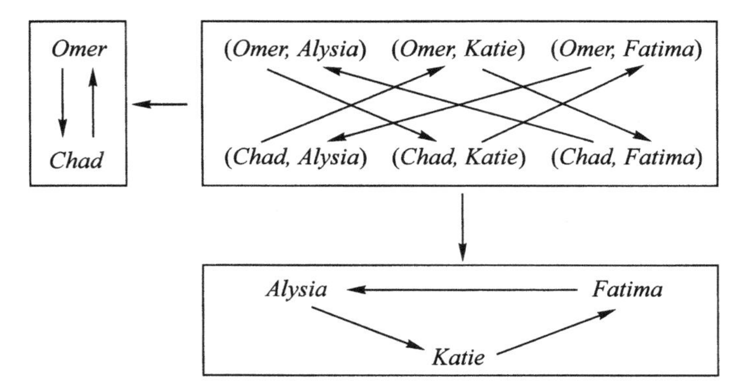

This shows that multiplying a 2-cycle by a 3-cycle we get a 6-cycle. But don't be fooled by this apparent simplicity. Try multiplying these cycles:

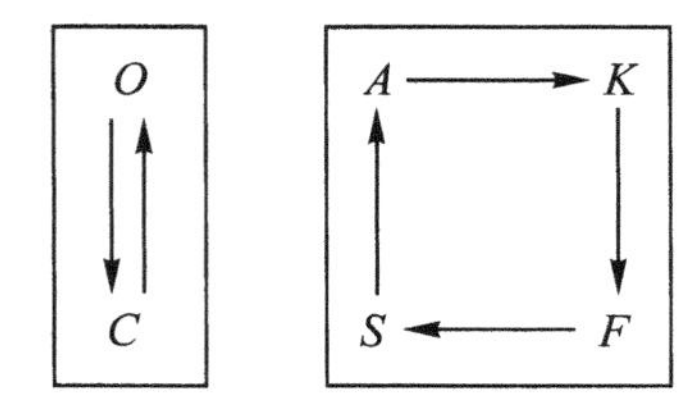

You won't get an 8-cycle at all. What you get instead is two 4-cycles!

Exercise 2:
What is the product $C_m \times C_n$ of an m-cycle and an n-cycle? For example, what is the product $C_{12} \times C_8$? Hint: Start by investigating products of cycles of smaller sizes.

Exercise 3:
Return to Exercise 3 of Session 12. Show that the object which was called $\mathbf{G} \times \mathbf{C}$, when provided with appropriate projection maps, really is the product in the category $\boldsymbol{\mathcal{S}}^{\circlearrowleft\cdot\circlearrowright}$.

SESSION 22

Universal mapping properties Incidence relations

1. A special property of the category of sets

We want to discuss two related ideas:

1. universal mapping properties, and
2. detecting the structure of an object by means of figures and incidence relations.

An example of (1) is the property appearing in the definition of *terminal object*: to say that $\mathbf{1}$ is terminal means that *for each* object X, there is exactly one map $X \longrightarrow \mathbf{1}$. The 'for each', 'for every', or 'for all' is what makes us call this a *universal* property: the object $\mathbf{1}$ is described by its relation to every object in the 'universe', i.e. the category under consideration.

The idea of *figure* arises when, in investigating some category $\mathcal{C}$, we find a small class $\mathcal{A}$ of objects in $\mathcal{C}$ which we use to probe the more complicated objects X by means of maps $A \xrightarrow{x} X$ from objects in $\mathcal{A}$. We call the map x a *figure of shape A in X* (or sometimes a *singular figure of shape A in X*, if we want to emphasize that the map x may collapse A somewhat, so that the picture of A in X may not have all the features of A). This way of using maps is very well reflected in the German word for map, '*Abbildung*,' which means something like a picture of A in X.

If the category $\mathcal{C}$ has a terminal object, we can consider it as a basic shape for figures. Indeed, we have already given figures of that shape a special name: a figure of shape $\mathbf{1}$ in X, $\mathbf{1} \longrightarrow X$, is called a *point* of X. In sets, the points of X are in a sense all there is to X, so that we often use the words 'point' and 'element' interchangeably, whereas in dynamical systems points are fixed states, and in graphs they are loops.

The category of sets has a special property, roughly because the objects have no structure:

If two maps agree on points, they are the same map.

That is, suppose $X \xrightarrow{f} Y$ and $X \xrightarrow{g} Y$. If $fx = gx$ for every point $\mathbf{1} \xrightarrow{x} X$, we can conclude that $f = g$. This can also be expressed in the contrapositive form: if $f \neq g$, then there is at least one point $\mathbf{1} \xrightarrow{x} X$ for which $fx \neq gx$.

This special property of the category of sets is not true of $\mathcal{S}^{\circlearrowright}$, nor of $\mathcal{S}^{\downdownarrows}$. For example, in $\mathcal{S}^{\circlearrowright}$ the 2-cycle C_2 has no points at all, since 'points' are fixed points; any two maps from C_2 to any system agree at all points (since there are no points to disagree on) even though they may be different maps.

There are, of course, figures of other shapes. In the category of sets, a figure of shape **2** in X, where '**2**' indicates a two-element set like $\{me, you\}$, is just a pair of points of X, since it is a map $\mathbf{2} \xrightarrow{x} X$. For example:

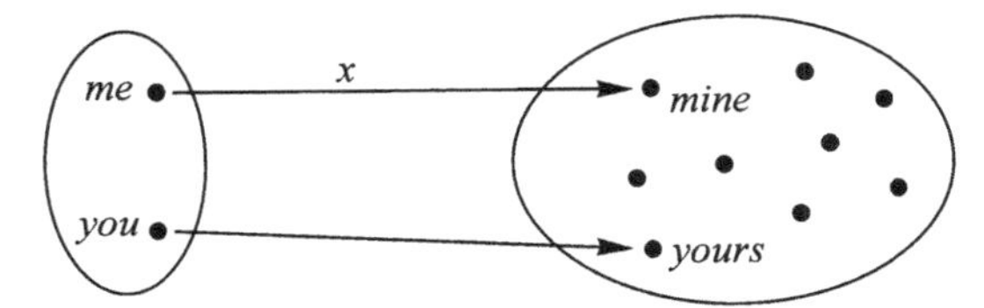

The two points will coincide if x is a constant map, so that a map x for which $mine = yours$ is also included as a figure of shape **2**. It is called a *singular* figure because the map 'collapses' the shape **2**. An example of a singular figure in graphs is the (unique) figure of shape A in **1**:

The special property of the category of sets can be viewed as saying that a very small class of shapes of figures (in fact just the shape **1**) suffices to test for equality of maps. Can we find such a small-and-yet-sufficient class of shapes in other categories?

2. A similar property in the category of endomaps of sets

What about the category of endomaps $\boldsymbol{\mathcal{S}}^{\circlearrowleft}$? Do we know some simple examples of objects that can be used as types of figures to probe other objects? Well, we had cycles such as

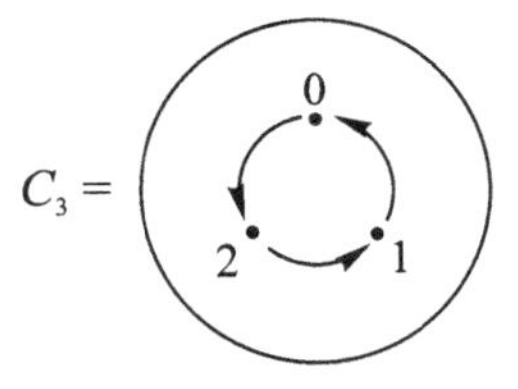

What is a figure of shape C_3? Imagine an endomap

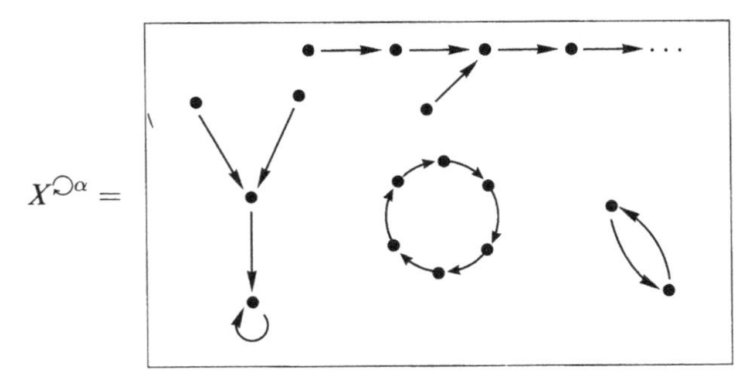

What's a figure of shape C_3 in $X^{\circlearrowright\alpha}$? It is a map $C_3 \longrightarrow X^{\circlearrowright\alpha}$. We should first look for a 3-cycle in $X^{\circlearrowright\alpha}$. In this example there aren't any, but we can map C_3 to a fixed point. This gives us a figure of shape C_3 in $X^{\circlearrowright\alpha}$, a 'totally singular' figure.

If instead of a figure of shape C_3, we look for a figure of shape

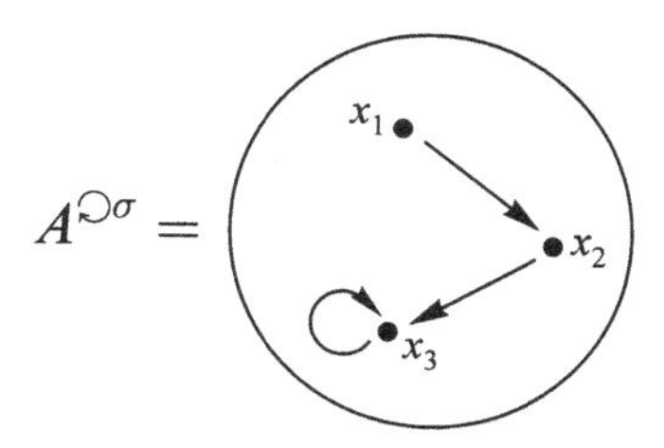

(the endomap that Katie came up with in Test 2 that satisfies $\sigma^3 = \sigma^2$), then we can find two non-singular figures in $X^{\circlearrowright\alpha}$. (Can you find any singular ones?) One special feature of the endomaps C_3 and $A^{\circlearrowright\sigma}$ is that they are generated by one single element: x_1 in the case of $A^{\circlearrowright\sigma}$, and any of its dots in the case of C_3. For example, if I want to map $A^{\circlearrowright\sigma}$ to $X^{\circlearrowright\alpha}$, it is sufficient to say where to map x_1. The images of the other dots are uniquely determined by the condition of preserving the structure of the endomap. Similarly, we can still consider that figures of shape C_3 'are' elements if we first specify a particular generator of C_3. The only restriction is that the point chosen as image must have the 'same positive properties' as the generator. This may give singular figures, for example we can map C_6 to C_2 this way:

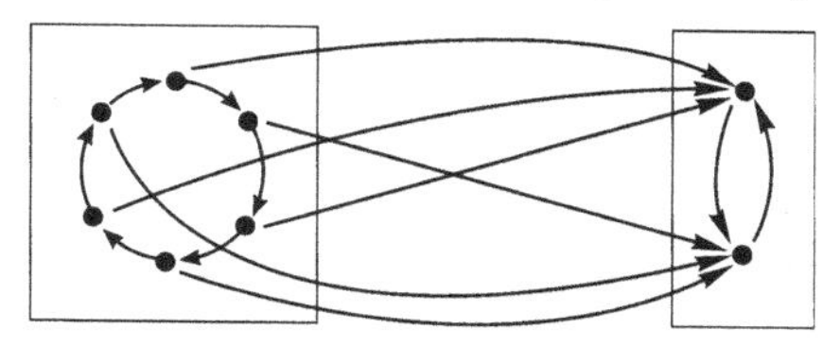

We can use this map to express a particular way in which figures of shape C_6 in other dynamical systems can be singular: a figure $C_6 \xrightarrow{x} X^{\circlearrowright\alpha}$ may factor through the above map $C_6 \longrightarrow C_2$ like this:

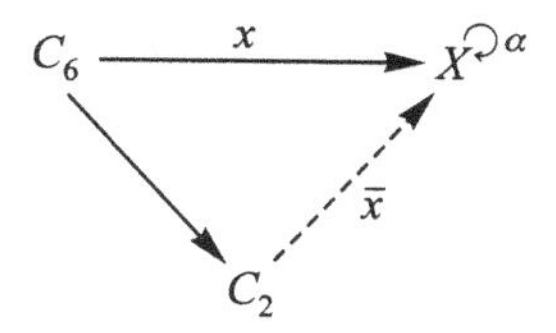

Now let us consider the 'successor' endomap $\sigma = (\,) + 1$ of the natural numbers $\mathbb{N}^{\circlearrowright\sigma}$ as a basic shape of figure. What is a figure of this shape? We saw before that any such figure $\mathbb{N}^{\circlearrowright\sigma} \xrightarrow{x} X^{\circlearrowright\alpha}$ is completely determined by the element of X to which the number 0 is mapped, and this without any conditions, so that every element of X determines one such figure. One can also say that each state of $X^{\circlearrowright\alpha}$ generates a figure in $X^{\circlearrowright\alpha}$ under the action of the dynamics, or endomap, and that all such figures are of shape $\mathbb{N}^{\circlearrowright\sigma}$, although possibly singular. For example, if a figure x of shape $\mathbb{N}^{\circlearrowright\sigma}$ in $X^{\circlearrowright\alpha}$ happens to factor through the cycle C_n

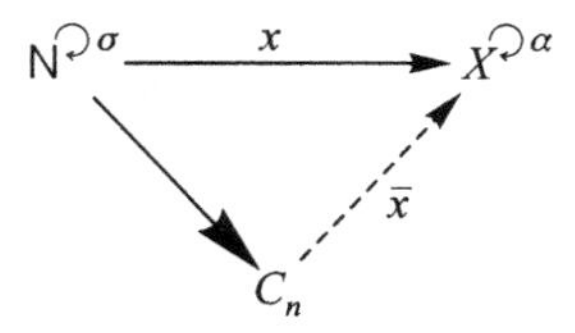

this means that the future of $x(0)$ in $X^{\circlearrowright\alpha}$ 'has the shape C_n.'

FATIMA: What does a heavy arrow-head mean?

It indicates that the map is an *epimorphism*, which is defined to mean that any problem of factoring a map through such a map has at most one solution. (For our map this follows from the fact that every element in C_n is an image of an element of $\mathbb{N}^{\circlearrowright\sigma}$.) When we see a diagram like this:

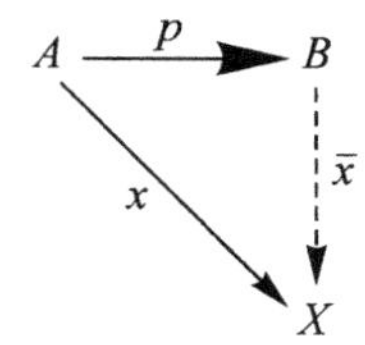

we know that there is at most one map $\bar{x}$ such that $\bar{x}p = x$. For example, all retractions are entitled to be drawn with a heavy arrow-head. An example of a map which doesn't have this property is the following:

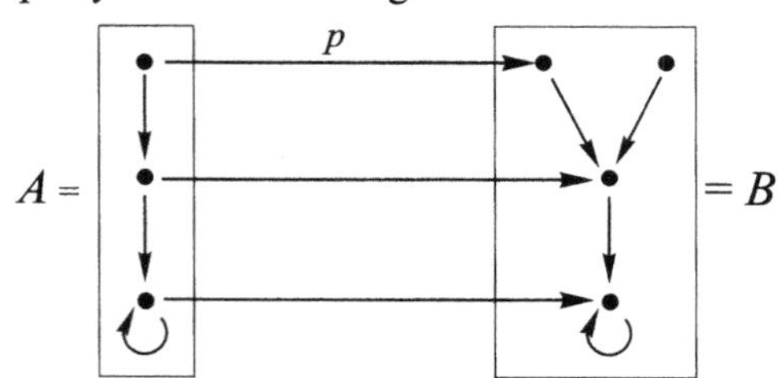

Some maps $A \longrightarrow X$ (for example p itself!) can be factored through p in several ways. Therefore, this p won't be drawn with a heavy arrow-head.

Going back to the natural numbers with the successor endomap, it turns out that it satisfies a property similar to that of the terminal object in the category of sets:

Given any pair of maps $X^{\circlearrowright\alpha} \overset{f}{\underset{g}{\rightrightarrows}} Y^{\circlearrowright\beta}$ *in* $\boldsymbol{\mathcal{S}}^{\circlearrowright}$, *if for all figures* $\mathbb{N}^{\circlearrowright\sigma} \xrightarrow{x} X^{\circlearrowright\alpha}$ *of shape* $\mathbb{N}^{\circlearrowright\sigma}$ *it is true that* $fx = gx$, *then* $f = g$.

The only difference between this and the case of sets is that there we were using a terminal object, while here we must use instead another figure type. Of course, in this category we also have a terminal object and the figures of its type are the fixed points.

But every fixed point $\mathbf{1} \longrightarrow X^{\circlearrowright\alpha}$ is also 'among' the figures of shape $\mathbb{N}^{\circlearrowright\sigma}$ (by composing the fixed point with the unique map $\mathbb{N}^{\circlearrowright\sigma} \longrightarrow \mathbf{1}$).

3. Incidence relations

Now we need to speak about incidence relations. Let's suppose that we have in X a figure x of shape A, and a figure y of shape B. We ask to what extent these figures are incident or to what extent they overlap, and what the structure of this overlap is. Well, we could have a map $u : A \longrightarrow B$ satisfying $yu = x$.

CHAD: Would B have to be smaller than A?

No. It could be smaller as in the example above with $A = \mathbb{N}^{\circlearrowright\sigma}$ and $B = \mathbf{1}$ where we had

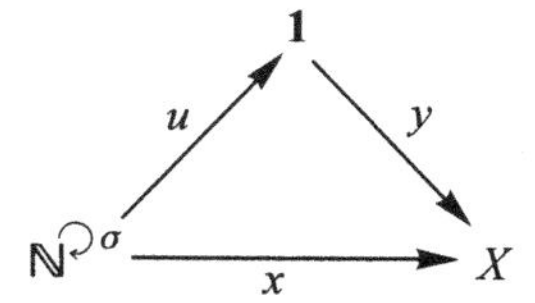

but it could also be bigger as in the case of

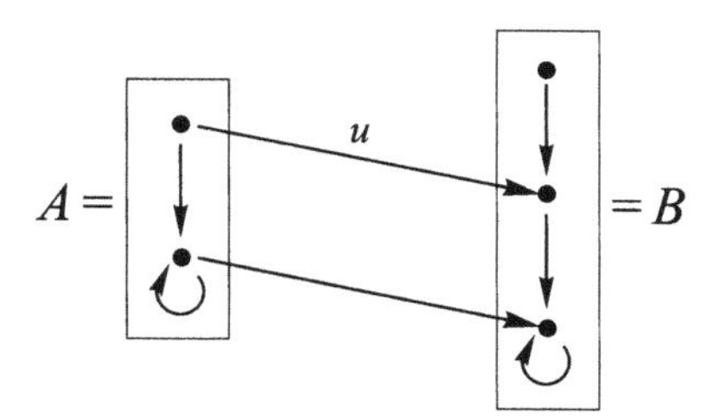

One way in which x may be incident to y is if there is a map u such that $yu = x$, i.e.

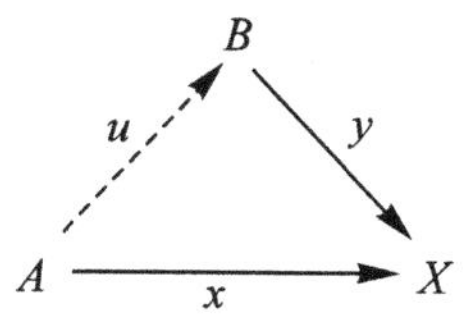

but another possibility is that we may have maps from an object T to A and to B so that

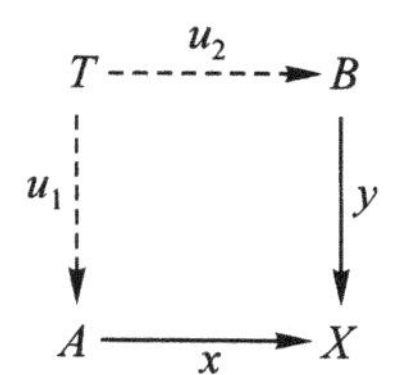

with $xu_1 = yu_2$. The second possibility means in effect that there is given a third figure $T \longrightarrow X$ together with incidences in the first sense to each of x and y.

4. Basic figure-types, singular figures, and incidence, in the category of graphs

Let's consider the case of the category of graphs $\mathbf{S}^{\downarrow\downarrow}$. In this category the two objects $D = \boxed{\bullet}$ and $A = \boxed{\bullet \longrightarrow \bullet}$ can serve as basic figure-types. What is a figure of shape A in an arbitrary graph?

DANILO: An arrow of the graph.

Right, and a figure of shape D is just a dot.

CHAD: Can the arrow be a loop?

Yes. Then we will have a singular figure of shape A. This happens when the map $A \longrightarrow X$ factors through 'the loop' or terminal object **1**.
In this category $\mathbf{S}^{\downarrow\downarrow}$ we see that:

Given any pair of maps $X \xrightarrow{f} Y$, $X \xrightarrow{g} Y$ in $\mathbf{S}^{\downarrow\downarrow}$, if $fx = gx$ for all figures $D \xrightarrow{x} X$ of shape D and for all figures $A \xrightarrow{x} X$ of shape A, then $f = g$.

(In the category of graphs, to test equality of maps we need two figure-types.)

Exercise 1:
Consider the diagram of graphs

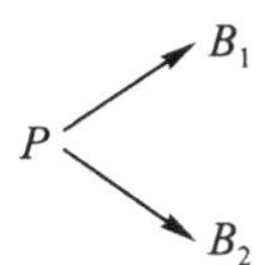

Suppose it satisfies the fragment of the definition of product in which we test against only the two figure-types $X = A$ and $X = D$. Prove that the diagram actually is a product, i.e. that the product property holds for all graphs X.

Another useful figure-type is that of the graph

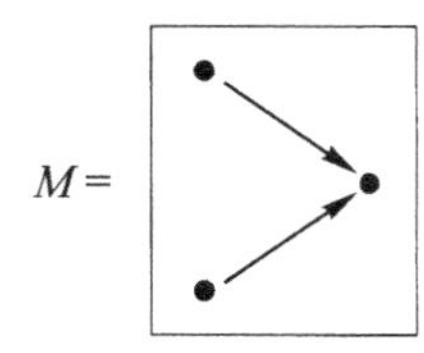

This graph has two arrows, which means that there are two different maps from A to M, namely the maps

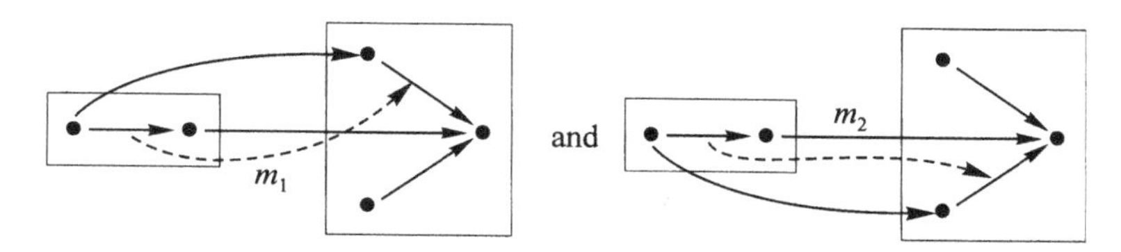

Are these two figures m_1 and m_2 incident in M?

FATIMA: Yes, they meet in one dot.

Right.

To express this incidence by maps, remember that in addition to the two fundamental objects D and A in the category of graphs, there are two important maps which we call 'source' and 'target.' They are the only two maps from D to A, namely

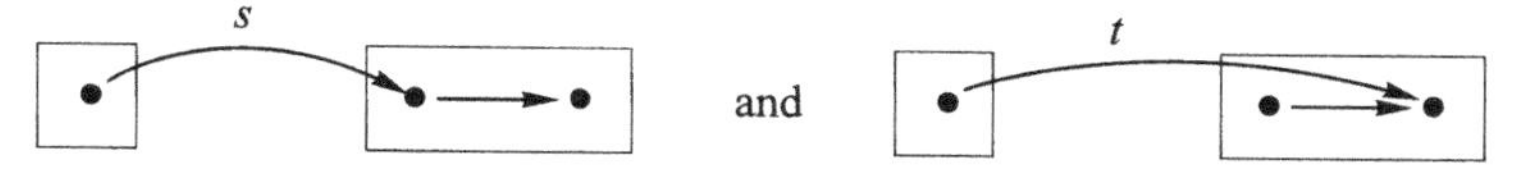

In terms of these two maps we can express the incidence of m_1 and m_2 by the commutativity of the diagram

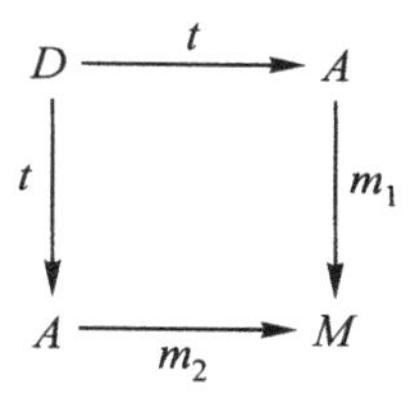

which means that $m_1t = m_2t$ or 'm_1 has the same target as m_2.' In fact, there is nothing else in the intersection of m_1 and m_2. (We can express that fact in terms of maps, too, but we don't need it now.) This graph M also has the property that for any graph X and any two arrows in X, $A \xrightarrow{x_1} X$ and $A \xrightarrow{x_2} X$ which have the same target, i.e. for which

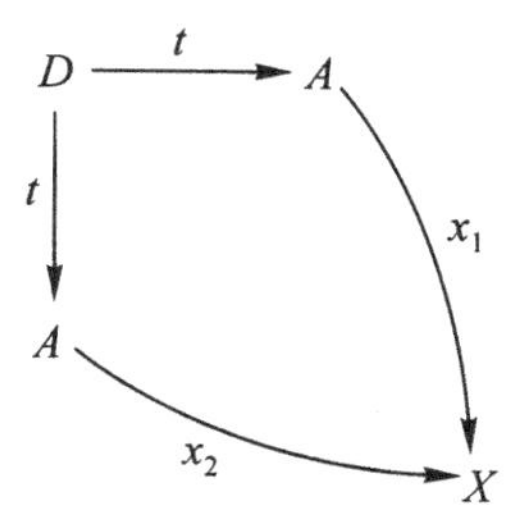

commutes, i.e. $x_1t = x_2t$, there is exactly one figure of shape M in X whose arrow m_1 matches with x_1 and whose arrow m_2 matches with x_2. In other words, there is exactly one solution x to the problem

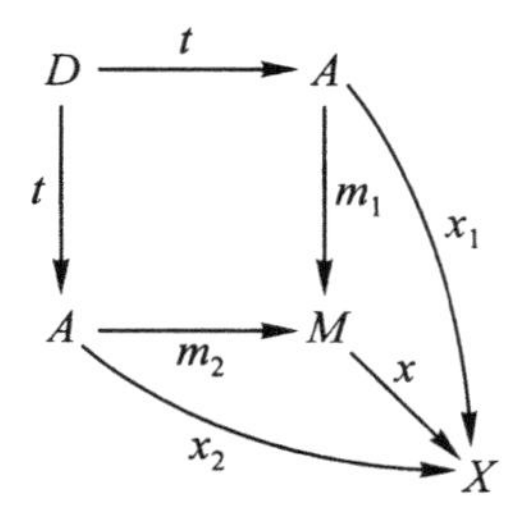

ALYSIA: Couldn't it be $xm_1 = x_2$?

Well, what we just said applies equally well to the figures x_2, x_1 (same as before, but taken in opposite order). Therefore there is also exactly one x' such that $x'm_1 = x_2$ and $x'm_2 = x_1$, but this x' will, in general, be different from x. They are equal only when the two arrows x_1, x_2 are themselves equal.

A picture of M in X might be singular, of course. In the graphs

and

there are figures of shape M in which two or more dots coincide; and in the graphs

and

there are figures of shape M in which the two arrows also coincide.

Exercise 2:
What is a figure of shape

$A_2 =$

in a graph X? What are the various ways in which it can be singular?

Notice that again we have two maps

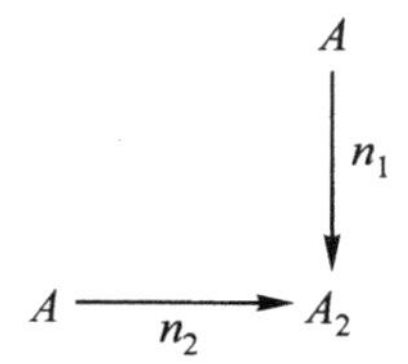

However, now their incidence relation is different: the source of n_2 is the target of n_1.

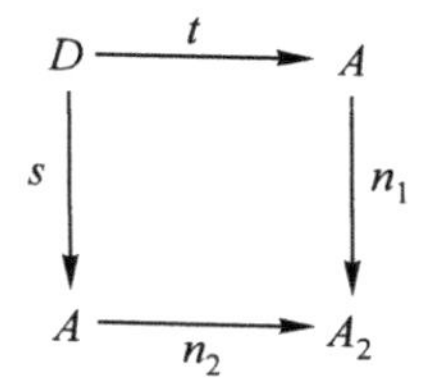

This object A_2 also has a universal mapping property: For any graph X with two arrows $A \xrightarrow{x_1} X$, $A \xrightarrow{x_2} X$, such that the source of x_2 is the target of x_1, there is exactly one figure of shape A_2 in X, $A_2 \xrightarrow{x} X$, such that $xn_1 = x_1$ and $xn_2 = x_2$.

In the discussion of presentations of dynamical systems at the end of Session 15, it was suggested that you think about presentations of graphs. You might want to try that again now, since figures and incidence relations are exactly what is needed. Suppose G is a graph. Label some (or none) of the arrows of G, say n of them, as

$$A \xrightarrow{a_1} G, A \xrightarrow{a_2} G, \ldots, A \xrightarrow{a_n} G$$

Also label some (or none) of the dots of G, say m of them, as

$$D \xrightarrow{d_1} G, D \xrightarrow{d_2} G, \ldots, D \xrightarrow{d_m} G$$

Now list some of the incidence relations that are true in G, of the forms $a_is = a_js$, $a_is = a_jt$, etc. and of the forms $a_is = d_j$, $a_it = d_j$. We call the two lists of labels, together with the list of equations a *presentation* of G if they have the property that for any graph G', and any n arrows $a'_1, a'_2, \ldots, a'_n$, and m dots $d'_1, d'_2, \ldots d'_m$ of G' satisfying the 'same' equations (with a_i and d_j replaced by a'_i and d'_j), there is exactly one map of graphs sending each a_i to a'_i and each d_j to d'_j. If you list all the arrows of G, all the dots, and all the true incidence relations, you will get a presentation. This is inefficient, though; for the graph M above, we found a presentation using labels m_1 and m_2 on the arrows, but no labelled dots and only one equation, $m_1t = m_2t$. Must every presentation of a graph label all the arrows? Can you find a 'minimal' presentation for any finite graph, with as few labels and equations as possible? You might want to try some small graphs first. The other problems at the end of Session 15 also can be considered for presentations of graphs.

SESSION 23

More on universal mapping properties

We will look at more illustrations of the use of figures to find objects having various universal mapping properties. There are many such properties and it will be helpful to put some of them in a list.

Universal mapping properties

Initial object	Terminal object
Sum of two objects	Product of two objects
...	Exponential, or power, or map space ...

The list is divided into two columns because universal mapping properties come in pairs; for each property in the right-hand column there is a corresponding one in the left, and vice versa. So far, we have studied only the first two properties on the right. The definition for a 'left column property' is similar to that of the corresponding 'right column property,' with the only difference that all the maps appearing in the definition are reversed – domain and codomain are interchanged. Let's clarify this with the simplest example.

The idea of initial object is similar to that of terminal object but 'opposite.' T is a terminal object if for each object X there is exactly one map from X to T, $X \longrightarrow T$. Correspondingly, I is an initial object if for each object X there is exactly one map from I to X, $I \longrightarrow X$.

In the category of abstract sets an initial object is an **empty set**: as we have seen, no matter what set X we choose, there is exactly one map

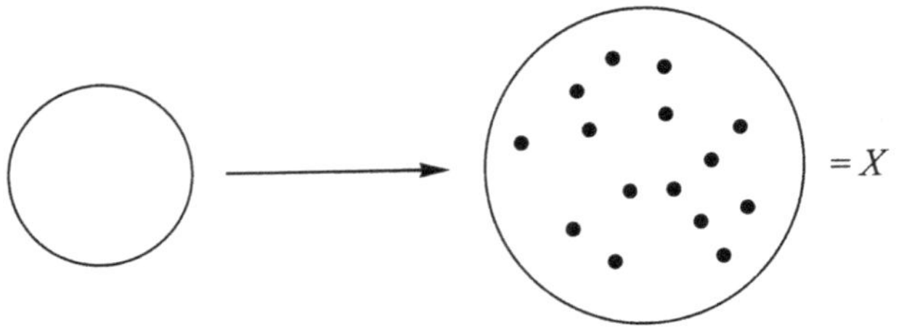

(It is not always the case that an initial object of a category deserves to be called 'empty,' although this description fits in all the categories we have studied so far.)

The second universal mapping property that we studied was *product of two objects*. The dual or opposite of this is *sum of two objects*, which we will study shortly. Another 'right column' universal mapping property is called *exponential*, or *power*, or *map space*, which we will study later.

1. A category of pairs of maps

It might occur to you that to study pairs of maps into two given objects B_1 and B_2 in $\mathcal{C}$, we could invent a new category, which we will call $\mathcal{C}_{B_1 B_2}$. An object of this category is an object of $\mathcal{C}$ equipped with a pair of maps to B_1 and B_2 respectively, i.e. a diagram of the type

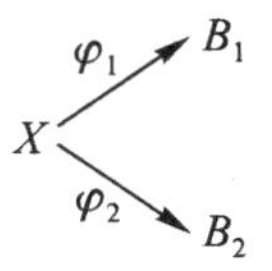

in $\mathcal{C}$, while a map between two objects in this category, for example a map

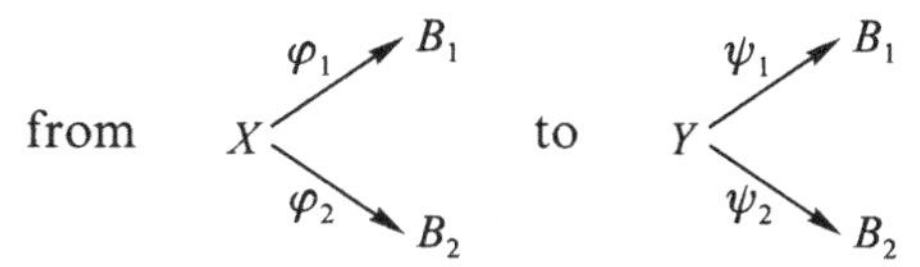

is simply a map $X \xrightarrow{f} Y$ in $\mathcal{C}$ which 'preserves the structure,' meaning that it satisfies the two obvious equations saying that this diagram commutes:

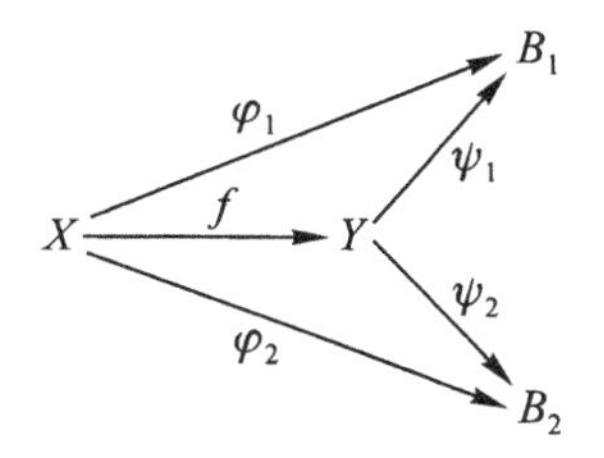

$$\psi_1 f = \varphi_1 \quad \text{and} \quad \psi_2 f = \varphi_2$$

Our main question about this category $\mathcal{C}_{B_1 B_2}$ is: What is its terminal object? The answer must depend only on B_1 and B_2 since these are the only ingredients used to construct this category. By the definition of terminal object we must find an object

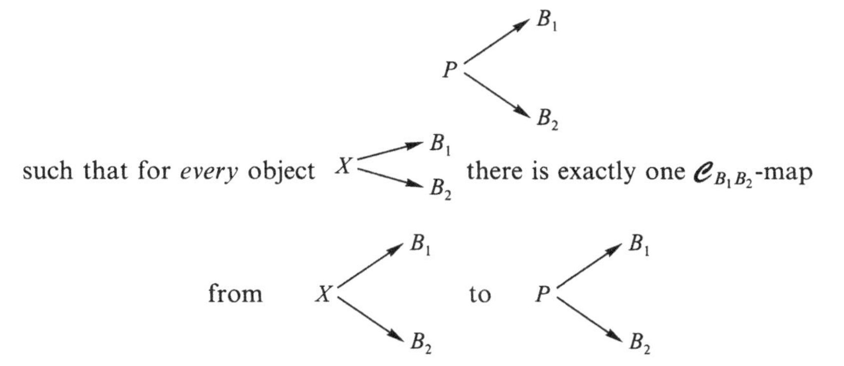

CHAD: So, it is the product of B_1 and B_2.

Exactly. The definition of a product of B_1 and B_2 in $\mathcal{C}$ says precisely the same thing as the definition of a terminal object in $\mathcal{C}_{B_1 B_2}$.

Why do we bother to invent a category in which a terminal object is the same as a product of B_1 and B_2 in $\mathcal{C}$? This construction, 'reducing' products in one category to terminal objects in another, in particular makes the uniqueness theorem for products a consequence of the corresponding theorem for terminal objects. Of course, it would seem to need a lot of effort to define the category $\mathcal{C}_{B_1 B_2}$ if our only purpose were to deduce that any two products of B_1 and B_2 are uniquely isomorphic from the uniqueness theorem for terminal objects. By the time we prove that $\mathcal{C}_{B_1 B_2}$ is actually a category we could have finished the direct proof of uniqueness of the product. However, after gaining some experience it becomes obvious that anything constructed as $\mathcal{C}_{B_1 B_2}$ was, is automatically a category; and there are many instances in which it is very helpful to think of a product of two objects as a terminal object in the appropriate category. The fact that this is always possible helps us to understand better the concept of product.

Exercise 1:
Formulate and prove in two ways the theorem of uniqueness of the product of two objects B_1 and B_2 of the category $\mathcal{C}$. (One way is the direct proof and the other way is to define the category $\mathcal{C}_{B_1 B_2}$, to prove that it is a category, to prove that its terminal object is the same as a product of B_1 and B_2 in $\mathcal{C}$, and to appeal to the theorem on uniqueness of terminal objects.)

2. How to calculate products

Just as we do not (rather, the category cannot) differentiate between any two terminal objects and so we refer to any of them as 'the' terminal object, we also refer to any product

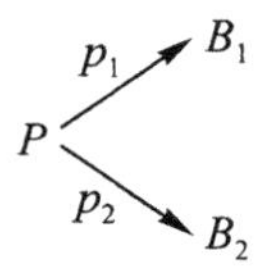

as 'the' product of B_1 and B_2 and we denote the object P by $B_1 \times B_2$, and we call the two maps p_1, p_2 'the projections of the product to its factors'. For any two maps from an object A to B_1 and B_2 respectively, i.e. for any

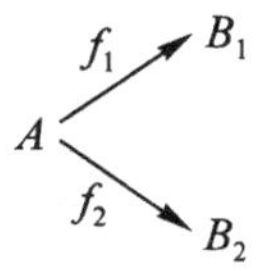

there is exactly one map $A \xrightarrow{f} B_1 \times B_2$ satisfying $p_1 f = f_1$ and $p_2 f = f_2$. This map is also denoted by a special symbol $\langle f_1, f_2 \rangle$ which includes the list of the maps that give rise to f.

Definition: *For any pair of maps*

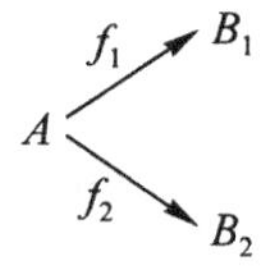

$\langle f_1, f_2 \rangle$ *is the unique map*

$$A \longrightarrow B_1 \times B_2$$

that satisfies the equations $p_1 \langle f_1, f_2 \rangle = f_1$ *and* $p_2 \langle f_1, f_2 \rangle = f_2$.

These equations can be read: 'the first component of the map $\langle f_1, f_2 \rangle$ is f_1' and 'the second component of the map $\langle f_1, f_2 \rangle$ is f_2.'

This means, in terms of figures, that the figures of shape A in the product $B_1 \times B_2$ are precisely the ordered pairs consisting of a figure of shape A in B_1 and a figure of shape A in B_2. On the one hand, given a figure of shape A in the product $B_1 \times B_2$, $A \longrightarrow B_1 \times B_2$, we obtain figures in B_1 and B_2 by composing it with the projections; on the other hand, the definition of product says that any two figures f_1 of shape A in B_1 and f_2 of shape A in B_2 arise this way from exactly one figure of shape A in $B_1 \times B_2$, which we called $\langle f_1, f_2 \rangle$.

This is precisely what was explained in the first session about Galileo's discovery. There, B_1 was the horizontal plane and B_2 was the vertical line, while $B_1 \times B_2$ was the space. The figures were motions, which can be considered as figures whose shape is *Time*, if by this we understand a time interval, *Time* = . Then a motion in the plane is a map

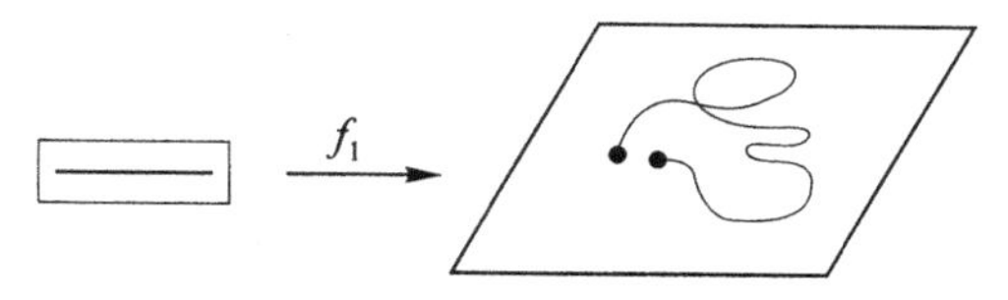

which is a figure of shape in this category.

Another quite compact way of expressing the relation between figures of any shape in a product of two objects and the corresponding figures in the factors is to write

$$\frac{A \longrightarrow B_1 \times B_2}{A \longrightarrow B_1, \quad A \longrightarrow B_2}$$

which is to be read: 'The maps $A \longrightarrow B_1 \times B_2$ correspond naturally to the pairs of maps $A \longrightarrow B_1, A \longrightarrow B_2$.'

In particular, we can consider figures whose shape is the terminal object. Since those figures are called 'points,' we see that the points of a product of two objects are the pairs of points, one from each factor, or with the notation just introduced:

$$\frac{\mathbf{1} \longrightarrow B_1 \times B_2}{\mathbf{1} \longrightarrow B_1, \ \mathbf{1} \longrightarrow B_2}$$

Because the category of sets has the special property explained in the last session (namely, that a map is completely determined by its values at points), the product of two sets is determined as soon as we know its points. Thus, this method tells us immediately the product of any two sets.

This method also tells us how to find the product of objects in other categories. Let me illustrate this with an example from the category of graphs. In this category we have two objects, $A = \boxed{\bullet \longrightarrow \bullet}$ and $D = \boxed{\bullet}$ such that the figures of shapes A together with the figures of shape D are sufficient to determine the maps of graphs. Therefore we can use these two graphs to calculate the product of any two graphs, as we used the terminal set to calculate the product of any two sets.

As an example let's calculate the product of the graph A with itself, i.e. $A \times A$. To do this we must determine its set of *arrows*, its set of *dots*, the relation between arrows and dots (which dots are the *source* and *target* of each arrow), and finally we must determine the two *projection maps* (without which there is no product).

The arrows of any graph X (including loops) are the graph maps $A \longrightarrow X$. The dots of X are the maps $D \longrightarrow X$, while the relation between arrows and dots is an instance of incidence relations that can be expressed in terms of those two special maps 'source' $D \xrightarrow{s} A$ and 'target' $D \xrightarrow{t} A$ that we introduced in the last session. For example, to say that a dot $D \xrightarrow{x} X$ is the source of an arrow $A \xrightarrow{p} X$ is the incidence relation

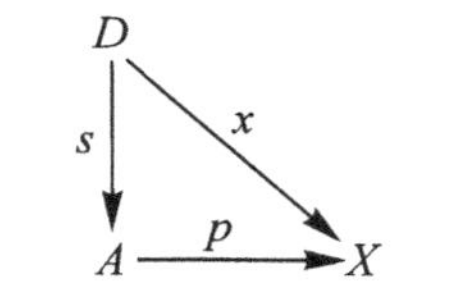

(*the source of* p *is* x, *or* $ps = x$)

In order to calculate $A \times A$ or A^2, we first find the set of dots of A^2:

$$\frac{D \longrightarrow A^2}{D \longrightarrow A, \ D \longrightarrow A}$$

This tells us that the dots of A^2 are the pairs of dots of A. Since A has two dots, there are four pairs and therefore A^2 has four dots. The arrows of A^2 are

$$\frac{A \longrightarrow A^2}{A \longrightarrow A, \ A \longrightarrow A}$$

the pairs of arrows of A. But A has only one arrow, thus we can form only one pair and therefore A^2 has only one arrow. At this point A^2 has been determined to be

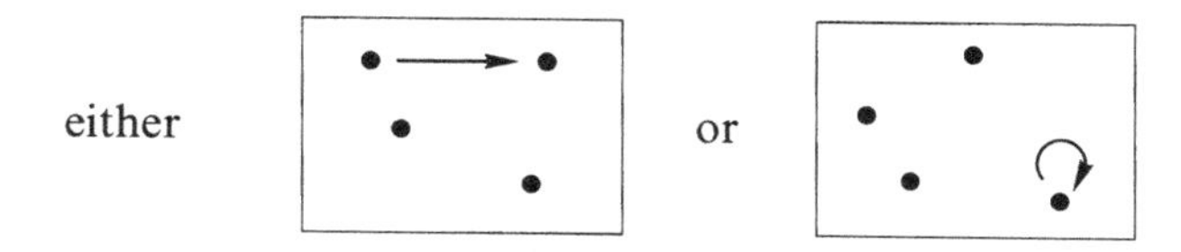

depending on whether the arrow of A^2 is a loop or not. This can be decided easily, since the loops of a graph X are the graph maps from the terminal graph $\mathbf{1} =$ [loop] to X, so that the loops of A^2 are

$$\frac{\mathbf{1} \longrightarrow A^2}{\mathbf{1} \longrightarrow A, \ \mathbf{1} \longrightarrow A}$$

the pairs of loops of A. But A has no loops. Hence A^2 doesn't have any either and therefore it must look like this

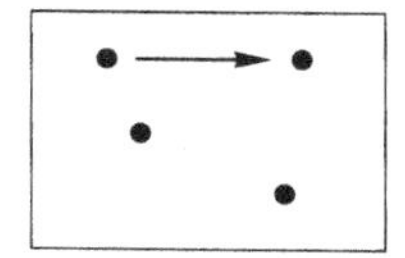

However, this only tells us how A^2 looks as a graph, not its structure as a product. To determine that, we need to know the projections $p_1 : A^2 \longrightarrow A$ and $p_2 : A^2 \longrightarrow A$. These are not hard to find if we label the dots and arrow of A, e.g. $A =$ $s \bullet \xrightarrow{a} \bullet\, t$ and accordingly we label A^2,

$(s,s) \bullet \xrightarrow{(a,a)} \bullet\, (t,t)$ $\quad (s,t)\bullet \quad \bullet\,(t,s)$

from which one easily figures out that the projections are the maps indicated in this 'sorting' diagram:

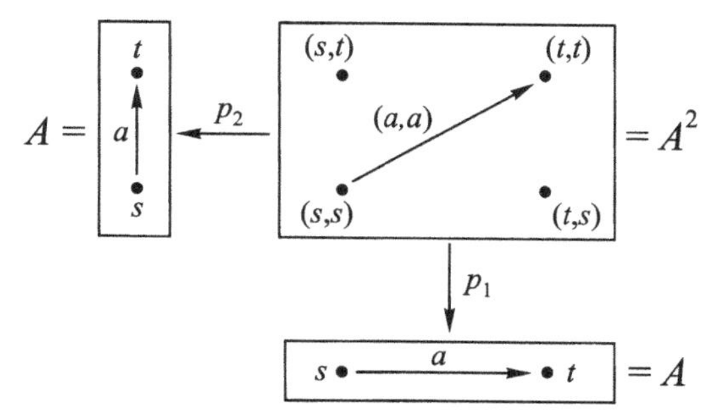

Notice that for any object X in any category having products there is a standard map $X \longrightarrow X \times X$, namely the one whose components (i.e. the composites with the two projections) are both the identity map 1_X. This standard map (which as we said before is denoted by $\langle 1_X, 1_X \rangle$) is often called the *diagonal* map. Since in our example there is only one map $A \longrightarrow A^2$, it must be the diagonal map. This is related to the fact that when A^2 is pictured internally with its standardized relation to the projections (as above), we get the picture

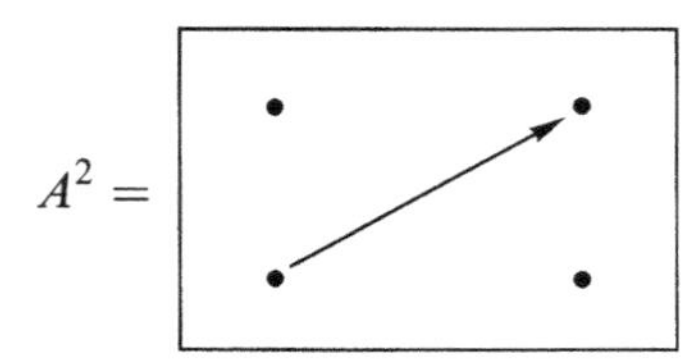

in which the arrow looks diagonal.

Notice that the graph A^2 consists of one arrow plus two naked dots. This can be expressed by the equation

$$A^2 = A + 2D$$

where by $2D$ is meant $D + D$, and the sum operation that appears here can actually be given a precise meaning as the opposite or dual of the product operation that is mentioned at the beginning of the lecture. We will explain more about this later. For now, try the following exercise.

Exercise 2:
Try to create the definition of 'sum' of two objects, in terms of a universal mapping property 'dual' to that of product, by reversing all maps in the definition of product. Then verify that in the category of sets and in the category of graphs, this property actually is satisfied by the intuitive idea of sum: 'Put together with no overlap and no interaction.'

SESSION 24

Uniqueness of products and definition of sum

In the last session we gave two exercises: one concerning the uniqueness of products and the other the definition of sum. One way of thinking about product and sum is that they combine two objects to get another object. In this session we will see that any product or sum also allows you to combine maps to get a new map. (Of course we already have one way of combining maps to get another map, namely *composition* of maps.)

1. The terminal object as an identity for multiplication

Let's start with an example to see how the uniqueness of products is useful. We saw that in the category of sets the number of elements of the product of two sets is precisely the product of the respective numbers of elements of the two sets, i.e. we had the formula

$$\#(A \times B) = \#A \times \#B$$

and therefore, as a particular case, if $\mathbf{1}$ is a terminal set,

$$\#(A \times \mathbf{1}) = (\#A) \times (\#\mathbf{1}) = (\#A) \times 1 = \#A$$

This suggests that we may have '$B \times \mathbf{1} = B$.' In fact, this 'equation' is 'true' in any category that has a terminal object, but we must say what it means!

To make B a product of B and $\mathbf{1}$ means that we must exhibit two maps

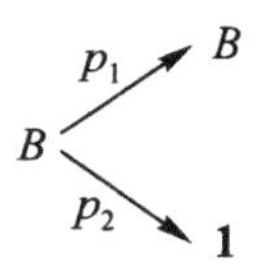

and prove that they satisfy the property of product projections. Actually there is only one choice for p_2, so we need only a map p_1 from B to B. There is an obvious choice: the identity of B. In fact this is the only thing one can think of, and therefore we hope it works. We want to see that

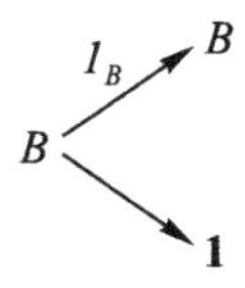

is a product. To prove it, let's suppose that we have two maps

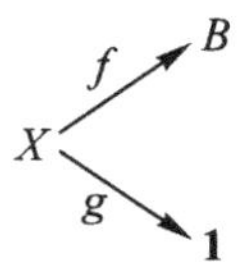

Is there exactly one map that makes this diagram commute?

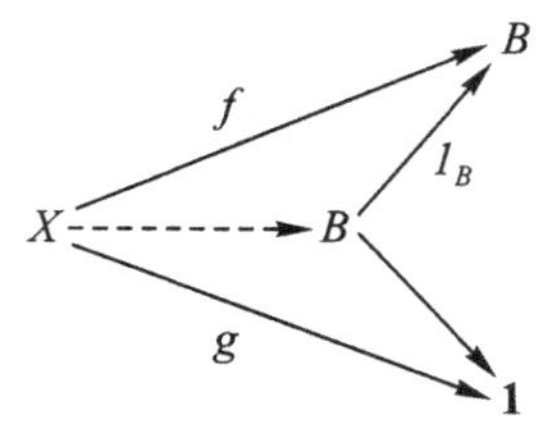

Well, there is only one possibility: it is to be a map which composed with 1_B gives f, so it can only be f itself. This works, because the other condition (that composing it with the map $B \longrightarrow \mathbf{1}$ gives g) is satisfied automatically by any map whatsoever $X \longrightarrow B$ (since $\mathbf{1}$ is terminal). Therefore we have proved that B is a product of B and the terminal object.

The reasoning is completely general and therefore the result holds in any category. Of course, a picture of $B \times \mathbf{1}$ (made in such a way as to make obvious the projection maps) may *look* different from B. For example, in the category of graphs, consider

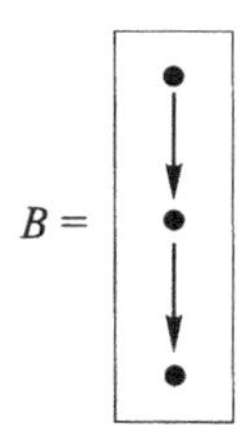

Since a terminal object in this category is a loop, $\mathbf{1} =$ [loop], we might draw the product $B \times \mathbf{1}$ as

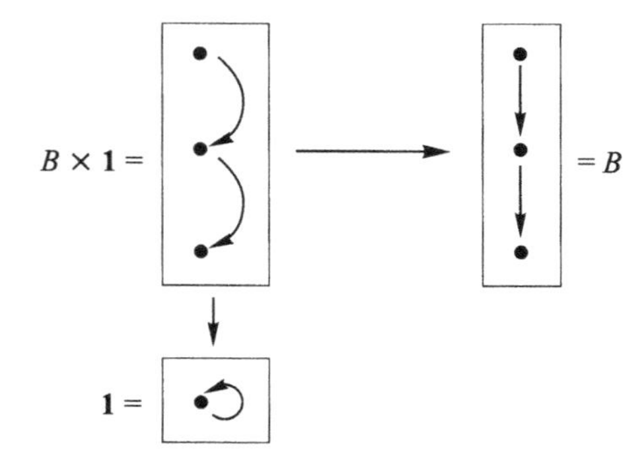

but it is obvious that $B \times \mathbf{1}$ is 'the same graph' as B, since the bending of the arrows that appears in the picture is not part of the graph, but an external device to make it obvious that the projection maps the two arrows to the loop.

We will recall now how a product in a category can be considered as giving rise to a way of combining two maps into one. Given two objects B_1 and B_2 in a category $\mathcal{C}$, a pair of product projections for B_1 and B_2 is a pair of maps

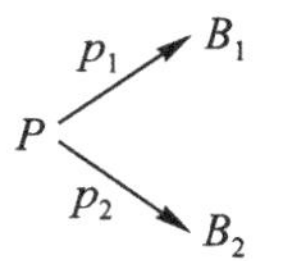

satisfying the following universal mapping property: For any two maps

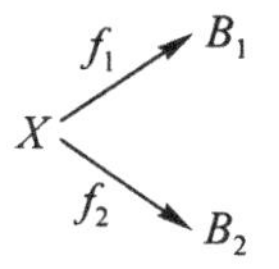

among all the maps $X \longrightarrow P$ there is exactly one $X \xrightarrow{f} P$ that satisfies both equations

$$f_1 = p_1 f \quad \text{and} \quad f_2 = p_2 f$$

As we said in the last session, that unique map f is denoted $\langle f_1, f_2 \rangle$. This means that a product P of B_1 and B_2 permits us to combine two maps

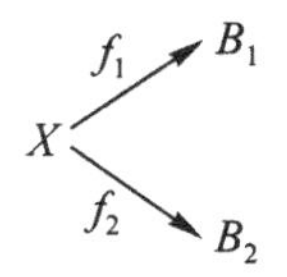

into one map $X \xrightarrow{\langle f_1, f_2 \rangle} P$.

The definition of 'product' is that this process of combining is inverse to the process of composing a map $X \longrightarrow P$ with the projections. If we are given a map $X \xrightarrow{g} P$ and compose it with the projections p_1 and p_2, we get two maps $g_1 (= p_1 g)$ and $g_2 (= p_2 g)$ which are 'the components of g' (relative to the product at hand). Indeed, if we now combine g_1 and g_2, the result must necessarily be the original map g.

Summing up: To say that two maps p_1, p_2 are product projections boils down to saying that this simple process of 'decomposing' a map (by composing it with each of p_1 and p_2) is invertible. In fact many universal mapping properties just state that a certain simple process is invertible.

2. The uniqueness theorem for products

Theorem: (Uniqueness of Products) *Suppose that both of*

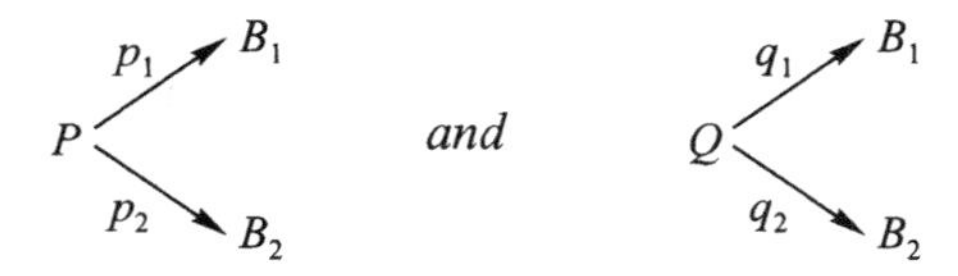

are product projection pairs (i.e. the ps as well as the qs satisfy the universal mapping property). Then there is exactly one map $P \xrightarrow{f} Q$ *for which*

$$q_1 f = p_1 \quad \textit{and} \quad q_2 f = p_2$$

This map f is in fact an isomorphism.

This theorem is sometimes crudely stated in the form (1) or (2) below:

1. Any two products of B_1 and B_2 are isomorphic objects.

More precisely:

2. Between any two products of B_1 and B_2 there is exactly one isomorphism compatible with the projections.

But the strongest and most precise statement is:

3. Between any two products of B_1 and B_2 there is exactly one map compatible with the projections, and that map is an isomorphism.

Proof of the uniqueness theorem: Suppose that

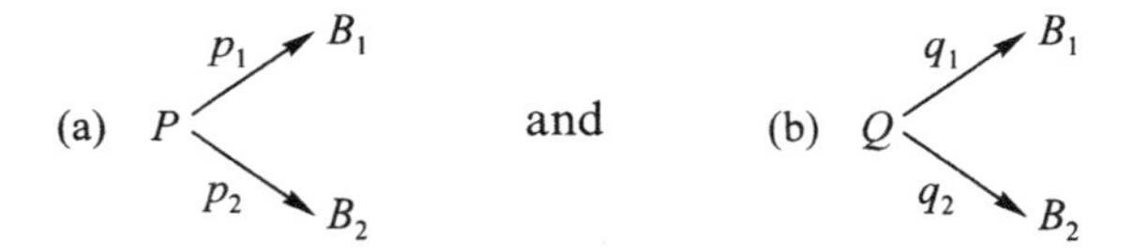

are two products of B_1 and B_2. Because (a) is a product, there is exactly one map $Q \xrightarrow{h} P$ for which $p_1 h = q_1$ and $p_2 h = q_2$; we would like to show that this map h is an isomorphism. For this, we should try to find its inverse. But there is an obvious thing to try: because (b) is a product, there is exactly one map $P \xrightarrow{k} Q$ for which $q_1 k = p_1$ and $q_2 k = p_2$. Is k really an inverse for h? Well, to prove that $hk = 1_P$, we calculate

$$p_1(hk) = (p_1 h)k = q_1 k = p_1$$

and

$$p_2(hk) = (p_2 h)k = q_2 k = p_2$$

This means that hk is the unique map which composed with p_1 gives p_1 and composed with p_2 gives p_2. But, isn't there another map with these properties?

DANILO: Yes. The identity of P.

Right. So the *exactly one* in the definition of product implies that hk is the identity of P. This proves half of the theorem. The other half is that kh is the identity of Q, which follows in the same way from

$$q_1(kh) = (q_1k)h = p_1h = q_1$$

and

$$q_2(kh) = (q_2k)h = p_2h = q_2$$

DANILO: In the case of sets I can picture the isomorphism between two different products of two sets. How does one picture it in more complicated cases?

It will be similar to the case of sets. Any two products will look like 'similar rectangles.'

3. Sum of two objects in a category

The challenge exercise that we gave in Session 23 was to invent the definition of *sum of two objects*. If we take the definition of product and reverse the maps that appear in it, we arrive at the following

Definition: *A* **sum of two objects** B_1, B_2 *is an object* S *and a pair of maps*

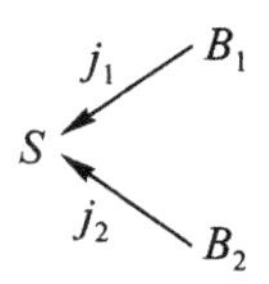

having the following universal mapping property: For any two maps

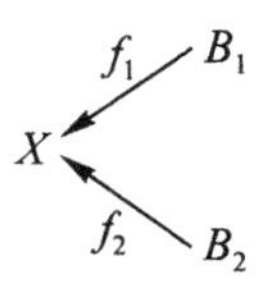

among all the maps $X \xleftarrow{f} S$ *there is exactly one that satisfies both*

$$f_1 = fj_1 \quad \text{and} \quad f_2 = fj_2$$

i.e. the diagram below commutes

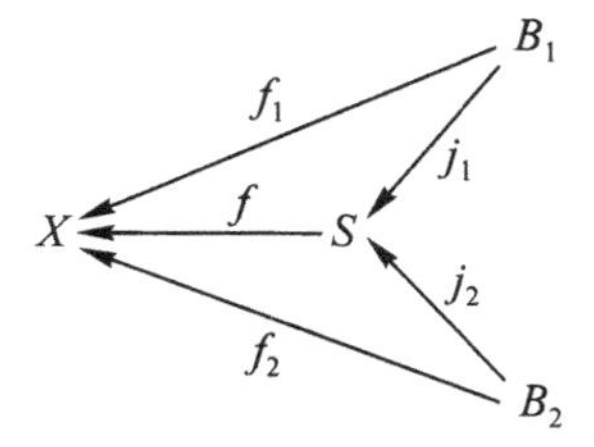

The maps j_1 and j_2 are called the *sum injections* of B_1 and B_2 into the sum S, and the notation for the unique map f is

$$f = \begin{cases} f_1 \\ f_2 \end{cases}$$

It should be a very good exercise for you to work out the following:

Exercise 1:
Formulate and prove the theorem of uniqueness of sums.

One answer is: 'Take everything said earlier in this session and reverse all the maps,' but you should work it out in detail.

What does the definition above have to do with *sum* as it is usually understood? Let's take the category of sets and see what our definition gives. Suppose that B_1 and B_2 are the sets

$$B_1 = \boxed{\bullet \quad \bullet \quad \bullet} \quad \text{and} \quad B_2 = \boxed{\bullet \quad \bullet}$$

What would you expect the sum of these two sets to be?

OMER: A five-element set.

That's right. The sum of B_1 and B_2 should be the set

$$S = \boxed{\bullet \quad \bullet \quad \bullet \quad \bullet \quad \bullet}$$

I will show you that this set does receive maps from B_1 and B_2 respectively, which satisfy the defining universal mapping property of sum. The two maps are

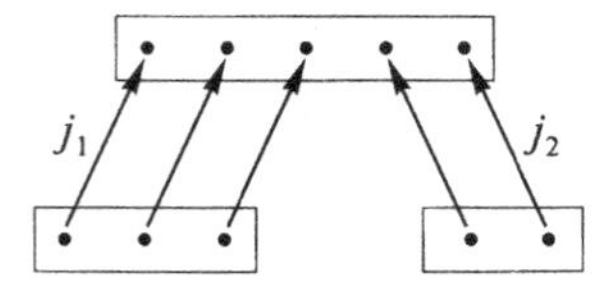

To prove the universal mapping property suppose that we are given any two maps from B_1 and B_2 to a set X, say

$$f_1 : B_1 \longrightarrow X \quad \text{and} \quad f_2 : B_2 \longrightarrow X$$

How can we 'combine' these two maps into a map $f = \begin{cases} f_1 \\ f_2 \end{cases}$? Well, we can define $f(s)$ 'by cases.' If s is 'in' B_1 (meaning that it is the j_1-image of some s' in B_1), we define $f(s)$ using f_1 (i.e. $f(s) = f_1(s')$), and if s is 'in' B_2 (i.e. $s = j_2(s'')$ for some s'' in B_2) we define $f(s)$ using f_2. In summary, we can define the map f by

$$f(s) = \begin{cases} f_1(s') \text{ if } s = j_1(s') \\ f_2(s'') \text{ if } s = j_2(s'') \end{cases}$$

(This expression is the origin of the notation $\begin{cases} f_1 \\ f_2 \end{cases}$ for the map f.)

This definition works because for each s in S, *either* $s = j_1(s')$ for *exactly one* s' in B_1 or $s = j_2(s'')$ for *exactly one* s'' in B_2, *but not both*. In other words, the maps j_1 and j_2 are *injective*, and they cover the whole of S (are *exhaustive*) and do not overlap (are *disjoint*).

Sums of objects in other categories may not look exactly as in this case of the category of sets, but this example justifies or motivates the definition of sum by the universal mapping property. As we shall see, now that we have a precise definition of 'sum,' we can *prove* equations such as the one that came up in Session 23, namely

$$A^2 = A + 2D$$

where A was the 'arrow' graph, and D was the 'naked-dot' graph.

If $\mathbf{1}$ is the terminal graph (the 'loop') we can define a whole sequence of graphs by summing $\mathbf{1}$s just as it is done with sets or numbers,

$$\mathbf{2} = \mathbf{1} + \mathbf{1}, \quad \mathbf{3} = \mathbf{1} + \mathbf{1} + \mathbf{1}$$

In this way we obtain the graphs

2 =  3 =

Even among graphs we have 'natural numbers,' while a graph such as the 'naked dot' D, neither $\mathbf{0}$ nor $\mathbf{1}$ but 'in between,' should be considered as a number in its own right, perhaps a different kind of number.

Having at our disposal multiplication and sum of objects, we can make all sorts of combinations and even write down *algebraic equations* among objects. (Compare with Exercise 19 in Article IV.)

Exercise 2:
Prove the following formulas:

(a) $D + D = 2 \times D$
(b) $D \times D = D$
(c) $A \times D = D + D$

Exercise 3:
Reread Section 5 of Session 15 and find a method, starting from presentations of $X^{\circlearrowleft \alpha}$ and $Y^{\circlearrowleft \beta}$, to construct presentations of
(a) $X^{\circlearrowleft \alpha} + Y^{\circlearrowleft \beta}$
(b) $X^{\circlearrowleft \alpha} \times Y^{\circlearrowleft \beta}$
Part (b) is harder than part (a).

SESSION 25

Labelings and products of graphs

Does anybody have any question about what has been explained so far?

OMER: I think I understand what a product is, but I don't quite understand what the X is that appears in the diagram

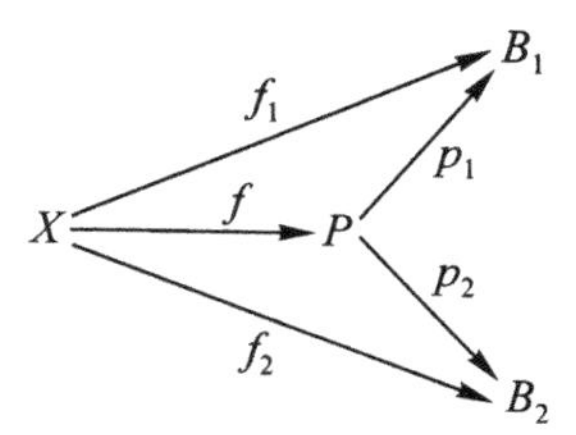

Well, in the *definition* of product of B_1 and B_2 in $\mathcal{C}$, there are infinitely many conditions the projections have to satisfy: one for *each* object X and *each* pair of maps f_1 and f_2. To say that Marco is the tallest person in the family means that for each person X in the family, X is at most as tall as Marco. The universal property of the product is like that: the product is the best thing of its type, and to say so requires comparing it with everything of its type, i.e. every object equipped with maps to B_1 and B_2.

Putting it another way: Suppose that you claim that the two maps

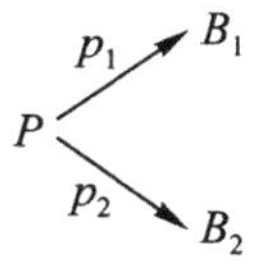

are product projections, and I claim that they are not. The definition of product means that if I want to prove to you that they are not, all I have to do is to select one particular object X and two particular maps,

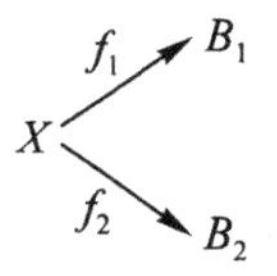

and show you that there isn't *exactly one* map f that makes this diagram commute:

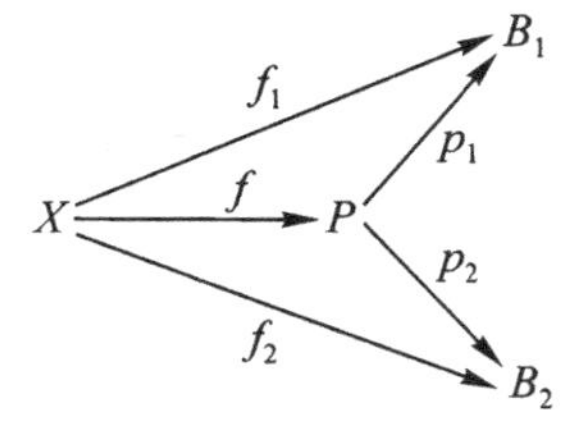

(so either I show that there is no such f, or I show at least two different ones).

Fortunately, in some categories to show that P (with its p_1 and p_2) is a product, we need only compare it with a few objects. For example, in the category of sets it suffices to test P against the terminal object; and in the category of graphs you were asked in Session 22 to prove that we need only test P against 'the naked dot' $X = \boxed{\bullet}$ and 'the arrow' $X = \boxed{\bullet \longrightarrow \bullet}$. Once we have tested P against these few, so that we know it is a product, we can take advantage of the fact that the universal property holds for each and every object X and every pair of maps f_1, f_2. We wrote this briefly as

$$\frac{X \longrightarrow B_1 \times B_2}{X \longrightarrow B_1,\ X \longrightarrow B_2}$$

meaning that to specify a morphism from X to the product is the same as to specify two morphisms, one from X to each of the factors.

1. Detecting the structure of a graph by means of labelings

To continue with the example of the category of graphs, we will see that the mere fact that the product of 'the arrow' A with any other graph Y has a map to A (the projection $p_1 : A \times Y \longrightarrow A$) gives us some information about the structure of $A \times Y$. To understand this more easily let's think first about the case of maps to the naked dot D. The question is: If a graph X has a map of graphs to D, what does this reveal about X itself? Well, a map of graphs $X \longrightarrow D$ takes every arrow in X to an arrow in D. But D doesn't have any arrows, therefore if X has at least one arrow, there can't be any maps from X to D. Also, a map $X \longrightarrow D$ must map all the dots to the unique dot of D. Thus we see that if X has a map of graphs to D, then X doesn't have any arrows and has exactly one map to D.

What is the meaning of a map to 'the arrow' $A = \boxed{s \bullet \xrightarrow{a} \bullet\, t}$? Well, a map $X \xrightarrow{f} A$ takes each dot of X either to s or to t. Therefore f divides the dots of X into two kinds: X_s and X_t. Here X_s is the set of those dots of X which are mapped to s, and X_t is the set of those dots of X which are mapped to t. Furthermore, every arrow of X must be mapped to a since this is the only arrow of A. This means that every arrow of X has its source in X_s and its target in X_t. The existence of a map of graphs from X to A means that the graph X is something like this:

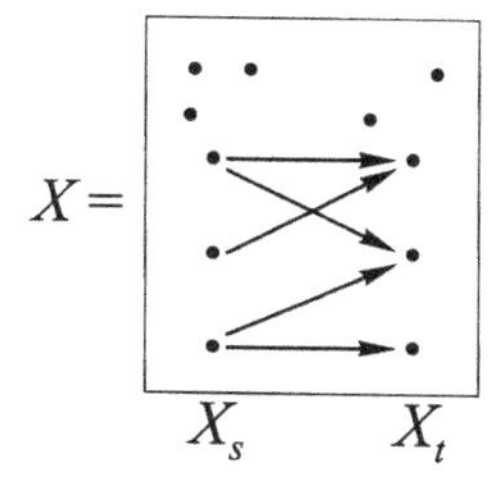

and the total number of maps from X to A depends only on the number of 'naked dots' that X has. There are $2^5 = 32$ maps from this graph X to 'the arrow' A.

Conversely, if we divide the dots of a graph X into two disjoint sets X_s and X_t in such a way that none of the dots in X_s is a target and none of the dots in X_t is a source (i.e. every arrow of X has its source in X_s and its target in X_t), then we have defined a map of graphs from X to A (the one that sends all the dots in X_s to s, all the dots in X_t to t, and all the arrows of X to a.)

Another interesting question of the same type is: What is the meaning of a sorting of the graph X by C_2, i.e. a map from X to the graph below?

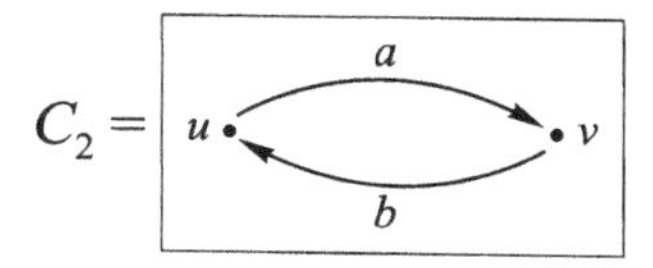

A map $X \longrightarrow C_2$ shows that X has no loops, and in general no odd cycles (cycles whose length is an odd number). Such a map divides the dots of X into two sorts: X_u = *dots mapped to u*, and X_v = *dots mapped to v*. Every arrow of X has either source in X_u and target in X_v or the other way around, i.e. no arrow has both source and target in X_u, nor both in X_v.

Another similar question is: What is the content of a map from any graph into the graph

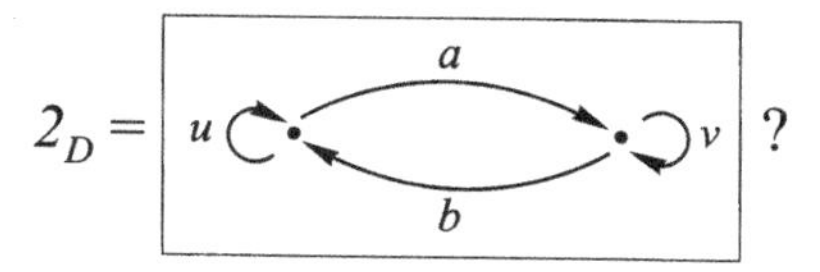

In this case a labeling of X by 2_D (i.e. a map $X \longrightarrow 2_D$) represents no restriction on X. The map itself is a choice of an arbitrary division of the dots of X into two sorts.

Exercise 1:
Find a graph 2_A such that a map $X \longrightarrow 2_A$ amounts to a division of the arrows of X into two sorts.

We have seen that the specific nature of a product can be successfully determined by the use of figures. What about trying to determine the structure of a sum? Could we do this with figures? If not, is there something analogous to figures that can be used in the case of sums? Remember that sums are defined 'like products,' but with all the maps reversed. The first implication of this is that the rule for defining maps from an object X into a product ('use a map from X into each factor') is converted into a rule for defining maps *from a sum* to an object Y ('use a map *from* each summand to Y').

When a map $X \longrightarrow Y$ is regarded as a figure of shape X in Y we think of Y as a fixed object and of X as variable, so as to give all possible shapes of figures in Y. But we can also take the opposite view and think of X as a fixed object and of Y as variable. Then the maps $X \longrightarrow Y$ would be considered as different 'labelings' or 'sortings' of X by Y. Other words that are used with the same meaning as 'Y-labeling' are 'Y-valued functions' and 'cofigures.' (See Session 6. The prefix 'co' meaning 'dual of' is used very often.) Exercise 2(b) is dual to Exercise 1 of Session 22.

Exercise 2:
(a) Show that if a diagram of *sets*

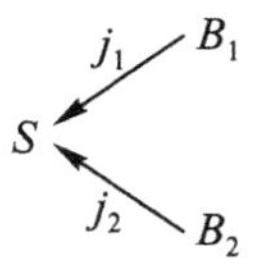

has the property of a coproduct, but restricted to testing against only the one cofigure-type $Y = \mathbf{2}$, then it is actually a coproduct, i.e. has that property for each object Y.
(b) Show that if a diagram of *graphs*

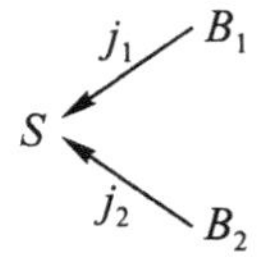

has the property of a coproduct, but restricted to testing against only the two cofigure-types $Y = 2_A$ and $Y = 2_D$, then it is actually a coproduct, i.e. has that property for each object Y.

Exercise 3:
Tricoloring a graph means assigning to each dot one of the three colors white, red, or green, in such a way that for each arrow, the source and target have different colors. If you fix a tricoloring of a graph X, and you have a map of graphs $Y \xrightarrow{f} X$, then you can color the dots of Y also: just color each dot $D \xrightarrow{y} Y$ the same color as fy. This is called the 'tricoloring of Y *induced* by f.'

(a) Show that this induced coloring is a tricoloring; i.e. no arrow of Y has source and target the same color.
(b) Find *Fatima's tricolored graph F*. It is the best tricolored graph: For any graph Y, each tricoloring of Y is induced by exactly one map $Y \longrightarrow F$.

Exercise 4:
In this exercise, $\mathbf{0}$ is the initial graph, with no dots (and, of course, no arrows) and A_2 is the graph $\boxed{\bullet \longrightarrow \bullet \longrightarrow \bullet}$.
Show that for each graph X:
(a) there is either a map $X \longrightarrow \mathbf{0}$ or a map $D \longrightarrow X$, but not both; and
(b) there is either a map $X \longrightarrow D$ or $A \longrightarrow X$, but not both; and
(c) there is either a map $X \longrightarrow A$ or $A_2 \longrightarrow X$, but not both.

Can the sequence $\mathbf{0}$, D, A, A_2 be continued? That is, is there a graph C such that for each graph X
(d) there is either a map $X \longrightarrow A_2$ or $C \longrightarrow X$, but not both?

2. Calculating the graphs $A \times Y$

As we saw earlier in this session, a graph of the form $A \times Y$ has a structure similar to that of the graph X pictured earlier. In other words, the dots of $A \times Y$ are divided into two sorts so that in one of them there are no targets and in the other there are no sources. Furthermore,

$$\frac{A \longrightarrow A \times Y}{A \longrightarrow A,\ A \longrightarrow Y}$$

i.e. the arrows of $A \times Y$ are the pairs of arrows $\langle a, y \rangle$ where a is an arrow of A and y is an arrow of Y. Since A has only one arrow, we conclude that $A \times Y$ has precisely as many arrows as Y has. Also,

$$\frac{D \longrightarrow A \times Y}{D \longrightarrow A,\ D \longrightarrow Y}$$

implies that the number of dots of $A \times Y$ is twice the number of dots of Y, since A has two dots.

To determine the source target relation in $A \times Y$, we compose each arrow of $A \times Y$, seen as a map $A \longrightarrow A \times Y$, with the maps 'source' and 'target'

$$D \underset{t}{\overset{s}{\rightrightarrows}} A$$

Recall that for an arrow $A \xrightarrow{y} Y$ of Y, the source and target are

$$y \circ s \quad \text{and} \quad y \circ t$$

$$D \underset{t}{\overset{s}{\rightrightarrows}} A \xrightarrow{y} Y$$

The source of the arrow $\langle a, y\rangle$ of $A \times Y$ is $\langle a, y\rangle \circ s = \langle s, y \circ s\rangle$, since the commutativity of the diagram

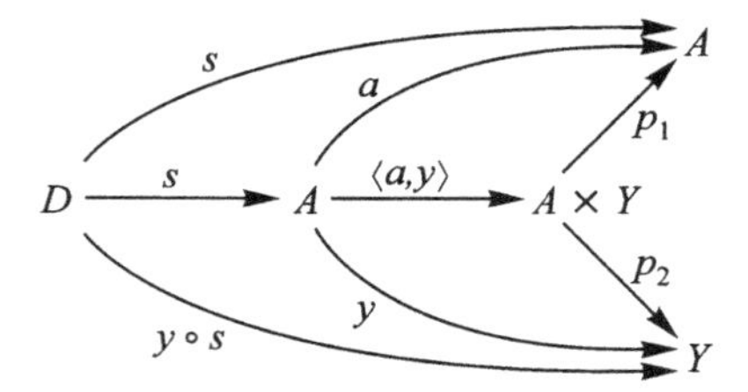

shows that $p_1 \circ (\langle a, y\rangle \circ s) = s$ and $p_2 \circ (\langle a, y\rangle \circ s) = y \circ s$. In a similar way we see that the target of $\langle a, y\rangle$ is $\langle a, y\rangle \circ t = \langle t, y \circ t\rangle$ since $p_1 \circ (\langle a, y\rangle \circ t) = t$ and $p_2 \circ (\langle a, y\rangle \circ t) = y \circ t$.

These rather long calculations were done in detail to give you some practice in such things, and to illustrate the general principle that $X \times Y$ looks like a rectangle with base X and height Y. For example,

Exercise 5:
In this exercise, $B = \boxed{\bullet \rightarrow \bullet \rightarrow \bullet}$ and $C = \boxed{\bullet \rightleftarrows \bullet \quad \bullet}$. Show that B is not isomorphic to C, but that $A \times B$ is isomorphic to $A \times C$. (We already know examples of the 'failure of cancellation': $\mathbf{0} \times X$ and $\mathbf{0} \times Y$ are isomorphic for every X and Y; we also saw that $D \times A$ is isomorphic to $D \times \mathbf{2}$. This exercise shows that cancellation can fail even when the factor we want to cancel is more 'substantial.')

3. The distributive law

In all the categories which we have studied, sums and products are related by the *distributive law*. As an application of the rule for defining maps into a product (and of the 'dual' rule for defining maps *on* a sum) try to do the following:

Exercise 6:
Assuming that X, B_1 and B_2 are objects of a category with sums and products, construct a map from the sum of $X \times B_1$ and $X \times B_2$ to the product of X with $B_1 + B_2$, i.e. construct a map

$$(X \times B_1) + (X \times B_2) \longrightarrow X \times (B_1 + B_2)$$

Hint: Use the universal mapping properties of sum and product, and combine appropriate injections and projections.

Some categories obey the 'law' that the map constructed in the exercise always has an inverse. Such 'distributivity' is structural and not merely quantitative, but a useful rough way of thinking about this distributive law of products with respect to sums is to consider that the area of a rectangle made up of two rectangles

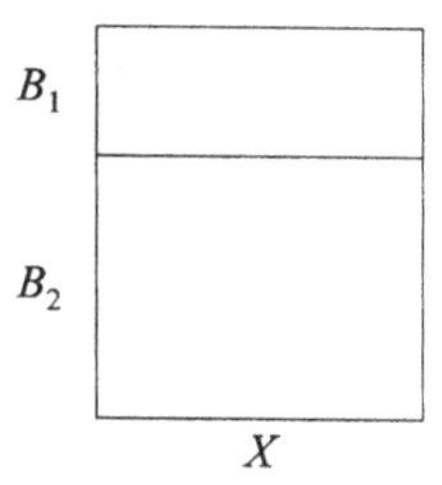

is equal to the sum of the areas of the two small rectangles:

$$\textit{Area of } X \times (B_1 + B_2) = \textit{Area of } X \times B_1 + \textit{Area of } X \times B_2$$

SESSION 26

Distributive categories and linear categories

1. The standard map $A \times B_1 + A \times B_2 \longrightarrow A \times (B_1 + B_2)$

An exercise in the last session asked you to find, for any three objects A, B_1, and B_2 of a category $\mathcal{C}$ that has sums and products, a 'standard' map

$$A \times B_1 + A \times B_2 \longrightarrow A \times (B_1 + B_2)$$

In many categories, this map and the standard (only) map $\mathbf{0} \longrightarrow A \times \mathbf{0}$ have inverses; when this happens we say that the *distributive law* holds in $\mathcal{C}$, or that the category $\mathcal{C}$ *is distributive*. This is the case in all the categories that we have discussed in the sessions.

In categories in which the distributive law doesn't hold, the use of 'sum' for that construction is often avoided; it is instead called 'coproduct,' which means (as mentioned in the last session) 'dual of product.' One of the fundamental ways in which one category differs from another is the relation between the concepts and the coconcepts. In many categories the distributive law is valid, but in other categories there are instead quite different, but equally interesting, relationships between product and coproduct.

The construction of the standard map mentioned above will be an application of a general fact which follows by combining the universal mapping property of products with that of coproducts: *A map from a coproduct of two objects to a product of two objects is 'equivalent' to four maps, one from each summand to each factor.* Since its domain is a coproduct, we know that a map f from $C_1 + C_2$ to $A \times B$ is determined by its composites with the injections of C_1 and C_2, and can be denoted

$$f = \begin{Bmatrix} f_1 \\ f_2 \end{Bmatrix}$$

where f_1 and f_2 are the result of composing f with the injections of C_1 and C_2 into $C_1 + C_2$.

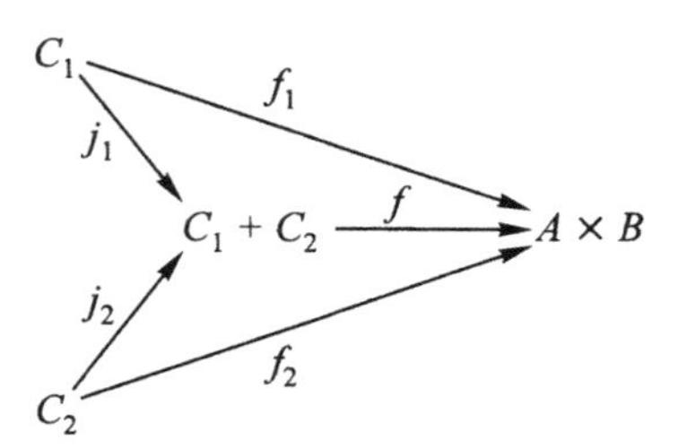

Furthermore, each of the maps f_1 and f_2 is a map into a product, and thus has two components, so that $f_1 = \langle f_{1A}, f_{1B} \rangle$ and $f_2 = \langle f_{2A}, f_{2B} \rangle$

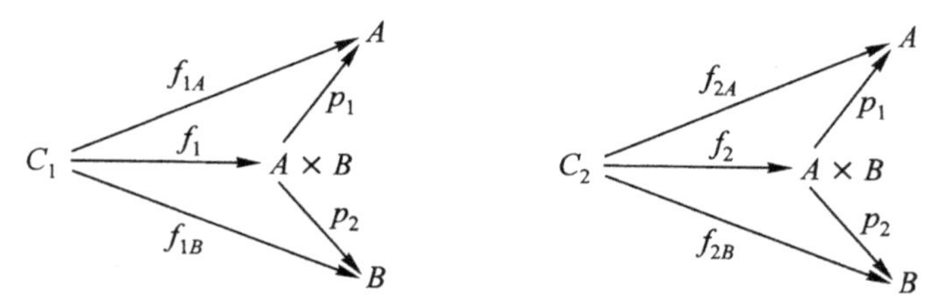

The end result is that f can be analyzed into the four maps

$$C_1 \xrightarrow{f_{1A}} A \quad C_1 \xrightarrow{f_{1B}} B$$
$$C_2 \xrightarrow{f_{2A}} A \quad C_2 \xrightarrow{f_{2B}} B$$

and, conversely, any four such maps determine a map from $C_1 + C_2$ to $A \times B$ given by

$$f = \begin{cases} \langle f_{1A}, f_{1B} \rangle \\ \langle f_{2A}, f_{2B} \rangle \end{cases}$$

which is more often denoted by the *matrix* (using a rectangular array of maps enclosed in brackets)

$$f = \begin{bmatrix} f_{1A} f_{1B} \\ f_{2A} f_{2B} \end{bmatrix}$$

This analysis can be carried out more generally, for coproducts and products of any number of objects. For any objects $C_1, \ldots, C_m$, and $A_1, \ldots, A_n$, denote product projections by $A_1 \times \ldots \times A_n \xrightarrow{p_\nu} A_\nu$ and sum injections by $C_\mu \xrightarrow{j_\mu} C_1 + \ldots + C_m$. Then for any matrix

$$\begin{bmatrix} f_{11} f_{12} \cdots f_{1n} \\ \vdots \qquad \vdots \\ f_{m1} \quad \cdots f_{mn} \end{bmatrix}$$

where $f_{\mu\nu} : C_\mu \longrightarrow A_\nu$, there is *exactly one* map

$$C_1 + \ldots + C_m \xrightarrow{f} A_1 \times \ldots \times A_n$$

satisfying all the $m \times n$ equations

$$f_{\mu\nu} = p_\nu f j_\mu$$

This way of stating the result, which gives the matrix for f by *formulas*, also makes it clear that if we analyze the map f in the opposite way – first using that $A \times B$ is a product, then that $C_1 + C_2$ is a coproduct – we obtain the same matrix.

To apply this to the problem of defining a map

$$A \times B_1 + A \times B_2 \longrightarrow A \times (B_1 + B_2)$$

we must define four maps as follows:

$$\begin{array}{ll} A \times B_1 \longrightarrow A & A \times B_1 \longrightarrow B_1 + B_2 \\ A \times B_2 \longrightarrow A & A \times B_2 \longrightarrow B_1 + B_2 \end{array}$$

and they are to be defined using only the standard product projections and sum injections. What maps can we choose? The two on the left don't require much thought; we can choose the product projections to A. Even the maps on the right are not too difficult, since we can use for each a product projection (to B_1 and B_2 respectively), followed by a sum injection. For example, for the first map on the right we take the composite

$$A \times B_1 \xrightarrow{proj} B_1 \xrightarrow{inj} B_1 + B_2$$

These choices provide the standard map $A \times B_1 + A \times B_2 \xrightarrow{f} A \times (B_1 + B_2)$. This map can be visualized by means of the diagram

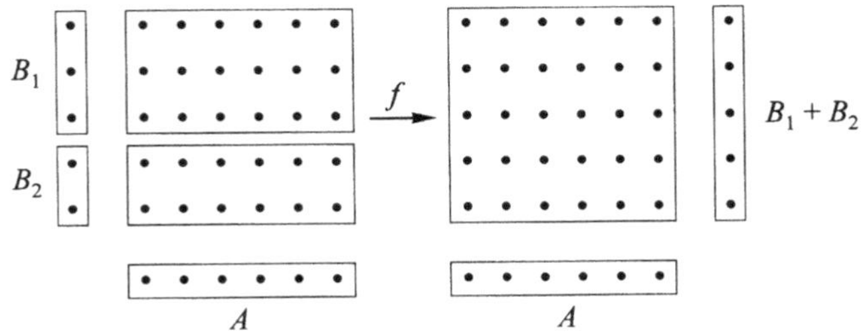

There is a **general distributive law** which is valid in all distributive categories. If $B_1, B_2, \ldots, B_n$ and A are objects in any category with sums and products there is a standard map

$$A \times B_1 + A \times B_2 + \ldots + A \times B_n \longrightarrow A \times (B_1 + B_2 + \ldots + B_n)$$

The general distributive law says that this standard map is an isomorphism. In the case of $n = 0$ (sum of no objects) the domain of this map is an initial object, and the map itself is the unique map

$$\mathbf{0} \longrightarrow A \times \mathbf{0}$$

which is obviously a section for the product projection $A \times \mathbf{0} \longrightarrow \mathbf{0}$. The general distributive law implies that this is actually an isomorphism, so that the 'identity' $A \times \mathbf{0} = \mathbf{0}$ can be seen as a consequence of the distributive law. Conversely, it can be shown that the two special cases $n = 0$, and $n = 2$ of distributivity imply the general distributive law. In Part V, we will study 'exponential objects,' and will prove that any category with these is distributive.

DANILO: What sort of categories would not satisfy the distributive law?

Exercise 20 of Article IV says that the category of pointed sets is not distributive. There is also an important class of categories, which we call *linear categories*, in which $A \times B$ is always isomorphic to $A + B$; and only trivial linear categories satisfy the distributive law.

2. Matrix multiplication in linear categories

Let me make a brief departure from our main topic to say something about these linear categories. First, linear categories have zero maps. By this we mean that for any two objects X, Y there is a special map from X to Y called *the zero from X to Y*, and we denote it by 0_{XY}. The fundamental property of a zero map is that composed with any other map it gives another zero map. Thus for any map $Y \xrightarrow{g} Z$, the composite $g0_{XY}$ is the zero map 0_{XZ}. Similarly, for any map $W \xrightarrow{f} X$, $0_{XY}f = 0_{WY}$. As usual, it is a good idea to draw the external diagrams for these composites, to see how the domains and codomains match.

The existence of zero maps has as a consequence that we can define a preferred map from the coproduct $X + Y$ to the product $X \times Y$,

$$f = \begin{bmatrix} 1_X & 0_{XY} \\ 0_{YX} & 1_Y \end{bmatrix} : X + Y \longrightarrow X \times Y$$

This map is called the 'identity matrix.'

Definition: *A category with zero maps in which every 'identity matrix' (as defined above) is an isomorphism is called a* **linear category**.

In a linear category, since every identity matrix is an isomorphism, we can 'multiply' any matrices $A + B \xrightarrow{f} X \times Y$ and $X + Y \xrightarrow{g} U \times V$. We simply define their 'product' as

$$\begin{bmatrix} f_{AX} & f_{AY} \\ f_{BX} & f_{BY} \end{bmatrix} \cdot \begin{bmatrix} g_{XU} & g_{XV} \\ g_{YU} & g_{YV} \end{bmatrix} = \begin{bmatrix} g_{XU} & g_{XV} \\ g_{YU} & g_{YV} \end{bmatrix} \circ \begin{bmatrix} 1_X & 0_{XY} \\ 0_{YX} & 1_Y \end{bmatrix}^{-1} \circ \begin{bmatrix} f_{AX} & f_{AY} \\ f_{BX} & f_{BY} \end{bmatrix}$$

This 'product' is another matrix (but now from $A + B$ to $U \times V$) since it is nothing but the composite

$$A + B \xrightarrow{f} X \times Y \xrightarrow{\alpha} X + Y \xrightarrow{g} U \times V$$

where α is the assumed inverse of the identity matrix.

3. Sum of maps in a linear category

This matrix multiplication has a very interesting consequence. If A and B are any two objects in a linear category, we can *add* any two maps from A to B and get another map from A to B. We use the following particular case of the above matrix multiplication (denoting the two maps that we are going to add by $A \xrightarrow{f} B$ and $A \xrightarrow{g} B$): Take $X = U = A$, $Y = V = B$ and the matrices

$$\begin{bmatrix} g_{XU} & g_{XV} \\ g_{YU} & g_{YV} \end{bmatrix} = \begin{bmatrix} 1_{AA} & g \\ 0_{BA} & 1_{BB} \end{bmatrix} \quad \text{and} \quad \begin{bmatrix} f_{AX} & f_{AY} \\ f_{BX} & f_{BY} \end{bmatrix} = \begin{bmatrix} 1_{AA} & f \\ 0_{BA} & 1_{BB} \end{bmatrix}$$

One can show that the 'product' of these two matrices must be of the form

$$\begin{bmatrix} 1_{AA} & f \\ 0_{BA} & 1_{BB} \end{bmatrix} \cdot \begin{bmatrix} 1_{AA} & g \\ 0_{BA} & 1_{BB} \end{bmatrix} = \begin{bmatrix} 1_{AA} & h \\ 0_{BA} & 1_{BB} \end{bmatrix}$$

for exactly one map $h : A \longrightarrow B$. The sum of f and g is now *defined* to be this map h, so that $f + g$ is uniquely determined by the equation

$$\begin{bmatrix} 1_{AA} & f \\ 0_{BA} & 1_{BB} \end{bmatrix} \cdot \begin{bmatrix} 1_{AA} & g \\ 0_{BA} & 1_{BB} \end{bmatrix} = \begin{bmatrix} 1_{AA} & f+g \\ 0_{BA} & 1_{BB} \end{bmatrix}$$

Even more interesting is that we now get a formula for multiplication of matrices in terms of this addition of maps:

Exercise 1:
Using the above definitions of matrix multiplication and addition of maps, prove the following *formula for matrix multiplication*:

$$\begin{bmatrix} f_{AX} & f_{AY} \\ f_{BX} & f_{BY} \end{bmatrix} \cdot \begin{bmatrix} g_{XU} & g_{XV} \\ g_{YU} & g_{YV} \end{bmatrix} =$$
$$\begin{bmatrix} g_{XU} \circ f_{AX} + g_{YU} \circ f_{AY} & g_{XV} \circ f_{AX} + g_{YV} \circ f_{AY} \\ g_{XU} \circ f_{BX} + g_{YU} \circ f_{BY} & g_{XV} \circ f_{BX} + g_{YV} \circ f_{BY} \end{bmatrix}$$

It is worth mentioning where the zero maps come from. In a linear category, the product of a finite family of objects is isomorphic to the coproduct. For an empty family, this says that the terminal object is isomorphic to the initial object. This isomorphism allows us to define 'the zero map' from an object X to an object Y by composing the unique maps

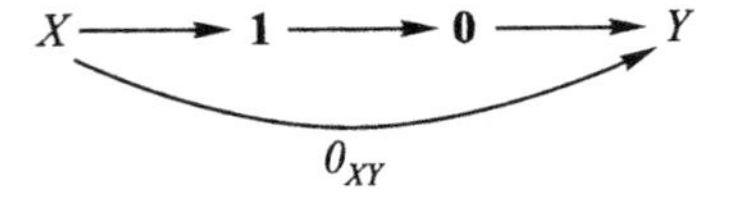

Exercise 2:
Prove that a category with initial and terminal objects has zero maps if and only if an initial object is isomorphic to a terminal object.

Saying that an initial object is isomorphic to a terminal object is equivalent to saying that there exists a map from the terminal object to the initial object; such a map is necessarily an isomorphism. (Why? What is its inverse?) *Warning*! In order to compare distributive categories with linear categories, we have written the matrices in a different ('transpose') way than they are usually written in linear categories.

4. The associative law for sums and products

The associative law for multiplication of objects is true in *any* category with products. (This is the subject of Exercise 16 in Article IV.) Just for practice in dualizing, we will discuss instead the corresponding problem for sums.

The sum of three objects can be defined much in the same way as the sum of two objects. The only difference is that the universal mapping property will now involve three injection maps:

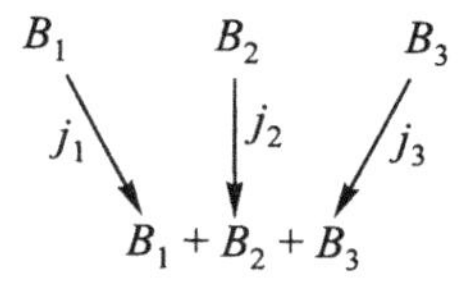

and the defining universal mapping property is that for any three maps from B_1, B_2, and B_3 to any object X,

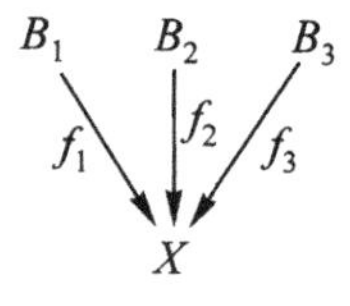

there is a unique map $B_1 + B_2 + B_3 \xrightarrow{f} X$, which can be denoted

$$f = \begin{cases} f_1 \\ f_2 \\ f_3 \end{cases}$$

such that $fj_1 = f_1, fj_2 = f_2$, and $fj_3 = f_3$. If a category has sums of two objects, then it also has sums of three objects: given B_1, B_2 and B_3 we first form the sum of B_1 and B_2, then the sum of $B_1 + B_2$ with B_3. We obtain injections from $B_1 + B_2$ and B_3, and composition with the injections from B_1 and B_2 to $B_1 + B_2$ yields the three injections required for a sum of three objects:

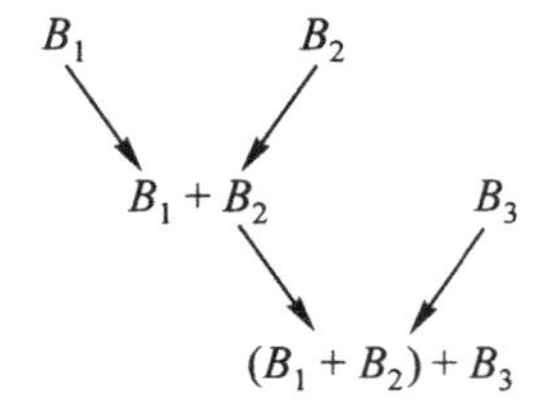

Check that the universal mapping property of a sum of three objects holds.

Let's see an example of this with sets. Let B_1, B_2 and B_3 be sets with 3, 2, and 4 elements, respectively. Then $B_1 + B_2$ is

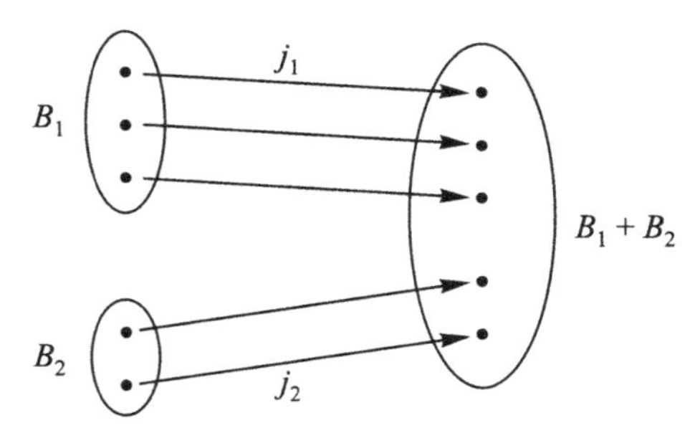

and if we sum $B_1 + B_2$ with B_3, we get a set with the following injections:

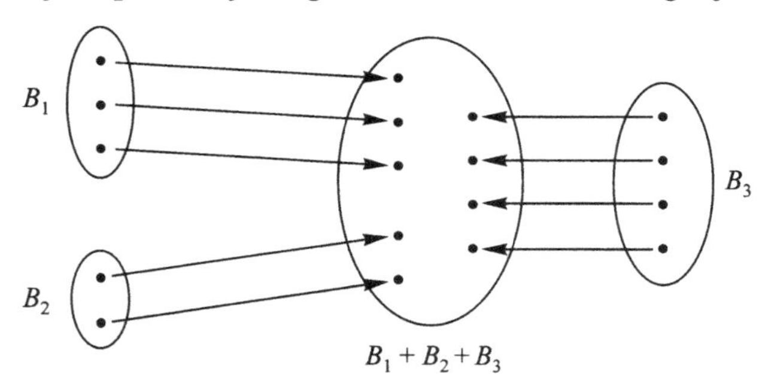

Our construction produced three-fold sums in terms of two-fold sums. Can you think of another construction?

IAN: Well, just B_1 plus $B_2 + B_3$.

Right. This construction is slightly different but you can verify in the same way that it also gives three injections which satisfy the correct universal mapping property. Then the uniqueness theorem for triple sums implies that we have an isomorphism

$$(B_1 + B_2) + B_3 \cong B_1 + (B_2 + B_3)$$

A similar reasoning applies to sums of four objects or more, and obviously all that we have said about sums applies also to products, so that one can find the triple product $A \times B \times C$ as $(A \times B) \times C$ or as $A \times (B \times C)$. In summary, if it is possible to form sums and products of two objects, then it is also possible to form sums and products of families of more than two objects. What would a sum or a product of a one-object family be? It should be just that object, right? And indeed it is. In order to prove it, one should first make clear the definition of sum or product of any family of objects and then use the fact that every object has an identity map. What about a sum or product of a family of no objects? If we sum no objects what is the result? Right. The result is zero, the initial object. On the other hand, if we multiply no objects, the result is one, the terminal object. These facts can be proved very easily, but for that you have to understand very well the universal mapping properties defining the sum and product of a family of objects. (See Article IV, Section 5.)

SESSION 27

Examples of universal constructions

1. Universal constructions

We have seen that there are two kinds of universal constructions: those similar to multiplication and terminal object – the technical term is 'limits' – and those similar to sum and initial object: 'colimits'. Let's summarize in a table all the universal constructions that we have studied.

Universal constructions

colimits	**limits**
Initial object (usually denoted **0**)	Terminal object (usually denoted by **1**)
Sum of two objects	Product of two objects
Sum of three objects, etc.	Product of three objects, etc.

Let's review what a terminal object is. To say that T is a terminal object in the category $\mathcal{C}$ means .... What?

CHAD: That there is only one map.

One map? From where to where?

CHAD: From the other object to T.

What other object?

CHAD: Any other object.

Right. From any other object. Start the sentence with that, don't leave it for the end, because then you are talking about something that nobody has introduced in the conversation. Now, what's an initial object?

FATIMA: An initial object is one that has exactly one map to any other object.

Right. But you should get used to starting with that other object: 'For each object X in $\mathcal{C}$ ...'. It is a curious definition because it refers to *all* objects of the category. That is the characteristic of definitions by universal mapping properties. What is a terminal object in the category of sets?

DANILO: A single element.

Right. Any set with exactly one element. What about in the category of dynamical systems?

FATIMA: A set with fixed point.

And anything else?

FATIMA: No.

Right. The terminal object in this category is just the identity map of any set with exactly one element, which we can picture as

or even better as

What about the category of graphs? What is a terminal object there?

CHAD: One element on top and one element on the bottom, and the only two maps as source and target.

Right. That is exactly the terminal graph. We used to draw it

where the solid arrow represents the map 'source,' and the dotted arrow represents the map 'target.' But we had a nicer way of picturing a graph, which was to draw all elements of the top (domain) set as arrows and to draw them together with the dots in one set, positioning the arrows with respect to the dots in a way that makes the 'source' and 'target' maps obvious. Thus the graph that Chad just described will be drawn as

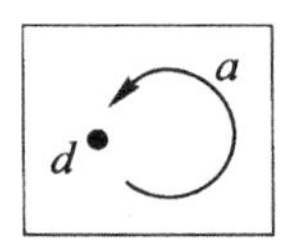

Notice the similarity to the terminal object of the category of dynamical systems.

Let's summarize the terminal objects in these various categories in a table.

Terminal object if $\mathcal{C}$ is the category ...

$\mathcal{S}$ = Sets and all maps	$\mathcal{S}^{\circlearrowright}$ = Dynamical systems	$\mathcal{S}^{\downdownarrows}$ = Graphs
$T =$	$T =$	$T =$

What about initial objects? What is an initial object in sets?

DANILO: An empty set.

Right. And we say that in the other categories the initial object was also 'empty,' so that we have:

Initial object if $\mathcal{C}$ is the category ...

$\mathcal{S}$ = Sets and all maps	$\mathcal{S}^{\circlearrowright}$ = Dynamical systems	$\mathcal{S}^{\downdownarrows}$ = Graphs
$I =$	$I =$	$I =$

Besides that, we also discovered some properties such as

$$\mathbf{0} + A = A \qquad | \qquad \mathbf{1} \times A = A$$

These equations look simple because they are familiar from numbers, but here they have more meaning. The product of two numbers is just a number, but the product of two objects R and Q is another object P *and* two 'projection' maps $P \longrightarrow R$ and $P \longrightarrow Q$. Thus when we said A is a product of A and $\mathbf{1}$, we had to *specify* the projection maps

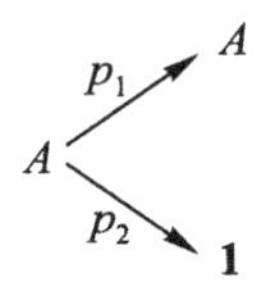

For p_2 there was exactly one possibility, and for p_1 we took the identity map on A, $1_A : A \longrightarrow A$. The statement that these choices make A a product of A and $\mathbf{1}$ *means* that for every object X and every pair of maps $f_1 : X \longrightarrow A$, $f_2 : X \longrightarrow \mathbf{1}$, there is exactly one map $f : X \longrightarrow A$ that we can use to fill in the picture and make the diagram commute:

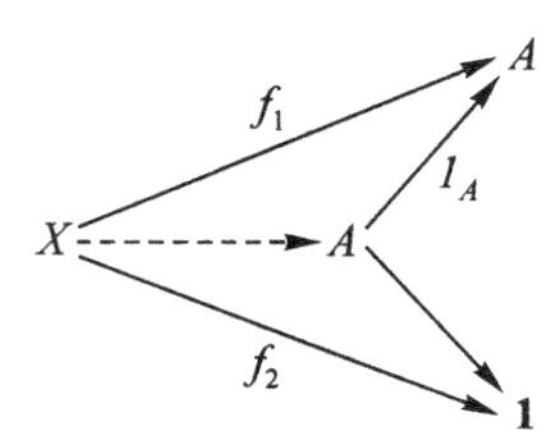

(The proof was easy since one of the two equations, $1_A \circ f = f_1$ forced us to choose the dotted arrow f to be f_1 itself; and the second equation was automatically satisfied, since f_2 and $p_2 \circ f$ are maps $X \longrightarrow \mathbf{1}$.)

2. Can objects have negatives?

For numbers, the *negative* of 3 is defined to be a solution of the equation $3 + x = 0$. Similarly, if A is an object of a category, a *negative of A* means an object B such that $A + B = \mathbf{0}$.

Each of the symbols '+,' '=' and '**0**' in that equation has a special meaning. '+' means coproduct of objects, '=' is here intended as 'is isomorphic to,' and '**0**' means 'initial object.' Similarly, A and B represent objects of the category, not numbers.

Can the initial object **0** serve as a coproduct of two objects A and B? Remember that a coproduct of A and B is an object C and a 'best' pair of maps $A \longrightarrow C \longleftarrow B$. We have to find maps $A \xrightarrow{j_1} \mathbf{0} \xleftarrow{j_2} B$, such that for every object X with maps $A \xrightarrow{f_1} X \xleftarrow{f_2} B$ there is exactly one map $\mathbf{0} \xrightarrow{f} X$ such that $f_1 = fj_1$, and $f_2 = fj_2$. What maps can one think of from an object A to **0**?

Let's pose the equation in the concrete case of the category of sets where we have a pretty good idea of what a coproduct is, since the coproduct of two sets is just 'all the elements of the two sets together,' as in this example:

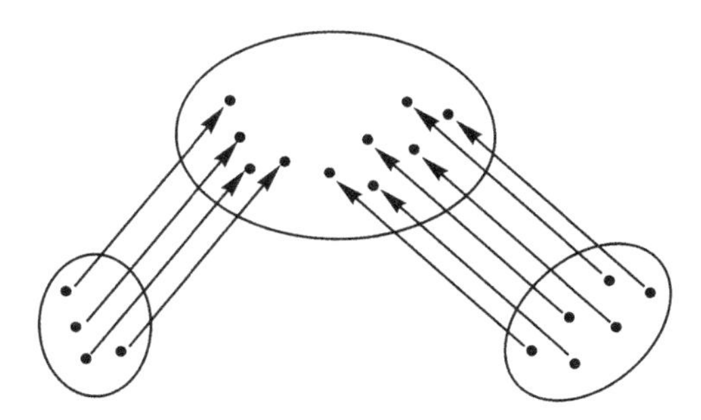

What would you say about the sets A and B if their coproduct is zero?

CHAD: Omer: Both have to be zero.

Right. We see that exactly one sort of set has a negative, namely an initial set, which is its own negative. This leads us to suspect that the same thing might be true in any category: can we prove that A and B must be **0** if $A + B = \mathbf{0}$?

We are assuming that there are injections $A \xrightarrow{j_1} \mathbf{0} \xleftarrow{j_2} B$, such that for *every* object X with maps $A \xrightarrow{f_1} X \xleftarrow{f_2} B$ there is exactly one map $\mathbf{0} \xrightarrow{f} X$ such that $f_1 = fj_1$ and $f_2 = fj_2$. How do we check whether A is an initial object?

OMER: There is one ...

Wrong. You must start ...

DANILO: For each object ...

Yes?

DANILO: For each object X in $\mathcal{C}$ there is exactly one map from A to X.

And how can we check it?

DANILO: To start with, we know that there is exactly one map $f : \mathbf{0} \longrightarrow X$.

Right, so fj_1 is one map from A to X; but we need to show that it is the only map from A to X, to prove that A is initial. Suppose $g : A \longrightarrow X$, and try to prove $g = fj_1$. Since we have $A \xrightarrow{g} X \xleftarrow{fj_2} B$, the universal property of coproduct gives us a map from $\mathbf{0}$ to X (which must be f) such that $fj_1 = g$ and $fj_2 = fj_2$. The second of these equations is uninteresting, but the first is what we needed.

There is another way of seeing the same thing, but treating simultaneously the objects A and B. To say that $\mathbf{0}$ is a coproduct of A and B means that for any object X, the pairs of maps $A \xrightarrow{f_1} X \xleftarrow{f_2} B$ are the same as the maps $\mathbf{0} \longrightarrow X$. There is only one map from $\mathbf{0}$ to X (since $\mathbf{0}$ is initial), therefore there is only one pair of maps $A \xrightarrow{f_1} X \xleftarrow{f_2} B$. Thus there is only one map $A \longrightarrow X$, and only one map $B \longrightarrow X$. This means that both A and B are initial objects. Now we have a complete answer to our question: an initial object has a negative, but only initial objects have negatives.

It is important to mention that, although in the categories we have studied it is trivial that $A + B = \mathbf{0}$ implies '$A = \mathbf{0}$' and '$B = \mathbf{0}$,' we have proved this also in other categories in which it is not nearly so evident. More strikingly, we can shift from the colimits column to the limits colum, thus 'dualizing' this theorem. The dualized statement and proof are obtained by reversing the direction of all the maps in the discussion above, which includes replacing each concept by its dual concept. By doing so we obtain a statement about a product of two objects and a terminal object, and we obtain also the proof of that statement. You should work this out yourself, so we will state it as an exercise.

Exercise 1:
Prove that if A and B are objects and $A \times B = \mathbf{1}$, then $A = B = \mathbf{1}$. More precisely, if $\mathbf{1}$ is terminal and

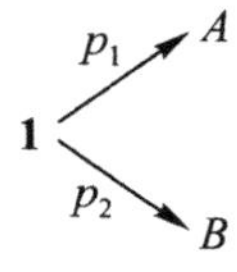

is a product, then A and B are terminal objects.

Our point is that the solution to this exercise is contained in the discussion of $A + B = \mathbf{0}$. You need only take everything that was said there, and

Where it says ...	write instead ...
coproduct	product
$+$	$\times$
$\mathbf{0}$	$\mathbf{1}$
$\longrightarrow$	$\longleftarrow$
$\longleftarrow$	$\longrightarrow$

After this 'translation' is completed you will have the solution to the exercise, so that the *logic* of the two results is the same. Yet, in some of the examples the second result is less obvious than the first. For example, in the category of graphs we found instances of products which were 'smaller' than one of the factors, as in the case of $A \times D = 2D$:

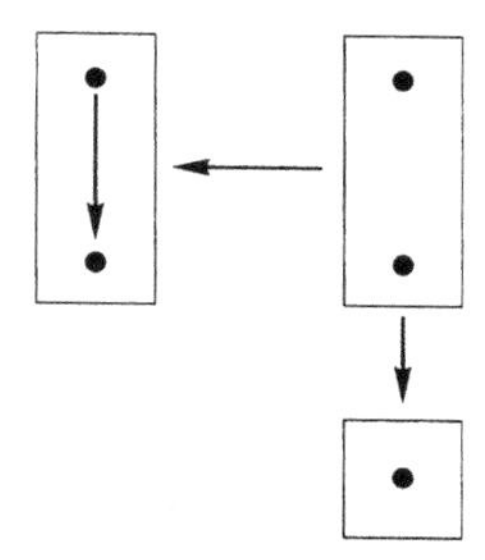

3. Idempotent objects

Let's look for objects C for which '$C \times C = C$.' This asks: Which objects C have maps

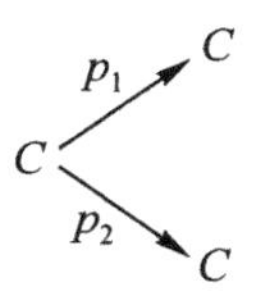

that are product projections? The question becomes more precise if the maps p_1 and p_2 are given. Let's ask for those objects C such that taking p_1 and p_2 both equal to the identity of C we get a product. That means that for any object X and any maps

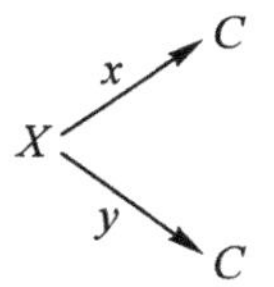

there is exactly one map $X \xrightarrow{f} C$ such that this diagram commutes:

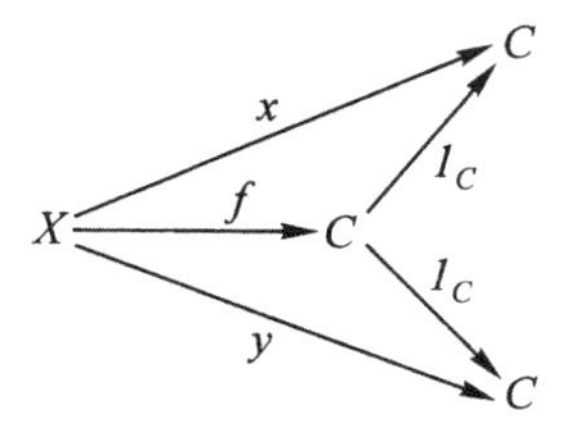

i.e. such that $1_C f = x$ and $1_C f = y$. This obviously implies $x = y$, so that any two maps from any object to C must be equal! That is: If

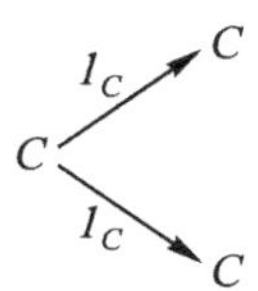

is a product, then for each X there is *at most one* map $X \longrightarrow C$. In fact, the converse is also true:

Exercise 2:

(a) Show that if C has the property that for each X there is at most one map $X \longrightarrow C$, then

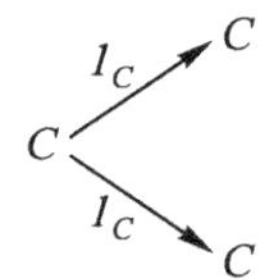

is a product.

(b) Show that the property above is also equivalent to the following property: The unique map $C \longrightarrow \mathbf{1}$ is a monomorphism.

Can you think of some examples?

DANILO: The empty set.

Yes. It seems that might work. Suppose that we have two maps from a set X to an empty set $\mathbf{0}$ ...

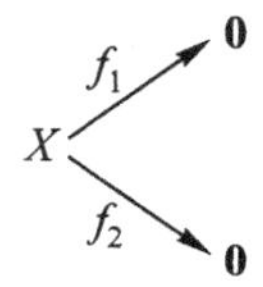

OMER: Then X is also empty.

Good. If there is any map at all from X to an empty set, then X must be empty.

CHAD: So there is at most one map $X \longrightarrow \mathbf{0}$, since if X is not empty there is none, and if X is empty there is exactly one.

Good. Here is an exercise about these objects.

Exercise 3:
Find all objects C in $\mathbf{S}$, $\mathbf{S}^{\circlearrowright}$, and $\mathbf{S}^{\downdownarrows}$ such that this is a product:

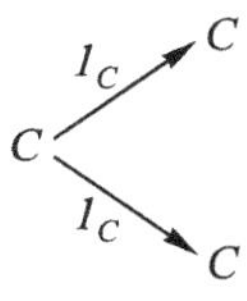

DANILO: Are there examples with $C \times C$ isomorphic to C, but for which the projections are not identity maps?

Yes. One of the most interesting examples was discussed by Cantor (about whom we will have more to say later.) the set $\mathbb{N}$ of natural numbers does have a pair of maps

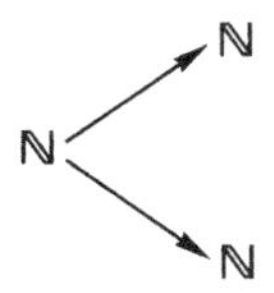

which form a product, but they are not the identity map. To find suitable 'projection' maps, picture $\mathbb{N} \times \mathbb{N}$ as a set of pairs in the way we usually have pictured products:

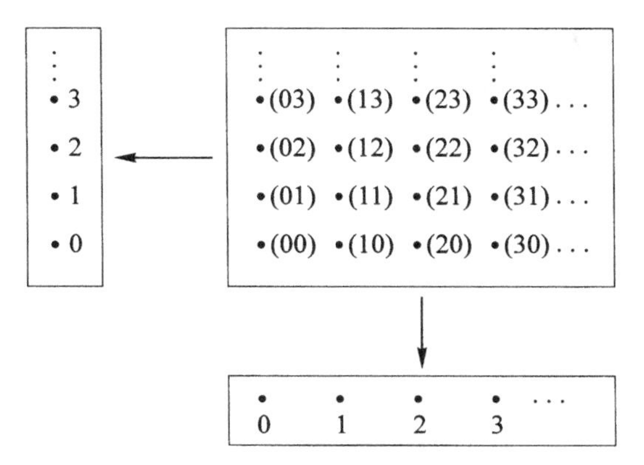

Now define an isomorphism $\mathbb{N} \xrightarrow{f} \mathbb{N} \times \mathbb{N}$ by making repeated 'northwest treks' through the elements of $\mathbb{N} \times \mathbb{N}$ as indicated in this figure:

9
5 8
2 4 7
0 1 3 6

That is $f(0)=(0,0)$, $f(1)=(1,0)$, $f(2)=(0,1)$, $f(3)=(2,0)$, $f(4)=(1,1)$, etc. Composing this isomorphism with the usual projection maps gives the two maps $\mathbb{N} \longrightarrow \mathbb{N}$ we wanted.

Exercise 4:
The inverse, call it g, of the isomorphism of sets $\mathbb{N} \xrightarrow{f} \mathbb{N} \times \mathbb{N}$ above is actually given by a quadratic polynomial, of the form

$$g(x,y) = \tfrac{1}{2}(ax^2 + bxy + cy^2 + dx + ey)$$

where a, b, c, d, and e are fixed natural numbers. Can you find them? Can you prove that the map g defined by your formula is an isomorphism of sets? You might expect that f would have a simpler formula than its inverse g, since a map $\mathbb{N} \xrightarrow{f} \mathbb{N} \times \mathbb{N}$ amounts to a pair of maps $f_1 = p_1 f$ and $f_2 = p_2 f$ from $\mathbb{N}$ to $\mathbb{N}$. But f_1 and f_2 are not so simple. In fact, no matter *what* isomorphism $\mathbb{N} \xrightarrow{f} \mathbb{N} \times \mathbb{N}$ you choose, f_1 cannot be given by a polynomial. Can you see why?

4. Solving equations and picturing maps

The general notions of limit and colimit are discussed in books on geometry, algebra, logic, etc., where category theory is explicitly used. While the special case of products extracts a single object from a given family of objects, the more general constructions extract a single object from a given diagram involving both objects and maps. An important example is a diagram of shape $\boxed{\bullet \rightrightarrows \bullet}$: two given objects and two given maps between them. (We call this a 'parallel pair' of maps.) To understand how the universal construction of limit applies to diagrams of that shape, consider first the notion of 'solution of an equation.' If $fx = gx$ in the diagram $T \xrightarrow{x} X \underset{g}{\overset{f}{\rightrightarrows}} Y$, we say that x is a solution of the equation $f \overset{?}{=} g$. It is not usually the case that $f = g$ (if it were, then all x would be solutions). Now we ask for a *universal* solution for a given pair f, g, meaning one which 'includes' all other solutions in a unique way.

Definition: $E \xrightarrow{p} X$ *is an* **equalizer** *of* f, g *if* $fp = gp$ *and for each* $T \xrightarrow{x} X$ *for which* $fx = gx$, *there is exactly one* $T \xrightarrow{e} E$ *for which* $x = pe$.

Exercise 5:
If both E, p and F, q are equalizers for the same pair f, g, then the unique map $F \xrightarrow{e} E$ for which $pe = q$ is an isomorphism.

Exercise 6:
Any map p which is an equalizer of some pair of maps is itself a monomorphism (i.e. injective).

Exercise 7:
If $B \xrightarrow{\alpha} A \xrightarrow{\beta} B$ compose to the identity $1_B = \beta\alpha$ and if f is the idempotent $\alpha\beta$, then α is an equalizer for the pair f, 1_A.

Exercise 8:
Any parallel pair $X \underset{g}{\overset{f}{\rightrightarrows}} Y$ of maps in sets, no matter how or why it occurred to us in the first place, can always be imagined as the source and target structure of a graph. In a graph, which are the arrows that are named by the equalizer of the source and target maps?

Another use of the word 'graph' which is very important in mathematics and elsewhere is to describe a certain kind of picture of the detailed behavior of a particular function, a picture that can be derived from the following in those cases where we can picture the cartesian product $X \times Y$ (e.g. as rectangle when X and Y separately are pictured as lines).

Consider the projection p_X to the first factor X from a product $X \times Y$. Any section of p_X will yield, by composition with the other projection, a map $X \longrightarrow Y$.

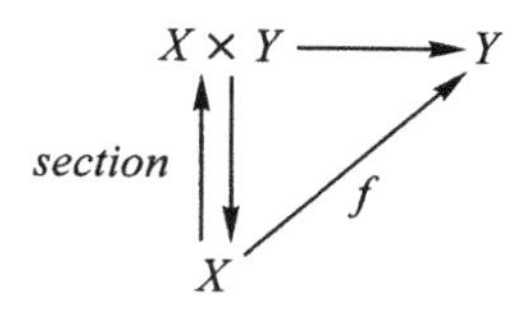

The universal property of products shows that this passage from sections of p_X to maps $X \longrightarrow Y$ can be inverted:

Exercise 9:
For any map $X \xrightarrow{f} Y$ there is a unique section Γ of p_X for which $f = p_Y\Gamma$, namely $\Gamma = \langle ?, f \rangle$.

This section Γ is called the *graph of f*. Like all sections, the graph of a map is a monomorphism, and hence can be pictured as a specific part of $X \times Y$, once we have a way of picturing the latter. 'Parts' are discussed in more detail in Part V, as is another important limit construction, known as 'intersection.'

Exercise 10:
Given two parallel maps $X \overset{f}{\underset{g}{\rightrightarrows}} Y$ in a category with products (such as $\mathcal{S}$), consider their graphs Γ_f and Γ_g. Explain pictorially why the equalizer of f, g is isomorphic to the intersection in $X \times Y$ of their graphs.

The internal diagrams of particular maps f which we frequently use in this book are pictures of the 'cograph' of f, rather than of the graph of f; for example they contain the sum $X + Y$ rather than being contained in $X \times Y$. Try to dualize the definition of graph of f to obtain the precise definition of 'cograph' of f. Try also to dualize the definition of equalizer to obtain the notion of 'coequalizer,' and explain why, when parallel maps in $\mathcal{S}$ are viewed as *source/target* structure, the coequalizer becomes the 'set of components' of the graph.

Exercise 11:
Say that $Y \xrightarrow{h} Z$ is a *cosolution* of the co-equation represented by a given source/target structure $X \overset{f}{\underset{g}{\rightrightarrows}} Y$ if $hf = hg$. Show that if such a cosolution h is universal, in the sense that any other cosolution $Y \xrightarrow{h'} Z'$ can be uniquely expressed as $h' = qh$, then h is an epimorphism. (A universal cosolution is called a *coequalizer* of the pair f, g; in many categories every epimorphism is a coequalizer of some pair.)

Exercise 12:
For a given map $Y \xrightarrow{h} Z$, consider all parallel pairs $X \overset{f}{\underset{g}{\rightrightarrows}} Y$ (for various X) such that $hf = hg$. Formulate the notion of a universal such; call it X_h. Show that $X_h \rightrightarrows Y$ is reflexive, symmetric, transitive, and jointly monomorphic. Here 'reflexive' you know from our discussion of directed graphs, 'symmetric' means there is an involution σ of X_h whose right action interchanges the universal f and g. 'Jointly monomorphic' means the map $X \longrightarrow Y \times Y$ with label $\langle f, g \rangle$ is injective. 'Transitivity' involves a trio of test maps $T \longrightarrow Y$. (An STM reflexive graph is called an *equivalence relation* on Y; in many categories every equivalence relation arises as the universal X_h for some h.)

SESSION 28

The category of pointed sets

1. An example of a non-distributive category

The various categories of dynamical systems and of graphs which we have discussed all satisfy the distributive law. A simple, frequently occurring, example of a category that is not distributive is $\mathbf{1}/\mathcal{S}$, the *category of pointed sets*. An *object* of this category is a set X together with a chosen *base point*, or *distinguished point*, $\mathbf{1} \xrightarrow{x_0} X$. We can picture an object $\mathbf{1} \xrightarrow{x_0} X$ of this category as

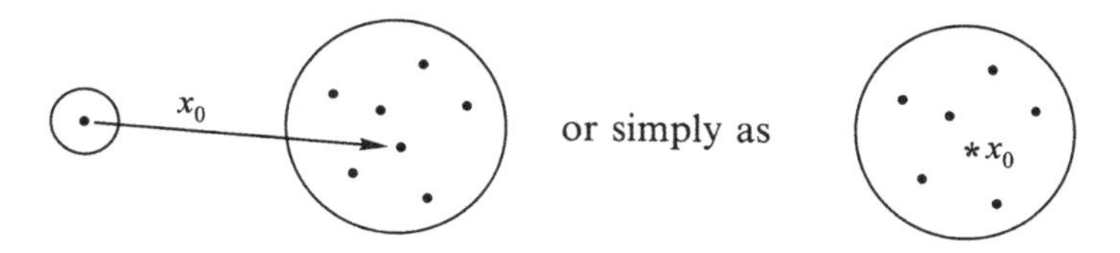

A distinguished point is a very simple kind of structure in a set. What should a map be in this category? A 'map that preserves the structure' seems to suggest just a map of sets that takes the base point of the domain into the base point of the codomain, so we take that as our definition. A *map* in $\mathbf{1}/\mathcal{S}$ from a set X with base point $\mathbf{1} \xrightarrow{x_0} X$ to a set Y with base point $\mathbf{1} \xrightarrow{y_0} Y$ is any map of sets $X \xrightarrow{f} Y$ such that $fx_0 = y_0$. This is the same as saying that the diagram below commutes:

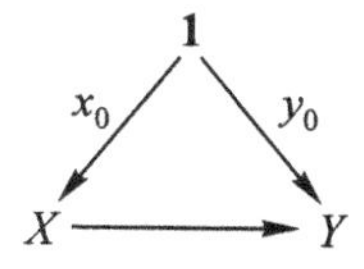

The internal diagram of such a map looks like this:

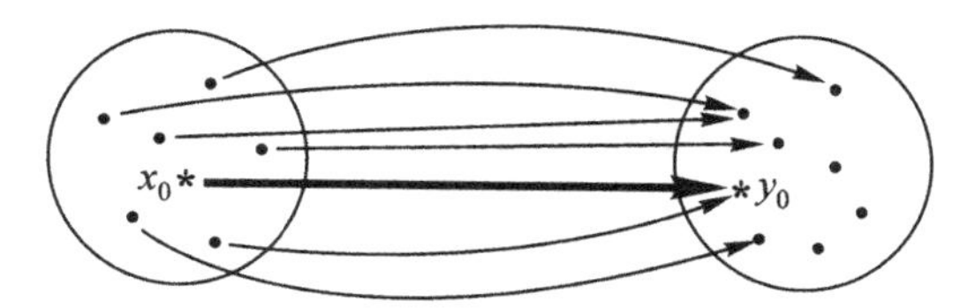

The base point of the domain is mapped to the base point of the codomain, while the other points can be mapped to any points of the codomain, including the base point.

Now that I have told you what the objects and the maps are, I hope you can complete the job of describing this category. Decide how to compose maps (being careful that the composite of maps is a map!), decide what the identity maps should be, and then check that the identity and associative laws are true.

The base point is sometimes called the 'origin' or 'preferred point', but in some applications in computer science, it is referred to as the 'garbage point.' Then, the fact that the maps in this category preserve the base point is expressed by the colorful phrase: 'Garbage in, garbage out.' Sometimes, even if the input isn't garbage, the result is garbage. The base point in the codomain serves as the recipient of all the garbage results of a particular map. This is useful because some processes for calculation have the property that for some inputs the process does not produce an output. In this category you won't have a problem because every codomain has a distinguished element where you can send any input whose image is undetermined.

Now, can anybody guess what the terminal object of this category is?

ALYSIA: One element?

Right. A set with just one element, in which that one element is the base point. It is easy to prove that this is really a terminal object, since for every set with base point, there is exactly one map to a one-element set, and obviously this map must preserve the base point so that it is indeed a map in this category.

What about an initial object?

DANILO: Also just a base point?

Yes! Every object in this category must have at least one point, otherwise it can hardly have a distinguished one. Now, a set with only one point (with that point taken as base point) is clearly initial, since to map it to any object you must send its only point to the base point of that object. Thus in this category we can write '$\mathbf{0} = \mathbf{1}$'! (This should not be too surprising, since we saw in Session 26 that all linear categories also have $\mathbf{0} = \mathbf{1}$.)

DANILO: So, the empty set is not an object of this category?

That's right. It doesn't have a point to be chosen as base point, so it cannot be made into a pointed set.

In this category the unique map $\mathbf{0} \longrightarrow \mathbf{1}$ is an isomorphism, and according to what was said in Session 26 this category has zero maps. What about products? Is there a product of the two pointed sets

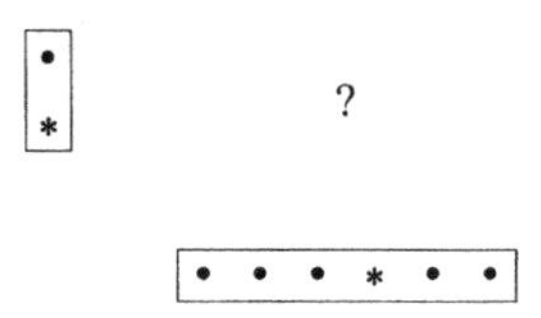

If we just calculate the product as sets we get

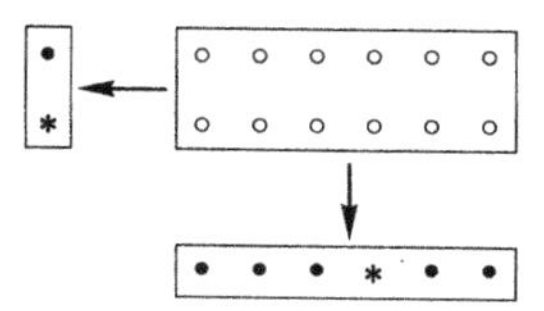

Is it possible to choose one point as base point in this product set so that it is preserved by the projections? For that it would have to be a point in the same row as the base point of the set on the left, and in the same column as the base point of the set on the bottom, so that the only choice is

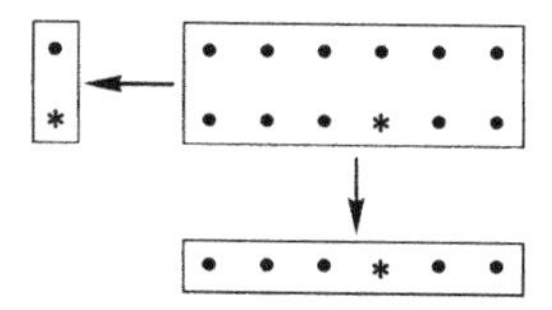

This indeed works as the product. The proof is not difficult since we are using as product a set that is a product in the category of sets.

What about sums? What pointed set can be used as the sum of the following two pointed sets?

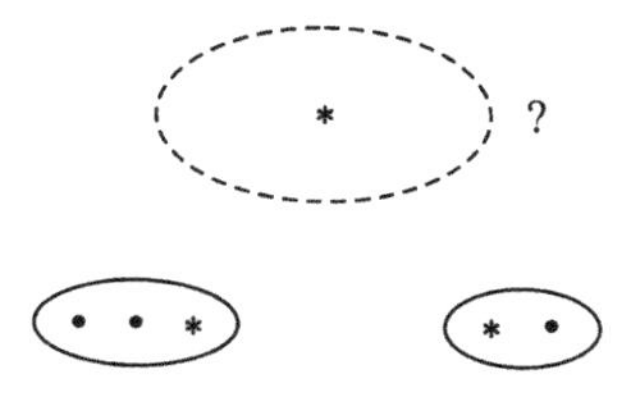

We have to choose maps that preserve the base point, so they have to be something like this:

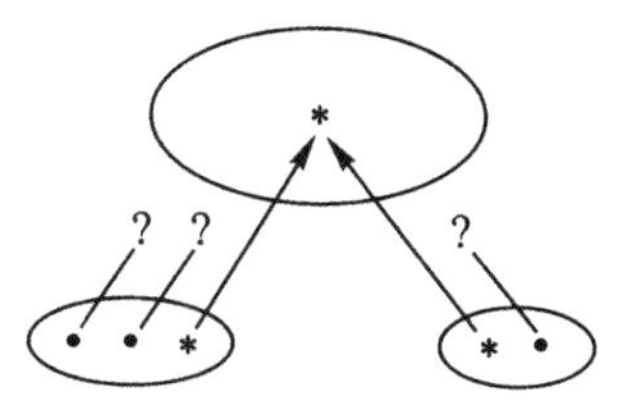

and I leave it for you to prove that the coproduct is this:

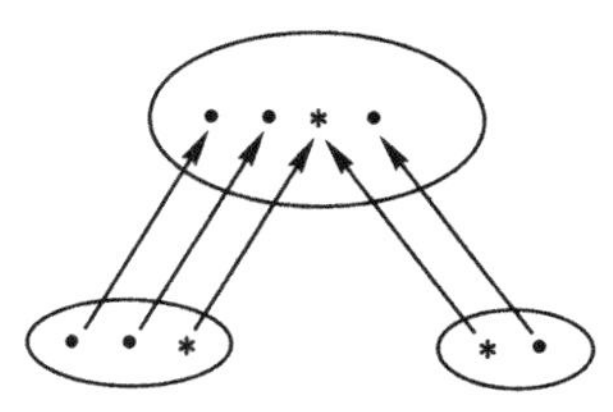

In this category the operation of coproduct consists in 'glueing by the base point.'

Exercise 1 below is closely related to Exercises 8, 9, and 20 of Article IV.

Exercise 1:
Both parts of the distributive law are false in the category of pointed sets:
(a) Find an object A for which the map

$$\mathbf{0} \longrightarrow \mathbf{0} \times A$$

is not an isomorphism.
(b) Find objects A, B_1, and B_2 for which the standard map

$$A \times B_1 + A \times B_2 \longrightarrow A \times (B_1 + B_2)$$

is not an isomorphism.

Exercise 2:
As we saw, in the category of pointed sets, the (only) map $\mathbf{0} \longrightarrow \mathbf{1}$ is an isomorphism. Show that the other clause in the definition of *linear category* fails, i.e. find objects A and B in $\mathbf{1}/\boldsymbol{S}$ for which the 'identity matrix'

$$A + B \longrightarrow A \times B$$

is not an isomorphism.

Test 3

1. Prove: If $\mathbf{1}$ is a terminal object, and X is any object, then *any* map $\mathbf{1}\xrightarrow{f} X$ is a *section* of a (the?) map $X \xrightarrow{g} \mathbf{1}$

2. Prove: If $\mathbf{1}$ is a terminal object and C is any object,then '$C \times \mathbf{1} = C$.' (First you should explain what '$C \times \mathbf{1} = C$' means. To get you started, you should *decide* what maps p_1 and p_2 should be the 'projection maps' in

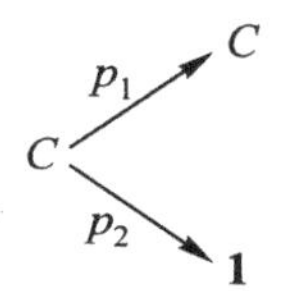

After you have *chosen* the particular maps p_1 and p_2, you must prove that they satisfy the correct 'universal property'.)

3. In $\boldsymbol{S}^{\downarrow\downarrow}$, the category of irreflexive graphs, 'find' $A \times A \times A$.
Express your answer in two ways:
   (a) draw a *picture* of $A \times A \times A$;
   (b) find numbers m and n such that $A \times A \times A \cong mD + nA$.
(The symbol $\cong$ means 'is isomorphic to'.)

Notes:
1. Recall that A is $\boxed{\bullet \longrightarrow \bullet}$
2. You may use the 'distributive law': $B \times C_1 + B \times C_2$ is isomorphic to $B \times (C_1 + C_2)$.

Test 4

1. Show that if $B \times C = \mathbf{1}$, then $B = \mathbf{1}$.
Your demonstration should work in any category.

Hint: First explain what '$B \times C = \mathbf{1}$' means!

2. All parts of this problem are in $\boldsymbol{S}^{\downarrow\downarrow}$, the category of irreflexive graphs.

$D = \boxed{\bullet}$ $A = \boxed{\bullet \longrightarrow \bullet}$ $B = \boxed{\circlearrowright\bullet \longrightarrow \bullet}$ $C = \boxed{\bullet \longleftarrow \bullet \longrightarrow \bullet}$

(a) Find the number of maps $\mathbf{1} \longrightarrow B + D$ and the number of maps $\mathbf{1} \longrightarrow C$.
(b) 'Calculate' $A \times B, A \times D$, and $A \times C$.
(Draw pictures – internal diagrams – of them.)
(c) Use the distributive law, and results from (b), to calculate

$$A \times (B + D)$$

(d) Show that $A \times (B + D)$ is isomorphic to $A \times C$.

Note: Comparing (a) and (d) illustrates the failure of 'cancellation':
From '$A \times (B + D) = A \times C$' we *cannot* cancel A and conclude that '$B + D = C$.'

Test 5

1. Find as many graphs with exactly 4 dots and 2 arrows as you can, with no two of your graphs isomorphic. (Draw an internal diagram of each of your graphs.)

Example: [• ⟶ • ⟵ •   •]

Hint: The number of such graphs is between 10 and 15.

2. $D =$ [•] $A =$ [• ⟶ •] $I =$ [• ⟶ • ⟶ •]

Find numbers a, b, c such that

$$I \times I = aD + bA + cI$$

Hint: First try to draw $I \times I$

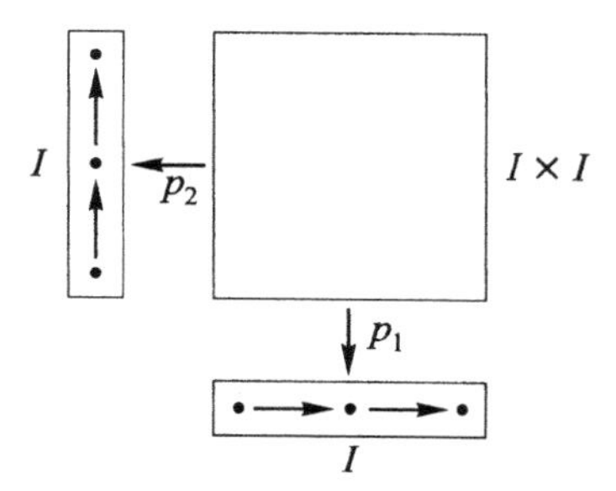

To *check* your picture, be sure that the two projection maps are *maps of graphs*!

SESSION 29

Binary operations and diagonal arguments

Objects satisfying universal mapping properties are in a sense trivial if you look at them from one side, but not trivial if you look at them from the other side. For example, maps from an object to the terminal object **1** are trivial; but if, after establishing that **1** is a terminal object, one counts the maps whose domain is **1**, $\mathbf{1} \longrightarrow X$, the answer gives us valuable information about X. A similar remark is valid about products. Mapping into a product $B_1 \times B_2$ is trivial in the sense that the maps $X \longrightarrow B_1 \times B_2$ are precisely determined by the pairs of maps $X \longrightarrow B_1$, $X \longrightarrow B_2$ which we could study without having the product. However, specifying a map $B_1 \times B_2 \longrightarrow X$ usually cannot be reduced to anything happening on B_1 and B_2 separately, since each of its values results from a specific 'interaction' of the two factors.

1. Binary operations and actions

In this session we will study two important cases of mapping a product to an object. The first case is that in which the three objects are the same, i.e. maps $B \times B \longrightarrow B$. Such a map is called a *binary operation* on the object B. The word 'binary' in this definition refers to the fact that an input of the map consists of two elements of B. (A map $B \times B \times B \longrightarrow B$, for which an input consists of three elements of B, is a *ternary operation* on B, and *unary operations* are the same as endomaps.)

Examples of binary operations are found among the operations of arithmetic. For example, if N is a number system (such as the natural numbers or the real numbers) the addition of numbers in N is a binary operation on N, that is, a map $N \times N \xrightarrow{+} N$. Given a pair of numbers, $\mathbf{1} \xrightarrow{\langle n,m \rangle} N \times N$, their sum is the composite

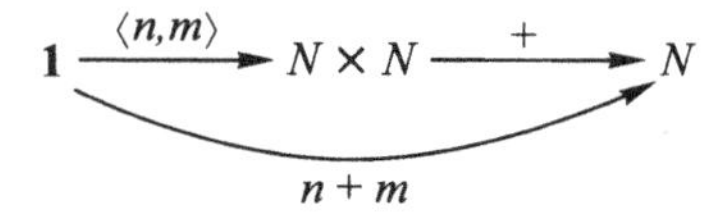

and the same can be said about multiplication $N \times N \xrightarrow{\cdot} N$. There would be no way of thinking of addition as one map if we could not form the cartesian product $N \times N$. An internal picture of the map 'addition' in the case of natural numbers is this:

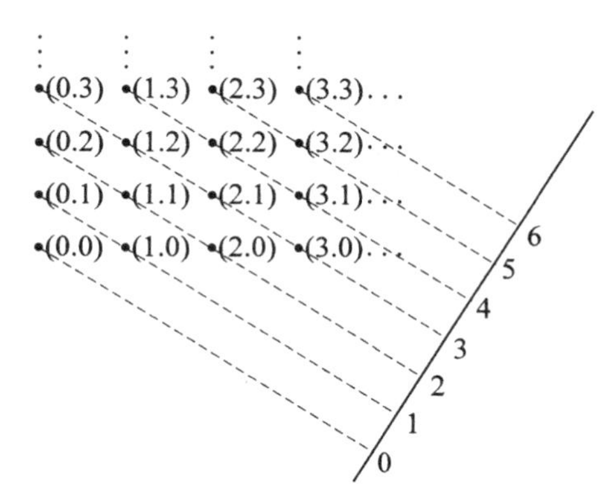

Of course, there is a lot to say about binary operations. They form a category in their own right, as we have seen in Session 4, and are the subject of much study.

Another important case of a mapping with domain a product is a map $X \times B \longrightarrow X$. Such a map is called an *action* of B on X. One can think of B as a set of available buttons that control the states in X, and of the given action $X \times B \xrightarrow{\alpha} X$ as an automaton; a particular button $\mathbf{1} \xrightarrow{b} B$ gives rise to an endomap of X, namely $\alpha(-,b)$. That is, for each element x of X its image is $\alpha(x,b)$, a new element of X. The endomap of X that is determined by $\mathbf{1} \xrightarrow{b} B$ can be understood as the composite of two maps

$$X \longrightarrow X \times B \xrightarrow{\alpha} X$$

of which the first is the graph of the 'constant map equal to b.' 'Pressing' the button b once changes a particular state x into the state $\alpha(x,b)$; pressing it twice changes x into $\alpha(\alpha(x,b),b))$, etc.

On the other hand, we can press a different button. Thus, an action involves not one endomap only, but many endomaps $\alpha(-,b_1)$, $\alpha(-,b_2),\ldots$, one for each element of B. Not only that, we can press one button and then press another; if the system is in state x and we press button b_1 and then button b_2, the resulting state will be $\alpha(\alpha(x,b_1),b_2)$ so that $\alpha(\alpha(-,b_1),b_2)$ is a new endomap of X. Similarly, any finite sequence of elements of B gives an endomap.

2. Cantor's diagonal argument

The most general case of a map whose domain is a product has all three objects different:

$$T \times X \xrightarrow{f} Y$$

Again each point $\mathbf{1} \xrightarrow{x} X$ yields a map

$$T \xrightarrow{f(-,x)} Y$$

so that f gives rise to a family of maps $T \longrightarrow Y$, one for each point of X, or as we often say, a family *parameterized* by (the points of) X, in this case a family of *maps* $T \longrightarrow Y$. As we will see in Part V, in the category of sets for each given pair T, Y of sets, there is a set X big enough so that for an appropriate *single* map f, the maps $f(-, x)$ give *all* maps $T \longrightarrow Y$, as x runs through the points of X. Such a set X tends to be rather large compared to T and Y; for example, if T has three elements and Y has five elements, then it would be necessary to take X with $5^3 = 125$ elements because that is the number of maps $T \longrightarrow Y$; we will later call an appropriate map f an 'evaluation' map. One might think that if T were infinite, we would not need to take X 'bigger'; however, that is wrong, as shown by a famous theorem proved over one hundred years ago by Georg Cantor†: T itself (infinite or not) is essentially *never* big enough to serve as the domain of a parameterization of *all* maps $T \longrightarrow Y$!

Diagonal Theorem: (In any category with products) *If Y is an object such that there exists an object T with enough points to parameterize all the maps $T \longrightarrow Y$ by means of some single map $T \times T \xrightarrow{f} Y$, then Y has the 'fixed point property': every endomap $Y \xrightarrow{\alpha} Y$ of Y has at least one point $\mathbf{1} \xrightarrow{y} Y$ for which $\alpha y = y$.*

Proof: Assume Y, T, f, and α given. Then there is the diagonal map $T \longrightarrow T \times T$ as always (which maps every element t to $\langle t, t\rangle$), so we can form the three-fold composite g:

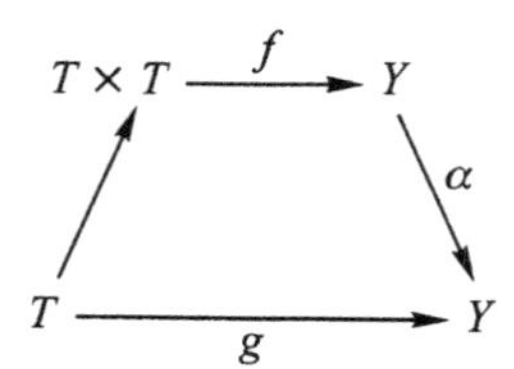

This new map by its construction satisfies

$$g(t) = \alpha(f(t, t))$$

for every point t of T. We have assumed that *every* map $T \longrightarrow Y$ is named as $f(-, x)$ for some point $\mathbf{1} \xrightarrow{x} T$, and g is such a map $T \longrightarrow Y$. So let $x = t_0$ be a parameter value corresponding to our g, i.e. $g = f(-, t_0)$, so that

$$g(t) = f(t, t_0)$$

for all t. Taking the special case $t = t_0$, we have

$$g(t_0) = f(t_0, t_0)$$

which by the definition of g says

†Historical note: Georg Cantor (1845–1918), German mathematician who founded set theory and influenced twentieth century topology. His diagonal argument is important in logic and computer science.

$$\alpha(f(t_0, t_0)) = f(t_0, t_0)$$

or, in other words, that $y_0 = f(t_0, t_0)$ defines a point of Y which is fixed by α:

$$\alpha(y_0) = y_0$$

Cantor's proof is called the 'diagonal argument' because of the role of the diagonal map; but the role of the endomap α is clearly equally important in the construction of g from f. That is especially evident if we state the theorem in the form of:

Cantor's Contrapositive Corollary: *If Y is an object known to have at least one endomap α which has no fixed points, then for every object T and for every attempt $f : T \times T \longrightarrow Y$ to parameterize maps $T \longrightarrow Y$ by points of T, there must be at least one map $T \longrightarrow Y$ which is left out of the family, i.e. does not occur as $f(-, x)$ for any point x in T.*

Proof: Use α and the diagonal as above to make f itself produce an example g which f leaves out.

In the category of sets, examples of Y without the 'fixed point property' abound. The simplest is a two-point set; if the points are called '*true*' and '*false*,' then the endomap α without fixed points is 'logical negation.' Applying Cantor's Theorem we can conclude that no map $T \times T \longrightarrow \mathbf{2}$ can parameterize all maps $T \longrightarrow \mathbf{2}$. That is often expressed: For all sets T,

$$T < \mathbf{2}^T$$

where $\mathbf{2}^T$ is a set which *does* parameterize all $T \longrightarrow \mathbf{2}$. Other important examples of such Y are the real numbers or the natural numbers $Y = \mathbb{N}$; if, for example, α is defined by $\alpha(y) = y + 1$ for all y, then α has no fixed points, so

$$T < \mathbb{N}^T$$

for all sets T, where $\mathbb{N}^T$ is a set for which there is a map $T \times \mathbb{N}^T \longrightarrow \mathbb{N}$ which parameterizes all maps $T \longrightarrow \mathbb{N}$. (Such a map is called an *evaluation* map and will be studied more in Part V.) Cantor drew the conclusion that for any infinite set T, there is a whole sequence

$$T < \mathbf{2}^T < \mathbf{2}^{(\mathbf{2}^T)} < \mathbf{2}^{\mathbf{2}^{(\mathbf{2}^T)}} \ldots$$

of genuinely 'more and more infinite' sets.

Exercise 1:

Cantor's proof, if you read it carefully, really tells us a bit more. Rewrite the proof to show that if $T \times T \xrightarrow{f} Y$ *weakly* parameterizes all maps $T \longrightarrow Y$, then Y has the fixed point property. To say that $T \times X \xrightarrow{f} Y$ 'weakly' parameterizes all maps $T \longrightarrow Y$ means that for each $T \xrightarrow{g} Y$ there is a point $\mathbf{1} \xrightarrow{x} X$ such that (letting ξ stand for the map whose components are the identity and the constant map with value x) the composite map $h = f \circ \xi$

$$T \xrightarrow{\xi} T \times X \xrightarrow{f} Y$$

agrees with $T \xrightarrow{g} Y$ on points; i.e. for each point $\mathbf{1} \xrightarrow{t} T$, $g \circ t = h \circ t$. (In the category of sets that says $g = h$; but as we have seen, in other categories it says much less.)

Cantor's Diagonal Theorem is closely related to the famous incompleteness theorem of Gödel, of which you may have heard. The setting for Gödel's theorem is

Subjective categories

Our proof of Cantor's Diagonal Theorem is clearly valid in any category with products. This fact was exploited by Russell around 1900 and by Gödel and Tarski in the 1930s to derive certain results (which are sometimes described in 'popular' books as 'paradoxes'). Gödel's work, in particular, went several steps beyond Cantor's. There is a frequent line of thought which does not begin by focussing on visualizing the possible dynamical systems or the possible graphs, etc. and then trying to understand these objects and their transformations. Instead, this line of thought:

1. starts with formulas and rules of proof and tries to
2. limit consideration only to those maps (or graphs or ...) which can be completely described by a formula, and
3. considers that two maps are equal only when the corresponding formulas can be proved equivalent on the basis of some given rules.

This part of the 'constructive' point of view has led to some advances in mathematics because, objectively, it leads to new examples of categories which are in some ways very similar to the categories of sets, of graphs, etc. but in some ways quite different.

As we will see in more detail in Part V, in most categories the relevant truth values are more than just the two $\{true, false\}$, and in many categories (both objective and subjective) the truth values actually constitute an object Ω in the category itself. In the subjective categories which are derived from formulas and rules of proof as alluded to above, this is due to the fact that truth values $\mathbf{1} \longrightarrow \Omega$ are themselves formulas. (They might, for example, arise by composing

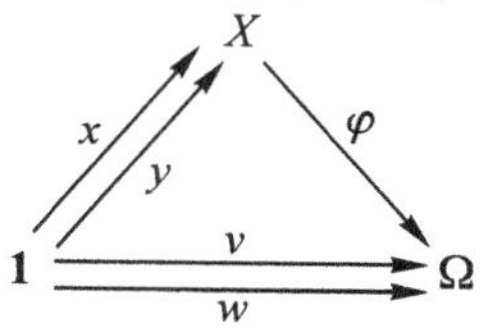

where x, y are formulas naming elements of type X and where φ is a formula naming a property of elements of type X; in linguistics x and y might be noun phrases, φ a predicate, and v and w the two resulting sentences.) If the rules of proof are not sufficient to prove v and w equivalent, then (because of (3) above) in this sort of category $v \neq w$; in particular, v may not be provably equivalent to either *true*: $\mathbf{1} \longrightarrow \Omega$ or *false*: $\mathbf{1} \longrightarrow \Omega$. The results of Gödel and Tarski show that this is quite often the case, i.e. the rules of proof permit four or more inequivalent truth values $\mathbf{1} \longrightarrow \Omega$ in such categories.

How does Cantor's Diagonal Theorem relate to these considerations of Russell, Gödel and Tarski? If T is an object whose elements are numerals, or words, or lists, or formulas, or proofs, or similar 'syntactical' elements, it is often possible to describe maps

$$T \times T \xrightarrow{f} \Omega$$

which do in a sense describe all *describable* properties $T \longrightarrow \Omega$. This is achieved by a device known as 'Gödel numbering' whereby the elements of T play a dual role: on the one hand they are names for the 'things' talked about (such as natural numbers or words syntactically considered as strings of letters) and on the other hand they are names for properties $T \longrightarrow \Omega$. Thus in

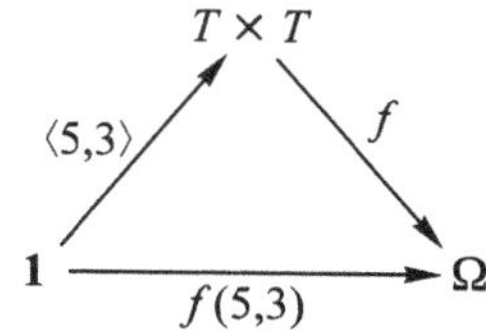

$f(5,3)$ is the sentence saying that the number 5 has property #3.
The key idealization in the latter example is that we imagine that all the properties (describable in a given syntactic scheme), can be listed in a fixed way so that we can speak of 'property #3,' etc. The list might look like

(0) $t^2 = t$
(1) $t^2 = 5t$
(2) $t = 0$
(3) $t + 2 = 7$
(4) $t^2 = t + 1$
(5) $t^3 = t^2 + t$
etc.

in which case we can say $f(5,3) = f(5,1)$ since both are *true*; $f(5,2) = f(5,4)$ also because both are *false*. But further down the list may be a more complicated property #13 for which the rules of proof are insufficient to prove either that 5 has or that it does not have the property. In this case

$$\mathbf{1} \xrightarrow{f(5,13)} \Omega$$

will be a point of Ω different from either *true* or *false*.

Now, since the maps $T \xrightarrow{\varphi} \Omega$ (in a category of the sort being described) are all supposed to be describable by formulas, it is frequently possible to choose the Gödel numbering and a single f such that *every* φ is in a sense representable by some $f(-,x)$. The new sense is crucial:

$$(S) \quad \begin{array}{l} f(t,x) = \textit{true} \text{ if and only if } \varphi(t) = \textit{true} \\ f(t,x) = \textit{false} \text{ if and only if } \varphi(t) = \textit{false} \end{array}$$

for each $t : \mathbf{1} \longrightarrow T$; this is the very weak sense in which φ 'is' property number (x) in the Gödel listing. Still, we cannot say that $f(-,x) = \varphi$ since $f(t,x) \neq \varphi(t)$ may happen for t for which $f(t,x)$ or $\varphi(t)$ is neither *true* nor *false*. Indeed, the Gödel/Tarski result is that there *must* be sentences $\mathbf{1} \longrightarrow \Omega$ which are neither (provably equivalent to) *true* nor *false* in *any* category of the sort we have just described. For if there were only the two points in Ω, then, since for every $T \xrightarrow{\varphi} \Omega$ there is a name $\mathbf{1} \xrightarrow{x} T$ for it in the sense (S), we would actually have $f(t,x) = \varphi(t)$ for all $\mathbf{1} \xrightarrow{t} T$. But that yields the conclusion of Cantor's diagonal theorem, which contradicts another feature of such categories: there is an endomap

$$\Omega \circlearrowleft^{not}$$

(which exchanges *true* and *false*) having *no* fixed points.

The 'constructivist' (also known as 'formalist,' 'intuitionist,' etc. in variants) idealization 'imagine a listing of all formulas and all proofs' is quite reasonable, provided one has already accepted the idealization 'imagine an object $\mathbb{N}$ whose elements $\mathbf{1} \longrightarrow \mathbb{N}$ are precisely all the natural numbers 0, 1, 2, ...' Like all serious idealizations, these lead to very interesting theories which might also be relevant some day. However, there is no evidence that anything exists in the real world which much resembles this particular idealization; all attempts to continue the '...' eventually *stop*, sometimes with the comment 'We could *imagine* going on ....' There is a widespread misconception that the lack of a real counterpart to the idea $\mathbb{N}$ is due to $\mathbb{N}$ being *infinite*. On the contrary, Cantor showed that the much 'larger' infinity $\mathbf{2}^{\mathbb{N}}$ is isomorphic as an abstract set with the idealization 'imagine the set of all points in this room.' The latter is an idealization of something that we regard as really there, though of course we can't 'list' all the points in this room by any syntactic process.

Regarding the scientific process of idealization, another great achievement of our old friend Galileo should be kept in mind. There are two equally important aspects. The idealization itself often consists of assuming that, of the many forces acting in a situation, one 'main' force is the only force. In Galileo's investigation of falling bodies, this one force is gravity. Such idealization can lead to very far-reaching development of theory; in the example of gravity, it led Galileo, Newton, Jacobi, Hamilton, Einstein, and others to theory which is constantly used in terrestrial and celestial navigation. In the cases discussed above, the one force idealized is (Brouwer) the urge to keep on counting. But the second equally important aspect of the scientific process of idealization is this: in applying the developed theory to new situations, one must constantly remain conscious of the likelihood that forces other than the 'main' idealized ones are also acting and sometimes becoming 'main' forces themselves.

Galileo knew quite well that if instead of a cannon ball or ball of wood, a dry leaf were dropped from Pisa's tower, friction and wind would be significant forces determining its fall; one might even observe the 'paradox' that sometimes the leaf falls upward, which does not mean that the theory of pure gravity would be wrong, but rather that a more comprehensive pure theory would be better applicable to the case. Since in the development and use of computers and software, forces other than the 'urge to keep on counting' are surely very significant, the beautiful theory of Russell, Brouwer, Tarski, Gödel, Turing (and of more recent logicians and computer scientists) in its pure form has had few applications. Counting is a subjective process, whereas gravity is an objective force. Even when the goal of applications involves a subjective component (such as computing the answer to an engineering problem), objective forces must also be taken into account.

PART V

Higher universal mapping properties

We find that the algebra of exponents comes from the notion of 'map object,' and we explore other universal mapping properties including that of 'truth-value' objects.

ARTICLE V

Map objects

Exponentiation

1. Definition of map object

In a category with products (including **1**) any map

$$T \times X \xrightarrow{f} Y$$

whose domain is a product may be considered as an X-parameterized family of maps $T \longrightarrow Y$. Namely, any point $x : \mathbf{1} \longrightarrow X$ gives rise via f to the map $T \xrightarrow{\langle 1_T, \bar{x} \rangle} T \times X \xrightarrow{f} Y$ (where $\bar{x}$ is the constant map $T \longrightarrow \mathbf{1} \xrightarrow{x} X$), which is often denoted for short by f_x. Thus $f_x(t) = f(t, x)$ for all t. For example, a calculator has a set $X = \{\sqrt{\ }, \log, \ldots\}$ of names of operations and a set T of possible numerical inputs, and a *pair* $\langle t, x \rangle$ must be entered before the calculation f can produce an output. For a given pair T, Y of objects, a randomly chosen X, f will fail to parameterize 'perfectly' the maps $T \longrightarrow Y$ in that

(a) there may be a map $T \longrightarrow Y$ which is not expressible by the given f, no matter what point x is chosen, and

(b) two different points $\mathbf{1} \overset{x}{\underset{x'}{\rightrightarrows}} X$ may name via f the same map $T \longrightarrow Y$.

However, a universal choice may be possible.

Definition: *Given two objects T, Y in a category with products, an object M together with a map $T \times M \xrightarrow{e} Y$, is an* **object of maps from T to Y with evaluation map**, *provided M and e satisfy: For each object X and each map $T \times X \xrightarrow{f} Y$, there is exactly one map, to be denoted $X \xrightarrow{\ulcorner f \urcorner} M$, for which $f = e(1_T \times \ulcorner f \urcorner)$*

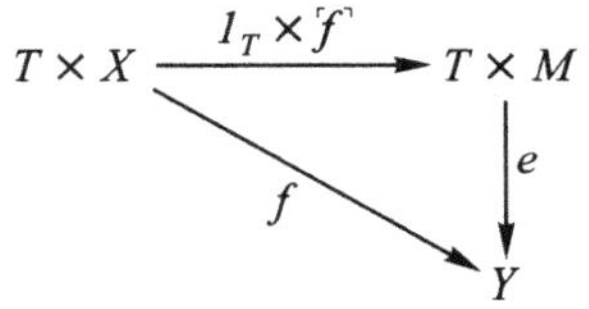

$$\begin{array}{ccc} T \times X & \xrightarrow{1_T \times \ulcorner f \urcorner} & T \times M \\ & \searrow^{f} & \downarrow e \\ & & Y \end{array}$$

i.e. for which $f(t, x) = e(t, \ulcorner f \urcorner (x))$ for all $S \xrightarrow{t} T$, $S \xrightarrow{x} X$.

Notation: The map $\ulcorner f \urcorner$, uniquely determined by f, is sometimes called the 'name of f.' The map e is called the *evaluation* map. Because of the uniqueness of map objects (Exercise 1 below) we can give M also a special symbol: call it Y^T with $T \times Y^T \xrightarrow{e} Y$. Now our 'exactly one' condition on e is abbreviated as:

$$T \times Y^T \xrightarrow{e} Y \quad \textit{induces} \quad \frac{X \longrightarrow Y^T}{T \times X \longrightarrow Y}$$

Map objects are also called 'function spaces.'

To master the idea of map objects, it is helpful to compare the definition with that of product. In both cases the universal property says that a certain simple process is invertible.

For products: Given any P with a pair of maps $P \xrightarrow{p_1} A$, $P \xrightarrow{p_2} B$, we can assign to each map $X \xrightarrow{f} P$ the pair $X \xrightarrow{p_1 f} A$, $X \xrightarrow{p_2 f} B$. Such a P with $P \xrightarrow{p_1} A$, $P \xrightarrow{p_2} B$ is a product with projection maps, if for each X this assignment process is invertible.
For map objects: Given any M with $T \times M \xrightarrow{e} Y$, we can assign to each map $X \xrightarrow{g} M$ the map $\hat{g}$ given by $T \times X \xrightarrow{1_T \times g} T \times M \xrightarrow{e} Y$. Such an M with $T \times M \xrightarrow{e} Y$ is a map object with evaluation map, if for each X this assignment process is invertible.

Exercise 1:
Formulate and prove a uniqueness proposition to the effect that if M_1, e_1 and M_2, e_2 both serve as map objects with evaluation map for maps $T \longrightarrow Y$, then there is a unique isomorphism between them which is compatible with the evaluation structures.

Exercise 2:
(Taking $X = \mathbf{1}$) The points of Y^T correspond to the maps $T \longrightarrow Y$.

Exercise 3:
Prove that

$$Y^{T \times S} \cong (Y^T)^S$$
$$Y^{\mathbf{1}} \cong Y$$

Exercise 4:
Prove that

$$Y^{T_1 + T_2} \cong Y^{T_1} \times Y^{T_2}$$
$$Y^0 \cong \mathbf{1}$$

if the category has sums and initial object, and if the indicated map objects exist. Therefore

$$Y^{\mathbf{1}+\mathbf{1}} \cong Y \times Y \text{ etc.}$$

Exercise 5:
Prove that

$$(Y_1 \times Y_2)^T \cong Y_1^T \times Y_2^T$$
$$\mathbf{1}^T \cong \mathbf{1}$$

Exercise 6:
In a category with products in which map objects exist for any two objects, there is for any three objects a standard map

$$B^A \times C^B \xrightarrow{\gamma} C^A$$

which represents composition in the sense that

$$\gamma\langle \ulcorner f \urcorner, \ulcorner g \urcorner \rangle = \ulcorner gf \urcorner$$

for any $A \xrightarrow{f} B \xrightarrow{g} C$.

2. Distributivity

Though many categories have products and 'sums,' only a fortunate few have map objects. Such categories are often called 'cartesian closed' categories, and automatically have further strong properties, some of which do not even refer directly to the map objects:

Proposition:
If sums exist in $\mathcal{C}$, and T is an object such that map objects Y^T exist for all objects Y, then $\mathcal{C}$ satisfies the distributive law for multiplication by T.

Sketch of proof: We need an inverse map $T \times (B_1 + B_2) \xrightarrow{?} T \times B_1 + T \times B_2$ for the standard map. The desired inverse can be found through the chain of invertible correspondences coming from universal mapping properties (UMP):

$$\begin{array}{c} T \times (B_1 + B_2) \longrightarrow T \times B_1 + T \times B_2 \\ \hline B_1 + B_2 \longrightarrow (T \times B_1 + T \times B_2)^T \\ \hline B_1 \longrightarrow (T \times B_1 + T \times B_2)^T, \; B_2 \longrightarrow (T \times B_1 + T \times B_2)^T \\ \hline T \times B_1 \longrightarrow T \times B_1 + T \times B_2, \; T \times B_2 \longrightarrow T \times B_1 + T \times B_2 \end{array} \quad \begin{array}{l} \text{UMP of map objects} \\ \text{UMP of the sum } B_1 + B_2 \\ \text{UMP of map objects (twice)} \end{array}$$

where in the last line we can choose the *injections* for the big sum. Feeding these injections in at the bottom and applying the three correspondences which are indicated by the horizontal lines, we get at the top a map with the desired domain and codomain. To show that it is really inverse to the standard distributivity map, one need only note that at each correspondence the map obtained is the only one satisfying appropriate equations involving injections, projections, and evaluations, and that both the identity map and the composition of the standard map with the 'inverse' satisfy the same equations. The other clause of the distributive law is proved similarly: To find an inverse for the map $\mathbf{0} \longrightarrow T \times \mathbf{0}$, run in reverse the correspondence

$$\frac{T \times \mathbf{0} \xrightarrow{?} \mathbf{0}}{\mathbf{0} \longrightarrow \mathbf{0}^T}$$

and verify that the result really is the desired inverse.

3. Map objects and the Diagonal Argument

Cantor's Diagonal Argument (see Session 29) is often used in comparing the 'sizes' of map objects; first note how the result itself can be slightly reformulated in the special case of a cartesian closed category.

Theorem (Cantor's Diagonal Argument) *Suppose Y is an object in a cartesian closed category, such that there exists an object T and a map $T \xrightarrow{f} Y^T$ which is 'onto' in the sense that for every map $T \xrightarrow{g} Y$ there exists a point of t of T such that $\ulcorner g \urcorner = ft$. Then every endomap of Y has a fixed point. Therefore, (contrapositive) if Y is known to have at least one endomap which has no fixed points, then for every object T, every map $T \longrightarrow Y^T$ fails to be onto ('$T < Y^T$').*

Proof: Suppose given T, f as described and let $Y \circlearrowleft^{\alpha}$ be any endomap. Consider the composite

$$T \xrightarrow{\langle 1_T, 1_T \rangle} T \times T \xrightarrow{\hat{f}} Y \xrightarrow{\alpha} Y, \qquad g = \alpha \hat{f} \langle 1_T, 1_T \rangle : T \longrightarrow Y$$

By the assumption that f is onto, there is a point t such that $\ulcorner g \urcorner = ft$, i.e. such that $g(s) = \hat{f}(s,t)$ for all s in T. But by definition of g, this means that $\alpha\hat{f}(s,s) = \hat{f}(s,t)$ for all s in T. In particular, if $s = t$, then $\alpha\hat{f}(t,t) = \hat{f}(t,t)$. This means that $\hat{f}(t,t)$ is a fixed point of α, as was to be shown.

4. Universal properties and 'observables'

The map object (or 'exponentiation') construction is used for constructing objects satisfying related universal properties, in categories of structured objects, e.g. in the category $\mathcal{S}^{\circlearrowleft}$ of discrete dynamical systems. If X is a discrete dynamical system with endomap α of states and if Y is just a set, then a map $X \xrightarrow{f} Y$ (from the set of states of X) may be considered as a definite process of observation or measurement (with values in Y) of some feature of states. Thus, if at a certain time the system X is in state x we will observe fx, one unit of time later we will observe $f\alpha x$, two units of time later we will observe $f\alpha\alpha x$, etc. so that x gives rise to a *sequence* of points of Y.

This can be made into a map of dynamical systems as follows: Given any set Y, consider $Y^{\mathbb{N}}$, the map set whose points correspond to sequences $\mathbb{N} \xrightarrow{y} Y$ in Y (here again $\mathbb{N} = \{0, 1, 2, \ldots\}$ is the set of natural numbers). On the set $Y^{\mathbb{N}}$ there is the 'shift' endomap β for which

$$(\beta y)(n) = y(n+1) \quad \text{for all } n \text{ and all } \mathbb{N} \xrightarrow{y} Y$$

Thus the set $Y^{\mathbb{N}}$ of sequences in Y is a dynamical system when equipped with the shift endomap. Now, returning to a given map $X \xrightarrow{f} Y$ where $X^{\circlearrowright\alpha}$ is a given dynamical system, we can define

$$X \xrightarrow{\bar{f}} Y^{\mathbb{N}}$$

by the formula

$$\bar{f}(x)(n) = f(\alpha^n x)$$

i.e. $\bar{f}$ assigns to any state x the *sequence* of all f-observations through its 'future.' The map $\bar{f}$ is actually a map of dynamical systems:

Exercise 7:
$\bar{f}$ is a map in the category $\mathcal{S}^{\circlearrowright}$, and the *only* such which, moreover, has $\bar{f}(x)(0) = f(x)$ for all x.

In applications, one often has only a limited stock of measurement instruments $X \xrightarrow{f} Y$ on the states of $X^{\circlearrowright\alpha}$; such an f can be called briefly an *observable*. One reason for introducing the map $\bar{f}$ of dynamical systems induced by an observable $X \xrightarrow{f} Y$ on a given dynamical system $X^{\circlearrowright\alpha}$ is that it permits a simple expression of some important properties that f may have, as in the following two definitions:

Definition: *An observable $X \xrightarrow{f} Y$ on a dynamical system $X^{\circlearrowright\alpha}$ is said to be* **chaotic** *if the induced $\mathcal{S}^{\circlearrowright}$-map*

$$X^{\circlearrowright\alpha} \xrightarrow{\bar{f}} (Y^{\mathbb{N}})^{\circlearrowright\beta}$$

is 'onto for states', i.e. if for every possible sequence $\mathbb{N} \xrightarrow{y} Y$ of future observations there is at least one state x of X for which $\bar{f}(x) = y$.

One interpretation of the chaotic nature of f is that (although $X^{\circlearrowright\alpha}$ itself is perfectly deterministic) f observes so little about the states that nothing can be predicted about the possible sequences of observation themselves. Often the 'remedy' for this is to observe more, i.e. to build $X \xrightarrow{f'} Y'$ (from which f might be recovered via a suitable $Y' \longrightarrow Y$) for which $\overline{f'}$ might not be onto.

Definition: *An observable* $X \xrightarrow{f} Y$ *on a dynamical system is an* **admissible notion of underlying configuration** *if* $\bar{f}$ *is 'faithful,' i.e. for any two states* x_1, x_2, *if the resulting sequences of future configurations are equal,* $\bar{f}(x_1) = \bar{f}(x_2)$, *then* $x_1 = x_2$.

The induced map $\bar{f}$ is often faithful even if f itself is not. The term 'state' in most applications means more precisely 'state of motion'; the state *of motion* usually involves more than merely the current position or 'configuration,' but for purely mechanical systems is determined by specifying additional quantities such as momentum which *are* determined by the *motion of the configuration.* (In common examples, Y itself is a map object E^B, where E is ordinary three dimensional physical space, B is the set of particles of a body such as a cloud, and the points of $Y = E^B$ correspond to *placements* $B \longrightarrow E$ of the body in space.)

Exercise 8:
Let $A \times A \xrightarrow{+} A$ be a binary operation such as addition of natural numbers or real numbers and let $X = A \times A$. The *Fibonacci*† *dynamics* α on X is defined by

$$\alpha(a, b) = \langle b, a + b\rangle$$

If $A = \mathbb{N}$ and $x = \langle 1, 1\rangle$ calculate $\alpha x, \alpha^2 x, \alpha^3 x, \alpha^4 x, \alpha^5 x$. Let $Y = A$ and let f be the projection: $f(a, b) = a$. Show that $X \xrightarrow{f} Y$ is an admissible notion of configuration for the Fibonacci dynamics.

Exercise 9: (more challenging)
Fix a point p on a circle C. Let $C \xrightarrow{\omega} C$ be the 'wrap twice around' map: the angle from p to $\omega(x)$ is twice the angle from p to x. Then $C^{\circlearrowright \omega}$ is a dynamical system. Let $C \xrightarrow{f} \{true, false\}$ answer the question 'Are we on the upper half-circle?' (Let's decide that 'upper half-circle' includes p but not its antipode.)

(a) Show that f is an admissible notion of underlying configuration.
(b) Show that f is not a chaotic observable, but is 'almost chaotic': Given any *finite* future (a list $y_0, y_1, \ldots, y_n$ of points of $\{true, false\}$), there is a state x for which $fx = y_0, f\omega x = y_1, \ldots,$ and $f\omega^n x = y_n$.

†Historical note: Fibonacci, also known as Leonardo of Pisa, lived from 1170 to 1250. He was sent by the merchants of Pisa to Africa to learn Arab mathematics. The sequence of numbers generated by the Fibonacci dynamics starting from the state $\langle 1, 1\rangle$ arose from a problem in his book *Liber Abaci*: 'A certain man put a pair of rabbits in a place surrounded on all sides by a wall. How many pairs of rabbits can be produced from that pair in a year, if it is supposed that every month each pair begets a new pair which from the second month on becomes productive?' In 1753 this dynamics was discovered to be intimately related to the golden section $\frac{1+\sqrt{5}}{2}$. It remains an important example in modern computer science.

Exercise 10:
For the generic arrow $A = \boxed{\underset{s}{\bullet} \longrightarrow \underset{t}{\bullet}}$ in $\boldsymbol{S}^{\downarrow\downarrow}$, the graph A^A exists; calculate it.

A syntactical scheme for calculating with map objects is often called a 'λ-calculus' because of a traditional use of the Greek letter lambda to denote the transformation involved in the universal property. In the exercise below, a closely related but not identical use of the same symbol occurs.

Exercise 11:
(a) For any map $W \xrightarrow{f} Y$ (in a category where $(\ )^T$ exists) there is an induced map $W^T \xrightarrow{f^T} Y^T$ for which $f^T(\ulcorner a \urcorner) = \ulcorner fa \urcorner$ for all $T \xrightarrow{a} W$.
(b) There is a standard map $X \xrightarrow{\lambda_T} (X \times T)^T$ (analogous to the diagonal map to a product.)
(c) For any $X \times T \xrightarrow{f} Y$, $\ulcorner f \urcorner = f^T \circ \lambda_T$ is the corresponding map $X \longrightarrow Y^T$.

5. Guide

Map object is a basic example of a higher universal mapping property. Session 29 treated some questions involving maps whose domain is a product without using map objects, but beginning in Session 30 map objects become crucial. The final two sessions introduce another universal mapping property, representing the logic of subobjects via truth-value objects.

SESSION 30

Exponentiation

1. Map objects, or function spaces

Map objects, or function spaces, are sometimes also called exponential objects because they satisfy laws of which the laws of exponents in arithmetic are special cases. They are used to study the way in which an output depends on a whole process, rather than just a single input. For example, the energy expended in walking from Buffalo to Rochester depends not only on the distance traveled, but on the whole 'motion' you perform. This motion is itself a map, say from an interval of time to 'space.'

We saw that a product of two objects X_1 and X_2 of a category $\mathcal{C}$ can be described as a terminal object in a certain category we constructed from $\mathcal{C}$, X_1, and X_2. In the same way, given two objects T and Y in $\mathcal{C}$, we can construct a category in which the corresponding map object Y^T may be described as a terminal object. It will be useful to introduce that category from the start because it will help us in future calculations.

Given two objects T and Y of a category $\mathcal{C}$ that has a terminal object and products, we define a category $\mathcal{C}/(T \longrightarrow Y)$ by saying that

1. an *object* in $\mathcal{C}/(T \longrightarrow Y)$ is an object X of $\mathcal{C}$ together with a map in $\mathcal{C}$ from $T \times X$ to Y, and
2. a *map* in $\mathcal{C}/(T \longrightarrow Y)$ from $T \times X' \xrightarrow{f'} Y$ to $T \times X \xrightarrow{f} Y$ is a $\mathcal{C}$-map $X' \xrightarrow{\xi} X$ such that $f' = f \circ (1_T \times \xi)$, i.e.

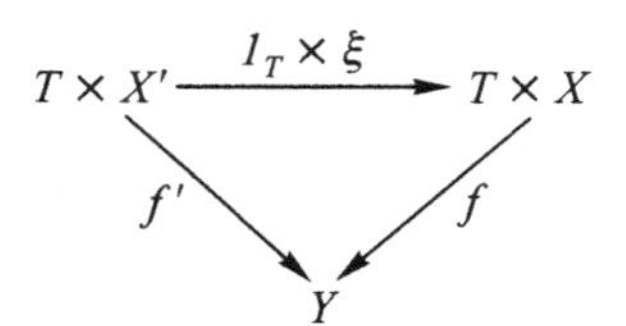

What is meant by $1_T \times \xi$? What is the product of two maps? If we have any two maps $A \xrightarrow{g} B$, $C \xrightarrow{h} D$ in a category with products, we can define a map $g \times h$ from $A \times C$ to $B \times D$ by first calculating the two composites

$$A \times C \xrightarrow{proj_1} A \xrightarrow{g} B, \quad A \times C \xrightarrow{proj_2} C \xrightarrow{h} D$$

and then forming the pair

$$\langle g \circ proj_1,\ h \circ proj_2 \rangle : A \times C \longrightarrow B \times D$$

which we take as the definition of $g \times h$. Thus our particular map $1_T \times \xi$ is defined by the diagram

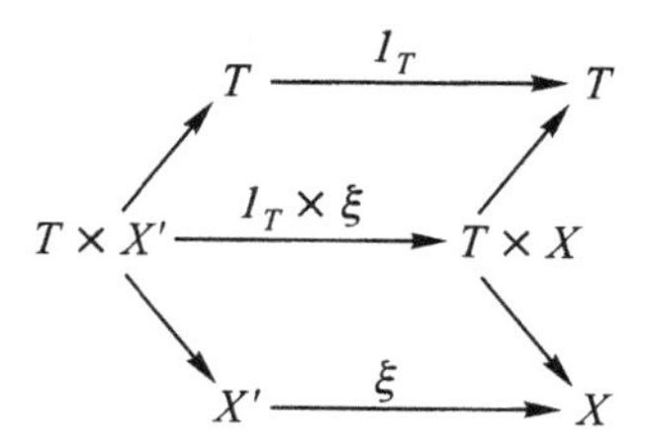

where the unlabeled maps are product projections.

We have defined the objects and the maps of the category $\mathcal{C}/(T \longrightarrow Y)$. It is necessary, of course, to define the identity maps and composition in such a way that the identity and associative laws hold. As is usually the case with categories defined from a previously given category, there is only one obvious way of defining identities and composition in order that the identity and associative laws hold, and the only question is whether these definitions indeed produce maps of the new category. This reduces to verifying that for any object $T \times X \xrightarrow{f} Y$ of $\mathcal{C}/(T \longrightarrow Y)$, the identity map of X in $\mathcal{C}$ is a map from $T \times X \xrightarrow{f} Y$ to itself; i.e. to verifying that

$$f \circ (1_T \times 1_X) = f$$

and that for any three objects $T \times X \xrightarrow{f} Y$, $T \times X' \xrightarrow{f'} Y$, and $T \times X'' \xrightarrow{f''} Y$, and any two maps in $\mathcal{C}/(T \longrightarrow Y)$, say ξ from $T \times X' \xrightarrow{f'} Y$ to $T \times X \xrightarrow{f} Y$ and η from $T \times X'' \xrightarrow{f''} Y$ to $T \times X' \xrightarrow{f'} Y$, the composite $\xi \circ \eta$ (a map of $\mathcal{C}$) is indeed a map in $\mathcal{C}/(T \longrightarrow Y)$ from $T \times X'' \xrightarrow{f''} Y$ to $T \times X \xrightarrow{f} Y$, i.e.

$$f \circ (1_T \times (\xi \circ \eta)) = f''$$

Both verifications come from simple properties of the product of maps explained above. The first follows from the fact that the product of identities is another identity – in particular, $1_T \times 1_X = 1_{T \times X}$ – and the second follows from a sort of 'distributivity of product with respect to composition,' which in our case takes the form $1_T \times (\xi \circ \eta) = (1_T \times \xi) \circ (1_T \times \eta)$.

How can we interpret the objects of this new category? The idea is that an object $T \times X \longrightarrow Y$ is a scheme for naming maps from T to Y in $\mathcal{C}$. An example of such an object is a calculator or processor, where X is the set of names of all the functions that the calculator can perform, and T and Y are the sets of all possible inputs and outputs. The map $T \times X \xrightarrow{f} Y$ describes the calculator itself: $f(t, x)$ is the result of applying the operation whose name is x to the input t. Therefore, for each element x of X, $f(-, x)$ represents a map $T \longrightarrow Y$. In particular, taking X to be the terminal object $\mathbf{1}$ of $\mathcal{C}$, an object $T \times X \longrightarrow Y$ amounts to just a single map $T \longrightarrow Y$ 'named' by $\mathbf{1}$, because $T \times \mathbf{1} \cong T$. Similarly, if $X = \mathbf{2}$, it names two maps $T \longrightarrow Y$, and so on. With larger X, the objects X, f can name more maps from T to Y.

Equally interesting is the interpretation of the maps in the category $\mathcal{C}/(T \longrightarrow Y)$. A map ξ from X',f' to X,f in $\mathcal{C}/(T \longrightarrow Y)$ is a way to correlate the names of the maps $T \longrightarrow Y$ as named by X',f' with the names of the maps $T \longrightarrow Y$ as named by X,f. It is a sort of dictionary. Of course, X',f' may not name all the maps from T to Y, but for those that are named, the map $X' \xrightarrow{\xi} X$ finds their corresponding names in the X, f 'language.' The condition for a map $X' \xrightarrow{\xi} X$ in $\mathcal{C}$ to belong to the category $\mathcal{C}/(T \longrightarrow Y)$ is that for any name x' in X', its image $\xi(x')$ names precisely the same map $T \longrightarrow Y$ which x' names. In other words, the map $f(-, \xi(x'))$ is the same as the map $f'(-, x')$, i.e. for every element t of T, $f(t, \xi(x')) = f'(t, x')$. This is what is meant by the condition $f \circ (1_T \times \xi) = f'$, since by the definition of product of maps $(1_T \times \xi)(t, x') = (t, \xi(x'))$.

For a given category $\mathcal{C}$, the category $\mathcal{C}/(T \longrightarrow Y)$ associated with some T and Y may have a terminal object. Then the corresponding object of $\mathcal{C}$, denoted by the exponential notation Y^T (Y raised to the power T) is called the map object from T to Y. The corresponding map of $\mathcal{C}$, $T \times Y^T \longrightarrow Y$, is denoted by e or ev and is called the evaluation map. Let's see what such a terminal object means. To say that $T \times Y^T \xrightarrow{e} Y$ is a terminal object in $\mathcal{C}/(T \longrightarrow Y)$ means that for every object $T \times X \xrightarrow{f} Y$ of this category there is exactly one map of $\mathcal{C}/(T \longrightarrow Y)$ from that object to $T \times Y^T \xrightarrow{e} Y$. By definition of $\mathcal{C}/(T \longrightarrow Y)$, this is a map $X \longrightarrow Y^T$ in $\mathcal{C}$ (to be denoted $\ulcorner f \urcorner$) such that

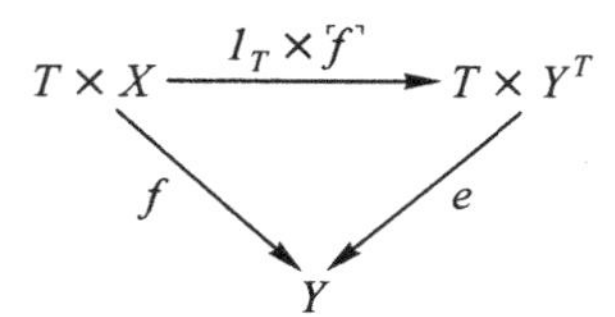

i.e. $e \circ (1_T \times \ulcorner f \urcorner) = f$. Thus, to say that $T \times Y^T \xrightarrow{e} Y$ is a terminal object in $\mathcal{C}/(T \longrightarrow Y)$ means, expressed in $\mathcal{C}$: For every map $T \times X \xrightarrow{f} Y$ in $\mathcal{C}$ there is exactly one map $\ulcorner f \urcorner : X \longrightarrow Y^T$ such that $e \circ (1_T \times \ulcorner f \urcorner) = \mathrm{f}$.

This correspondence between $\ulcorner f \urcorner$ and f is expressed as usual by

$$\frac{X \longrightarrow Y^T}{T \times X \longrightarrow Y}$$

Having map objects in a category is a strong condition from which we will deduce many consequences. In many categories $\mathcal{C}$, Y^T exists only for certain 'small' objects T. Best of all are the *cartesian closed categories*: those categories with products in which every pair of objects has a map object. (The word 'closed' refers to the fact that the maps from one object to another do not just form something outside of the category – a set – but form an object of $\mathcal{C}$ itself, while the word 'cartesian' refers to the French mathematician R. Descartes – in Latin *Cartesius* – who is usually credited with the idea of products although, as we have seen, these were already used by his older contemporary Galileo.)

We saw in Part IV that for any object T, $T \times \mathbf{1} = T$. Thus, applying the definition of the map object Y^T to the particular case $X = \mathbf{1}$, we deduce that

$$\frac{\mathbf{1} \longrightarrow Y^T}{T \longrightarrow Y}$$

In other words, the points of the map object Y^T correspond bijectively with the maps from T to Y. This is why Y^T is called a 'map object.'

2. A fundamental example of the transformation of map objects

For an important application of these ideas to the study of the motion of bodies in space we place ourselves in a category of smooth objects that includes among its objects a body B, a time interval I, and ordinary space E. We do not need to go into the details for defining such a category.

Imagine a body B moving in space, e.g. a cloud moving in the sky. The usual way of describing this motion during the time interval I is as a map $I \times B \longrightarrow E$ that associates to each particle of the body at each time a position in space. Thus, we have

1. $I \times B \xrightarrow{\textit{motion}} E$

But if the map object E^I exists, then this map is equivalent to a map

2. $B \xrightarrow{\ulcorner \textit{motion} \urcorner} E^I$

This map assigns to each particle of the body the whole of its motion. According to what was said before, the points of the object E^I are the 'paths' in space, i.e. maps $I \longrightarrow E$, and this object of paths is completely independent of the body B; once we have understood it, we can use it to study any motion of any body whatsoever.

There is a third way of looking at the motion of the body B. Composing with the isomorphism $I \times B \cong B \times I$, and using again the fundamental property of map objects, tells us that the motion of B can be viewed as a map

3. $I \longrightarrow E^B$

Since the points of E^B are the maps $B \longrightarrow E$ which represent the different possible positions or placements of the body in space, the above map assigns to each instant of time a particular position of the body as a whole. Again, the object E^B of placements of our body involves only the body B and space, and has nothing to do with time.

Each of the three different viewpoints about the motion of an object that we just presented has its own importance and application. We need the three ways of describing a motion to be able to calculate (by composition of maps) different quantities associated with that motion. For example, we may have a function $E \longrightarrow \mathbb{R}$, from the space E to real numbers which tells some property of space, say 'distance from the earth.' By composing this map with the motion in the form

$I \times B \longrightarrow E$, we obtain a map $I \times B \longrightarrow \mathbb{R}$ which tells us how the distance of the different parts of the body from the earth changes with time.

However, the *velocity* of a particle is not determined by just one position. The velocity is really a property of the paths through time. Indeed, corresponding to the object E, there is another object V of 'velocities' and differential calculus constructs a map

$$E^I \xrightarrow{\text{velocity}} V^I$$

which associates a 'velocity path' to each space path. Composing this map with the motion of the body in the form $B \longrightarrow E^I$ we obtain the velocity path for the particles of the body in that particular motion:

$$B \longrightarrow E^I \longrightarrow V^I$$

Having obtained this map $B \longrightarrow V^I$ we can go back and study it in the form $I \times B \longrightarrow V$ or in the form $I \longrightarrow V^B$. (Maps $B \longrightarrow V$ are called 'velocity fields.')

The third point of view is useful to calculate quantities that depend on the placement of the body in space. For example the *center of mass* or 'balance point' of the object depends only on the placement and therefore is given by a map (constructed by integral calculus)

$$E^B \xrightarrow{\text{center of mass}} E$$

By composing this map with the motion of the body as $I \longrightarrow E^B$ we obtain a path in space, $I \longrightarrow E$, which represents the motion of the center of mass of the body.

To summarize, a particular motion of a body can be described by any of three maps

$$I \times B \longrightarrow E \quad B \longrightarrow E^I \quad I \longrightarrow E^B$$

These contain the same information, but as maps they serve different purposes.

Maps such as *velocity* and *center of mass* above, whose *domain* is a map object, are often called *operators* or *functionals*. Functionals require much analysis, because there is no *generally valid* way of reducing them to something which does not involve map objects. (This 'non-trivial side' contrasts with the 'trivial side' treating maps whose *codomain* is a map object.)

3. Laws of exponents

Map objects exist in the category of sets and in the category of graphs, which therefore are both cartesian closed categories. Before we study these examples it is useful to know the laws of exponentiation. These are not additional assumptions; they follow from the definition.

If the base is a product, the relevant law is

We will now apply that method to the problem of finding map objects in the category of sets and in the category of graphs. We only work out the first part of the method (describing the solution) which, anyhow, is the hardest part. We will leave for you to prove that the objects we shall describe are indeed map objects. Let's start by recalling the universal mapping property defining map objects.

1. Definition of map object versus definition of product

If T and Y are objects in a category with products, the map object of maps from T to Y is two things: a new object, to be denoted Y^T, and a map $T \times Y^T \longrightarrow Y$, to be denoted e (for *evaluation*), satisfying the following universal mapping property. For every object X and every map $T \times X \xrightarrow{f} Y$, there is exactly one map from X to Y^T, to be denoted $X \xrightarrow{\ulcorner f \urcorner} Y^T$ which together with e determines f as the composite

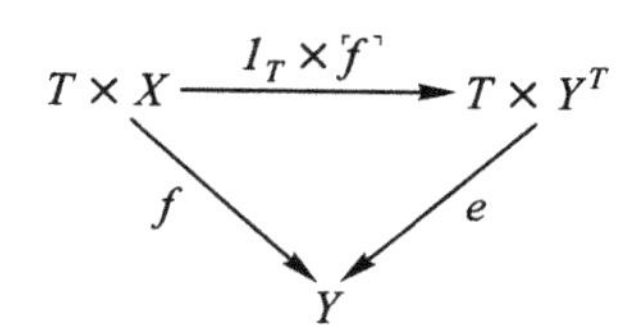

This definition is long. The best way to learn it is to apply it to solve the exercises. As soon as you get some practice, it won't seem so long. Besides, you should notice that this definition follows the same pattern as all the other definitions using universal mapping properties.

ALYSIA: I do not understand what the 'corners' mean.

The 'corners' are just a mark to make up a symbol for the new map. Since the new map is determined by f we want its symbol to remind us of the f, so we use an 'f with corners.' We could have used any other mark, but the corners are used for historical reasons; they were earlier used in logic. The map $\ulcorner f \urcorner$ is to be called 'the name of f.' The use of these corners in this definition is very similar to the use of the brackets $\langle , \rangle$ in the definition of product. In fact, the whole definition parallels that of products. It may be helpful to write the definitions side by side to see clearly the parallel:

Definition of map object

Given objects T, Y, a Map Object of maps from T to Y is two things
1. *an object denoted* Y^T *and*
2. *a map, called evaluation*

$$T \times Y^T \xrightarrow{e} Y$$

Definition of product

Given objects B_1, B_2, *a Product of* B_1 *and* B_2 *is two things*
1. *an object, denoted* $B_1 \times B_2$, *and*
2. *two maps, called projections*

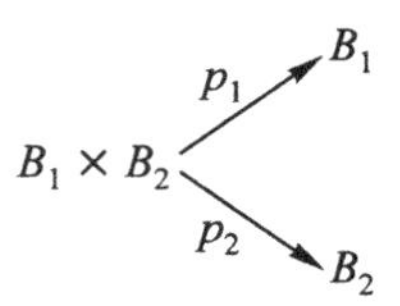

such that they satisfy the following

Universal mapping property defining a map object

For any object X and any map

$$T \times X \xrightarrow{f} Y$$

there is exactly one map from X to Y^T, *to be denoted*

$$X \xrightarrow{\ulcorner f \urcorner} Y^T$$

such that

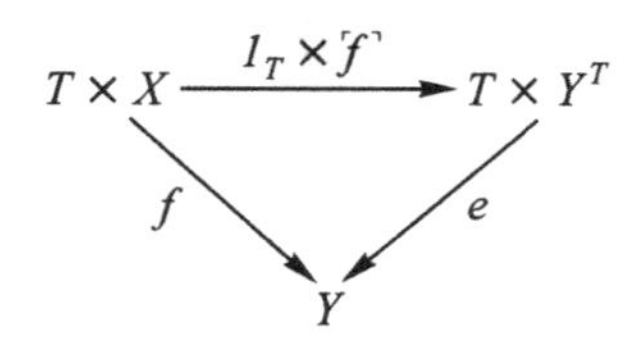

that is, the map f can be expressed as

$$f = e \circ (1_T \times \ulcorner f \urcorner).$$

Universal mapping property defining a product

For any object X and any maps

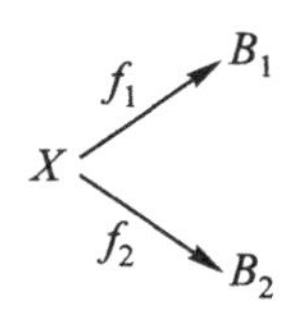

there is exactly one map from X to $B_1 \times B_2$, *to be denoted*

$$X \xrightarrow{\langle f_1, f_2 \rangle} B_1 \times B_2$$

such that

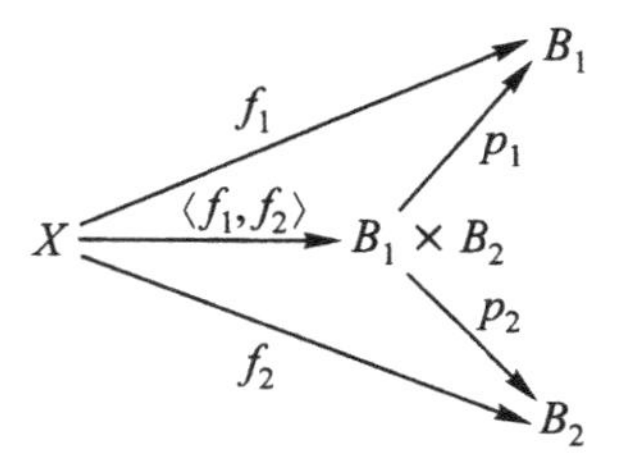

that is, the maps f_1, f_2 *can be expressed as*

$$f_1 = p_1 \circ \langle f_1, f_2 \rangle \quad \text{and} \quad f_2 = p_2 \circ \langle f_1, f_2 \rangle.$$

These universal mapping properties can be symbolically summarized as

$$\frac{X \longrightarrow Y^T}{T \times X \longrightarrow Y}$$ *which is briefly expressed as: the maps from any object X to the map object Y^T are the same as the maps from $T \times X$ to Y.*	$$\frac{X \longrightarrow B_1 \times B_2}{X \longrightarrow B_1, X \longrightarrow B_2}$$ *which is briefly expressed as: the maps from any object X to the product $B_1 \times B_2$ are the same as the pairs of maps from X to B_1 and B_2.*

CHAD: Can you interchange the X and the T?

Yes, as long as the category has the appropriate map objects. In some categories it may be that Y^T exists while Y^X does not exist. But as long as both map objects exist we can use the standard isomorphism $T \times X \cong X \times T$ to interchange the X and the T as follows:

$$\begin{array}{c} X \longrightarrow Y^T \\ \hline T \times X \longrightarrow Y \\ \hline X \times T \longrightarrow Y \\ \hline T \longrightarrow Y^X \end{array}$$

2. Calculating map objects

Let's now try to calculate some map objects, first in the category of sets. Suppose that T and Y are two sets. What set is Y^T? From all that we have said in this book everybody should guess that Y^T 'is' the set of all maps from T to Y. We can *deduce* that just from the universal mapping property. In order to know the *set* Y^T, all we need to know is what its points are. We can use the fact that $T \times \mathbf{1} = T$ to deduce immediately what the points of Y^T must be:

$$\begin{array}{c} \mathbf{1} \longrightarrow Y^T \\ \hline T \times \mathbf{1} \longrightarrow Y \\ \hline T \longrightarrow Y \end{array}$$

That is, *if* there are a set Y^T and a map e satisfying our universal property, then the points of Y^T 'are' the maps from T to Y.

For example, if Y has 5 points and T has 3 points, then Y^T can be any set with 125 points. It is the specification of the evaluation map which transforms this mere set of dots into a system of names for the detailed maps $T \longrightarrow Y$, somewhat as the circuitry and programming of a computer transform an empty memory bank into a system of useful meanings. Thus calculating map objects involves making a good choice of the map that is to play the role of the evaluation map, and somehow verifying that it has the universal mapping property.

Suppose we are given objects T and Y in the category of graphs. If the graph Y^T exists, its points (i.e. its loops) are the (graph) maps from T to Y. In particular, if $T = A$ (the arrow) and $Y = A$, this tells us that the loops of A^A are the maps $A \longrightarrow A$. Since there is only one map of graphs

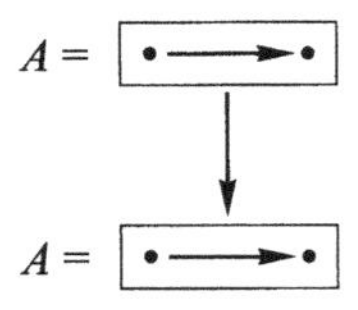

we conclude that A^A must have exactly one loop. Unfortunately, to know what the loops of a graph are is not to know very much about it. We need to know the arrows and the dots, and how they are interconnected. Now we must remember that the arrows of a graph X are the same as the graph maps from A to X, and that the dots of X are the graph maps from the 'naked dot' D to X, a fact that can be represented as

$$\frac{\textit{arrows of } X}{A \longrightarrow X} \quad \text{and} \quad \frac{\textit{dots of } X}{D \longrightarrow X}$$

Applying this to the graph Y^T, we deduce what its arrows are by using the universal mapping property, just as we used that property to find the loops:

$$\frac{\frac{\textit{arrows of } Y^T}{A \longrightarrow Y^T}}{T \times A \longrightarrow Y}$$

Thus the arrows of Y^T must be the graph maps from $T \times A$ to Y, and similarly for the dots:

$$\frac{\frac{\textit{dots of } Y^T}{D \longrightarrow Y^T}}{T \times D \longrightarrow Y}$$

the dots of Y^T are the graph maps from $T \times D$ to Y. We know what the set of arrows and the set of dots of Y^T must be, but in order really to apply this and do calculations, we must understand clearly what the graphs $T \times A$ and $T \times D$ are. The second one is easier: $T \times D$ has no arrows and its dots are the same as those of T. For example, if T is the graph

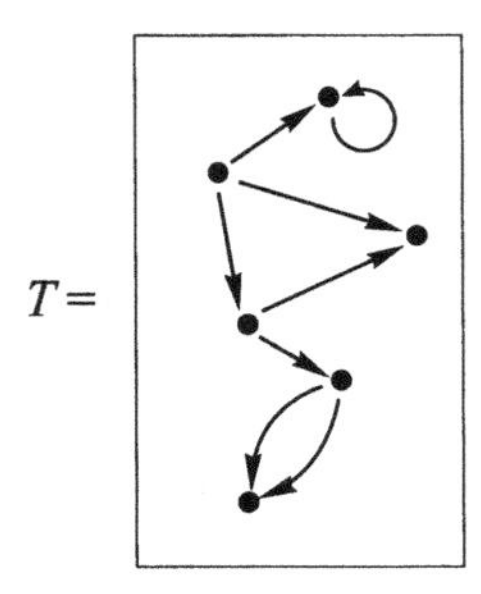

then

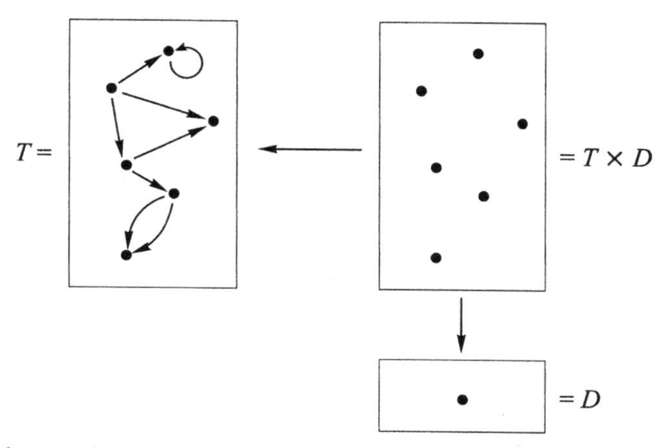

Since $T \times D$ doesn't have any arrows its graph maps to Y are the same as the maps of sets from the dots of $T \times D$ (which are the dots of T) to the dots of Y, namely: *the dots of* Y^T *are the set maps* $T_D \longrightarrow Y_D$.

In the case $T = A$ and $Y = A$ we know that the dots of A^A are the maps from $\{s, t\}$ to $\{s, t\}$ which are four in number. So, A^A *must have four dots* (one of which carries a loop, since we found that A^A has one loop).

In order to find out what the arrows of Y^T are, we need to understand the graph $T \times A$, since we have deduced that these arrows are precisely the graph maps from $T \times A$ to Y. Recall from Session 25 the nature of the product of a graph and the arrow. In the case of the graph T pictured above, the product $T \times A$ is

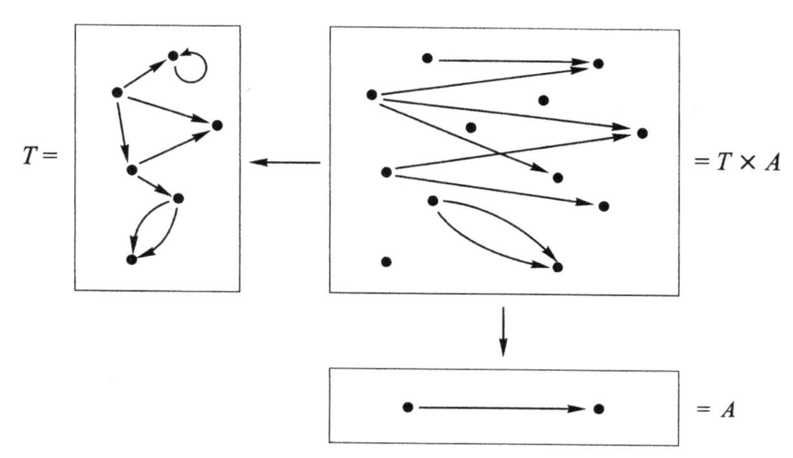

Now, try the following exercises.

Exercise 1:
Draw pictures of A^A, and the evaluation map $A \times A^A \xrightarrow{e} A$.

To do this you need first to draw $A \times A$, and then find all the graph maps from $A \times A$ to A. That is the set of arrows of A^A. Then you need to determine how the arrows fit with the dots. A^A exhibits nicely the distinction between being terminal and having exactly one point. The following exercise shows which objects X have X^X terminal.

Exercise 2:
Let X be an object in a cartesian closed category. Show that the following two properties are equivalent:

(a) $X \longrightarrow \mathbf{1}$ is a monomorphism;
(b) $X^X = \mathbf{1}$.

We saw in Session 27 that property (a) is equivalent to 'idempotence' of X (i.e. $X \xleftarrow{1_X} X \xrightarrow{1_X} X$ is a product).

Exercise 3:
'Primitive recursion data' for defining a sequence of functions $A \longrightarrow Y$ consist of an initial function $A \xrightarrow{f_0} Y$ and a rule $\mathbb{N} \times A \times Y \xrightarrow{h} Y$ for going from one function to the next. Show that for any primitive recursion data there is exactly one $\mathbb{N} \times A \xrightarrow{f} Y$ for which

$$f(0, a) = f_0(a) \qquad \text{for all } a$$
$$f(n+1, a) = h(n, a, f(n, a)) \quad \text{for all } n, a$$

(Hint: Consider a suitable dynamical system whose state space is $X = \mathbb{N} \times Y^A$; see Exercise 12, page 186.)

In part V we are studying three higher universal mapping properties: map spaces, the truth space, and the set of connected components; the next session 32 begins the discussion of what 'truth is good for'.

ARTICLE VI

The contravariant parts functor

1. Parts and stable conditions

The existence of map objects is already a very powerful (and very useful) property that a category $\mathcal{C}$ might have. But even stronger properties are realizable; for example, the categories $\mathcal{C}/X$ might themselves have map objects. The concept of *part* (or *subobject*) in $\mathcal{C}$ might be representable by a truth value object Ω, as is explained here and in Sessions 32 and 33.

We will investigate the relation between parts of X (monomorphisms with codomain X) and stable conditions on figures in X. A condition is called *stable* provided: for any figure x in X with shape A that satisfies the condition and for any $A' \xrightarrow{a} A$, the transformed figure $x' = xa$ also satisfies the condition. A fundamental kind of condition is one given by a map g as follows:

Definition 1: x **is in** g (or x **belongs to** g) *if and only if there exists* w *for which* x = gw.

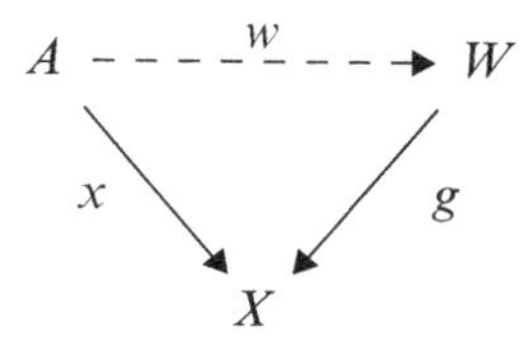

Exercise 1: The condition '. . . is in g' is stable.

Exercise 2: If g is a split epimorphism (i.e. has a section s) then every x (with the same codomain X as g) is in g.

Since most maps g are not split epimorphisms, the problem of which figures are in g (the same as the lifting or choice problem considered earlier) is more difficult unless some restriction is made. (One sort of restriction, not discussed here, is to limit the shapes A of the figures x to be 'projective' objects.) The most important restriction is to consider only those g which are parts of X (i.e. monomorphic mappings); then we use these as tools for investigating the general g via the notion of image (see Definition 2 below).

> **Exercise 3:** If g_1 and g_2 are parts of X such that for all x,
> x is in g_1 if and only if x is in g_2,
> then there is exactly one isomorphism h such that $g_1 = g_2h$.

When g_1 and g_2 are related as in Exercise 3, it is sometimes loosely said that g_1 and g_2 are the 'same part'. (Sometimes a condition on figures in X in $\mathcal{C}$ may be too complicated to determine a part in $\mathcal{C}$; but if the condition satisfies stability, it may determine a part of IX where $I:\mathcal{C} \longrightarrow \mathcal{C}'$ is an inclusion of $\mathcal{C}$ as a full subcategory of a larger category.)

Definition 2: *An* **image of a map** g *is a part* i *of the codomain of* g *for which*
(1) g *is in* i
(2) *for all parts* j, *if* g *is in* j *then* i is *in* j.
Of course any two images of the same g *are uniquely isomorphic as parts.*

> **Exercise 4:** Suppose g has image i (so obviously any figure in g is also in i). Suppose moreover that conversely every figure that is in i is also in g. Then the p proving that g is in i (i.e. $g = ip$) is a split epimorphism (i.e. there exists s such that $ps = 1$).

In Exercise 7, we will see that if $\mathcal{C}$ has equalizers, then a map p which proves that i is an image of g is itself special, in a sense dual to that in which parts are special, namely p is epimorphic ($up = vp$ implies $u = v$); but p is not usually a split epimorphism. Approaching the lifting problem gradually, it is appropriate to generalize the lifting relation by saying that x is *locally in* g if and only if there exists epimorphic a so that xa is in g. Then in categories with certain exactness properties, x is in the image of g if and only if x is locally in g itself.

2. Inverse Images and Truth

A more general way to specify a part of X is in terms of a map $X \xrightarrow{f} Y$ together with a part i of Y. Then the condition on x, that there exist t with $it = fx$, is stable. In many categories there is always a part of X that corresponds to this sort of condition.

Definition 3: *A part* j *such that for any* x,
x *is in j if only if* fx *is in* i
is called an **inverse image of** i **along** f.

(In terms of conditions on x, the inverse image is called the result of *substituting* f into the condition defined by i.)

Exercise 5: Of course, the most basic kind of condition on figures in X is that given by an equation as follows. Given two maps $f_1, f_2 : X \rightrightarrows Y$ we can consider the condition on x

$$f_1x = f_2x.$$

If this condition determines a part of X, that part is called the *equalizer* of f_1, f_2. If the category has products, equalizer is a special case of inverse image, as is seen by defining $f = \langle f_1, f_2 \rangle$, the induced map into $Y \times Y$, and considering the diagonal part i of $Y \times Y$.

Exercise 6: Very different conditions may correspond to the same part. For example, given $f_1, f_2 : X \rightrightarrows Y$, and also given $W \xrightarrow{g} X$, the condition (on figures x in X) to satisfy the equation $f_1x = f_2x$ may be equivalent to x being in the image of g in the sense of Definition 2. In that case we can say that g parameterizes (see section 2 of Session 6) the solutions of the equation $f_1 = f_2$, and the obvious diagram $W \longrightarrow X \rightrightarrows Y$ is said to be *exact*. If, moreover, there are no redundancies in the parameterization g, i.e. g is a part, then g is isomorphic to any equalizer of the pair f_1, f_2.

Exercise 7: In case the category has an equalizer for each parallel pair and an image for each map, then every map can be factored as an epimorphism followed by a monomorphism.

Hint: All that needs to be shown is that the map p connecting a given map to its image is an epimorphism (not necessarily split). The property of being epimorphic involves equality of maps, which can be tested using equalizers.

An important property that many categories of interest enjoy is the 'representability' of the general notion of part, as follows.

Definition 4: *An object* Ω *together with a given part* $\mathrm{T} \xrightarrow{\mathrm{v}} \Omega$ *is called a* **subobject classifier** *or* **truth value object** *for* $\mathcal{C}$ *if and only if for every part* g *of any* X *there is exactly one* $\mathrm{X} \xrightarrow{\mathrm{f}} \Omega$ *for which* g *is the inverse image of* v *along* f. *The map* v *with this remarkable property is often called simply* 'true' *(*v *for veritas). In general,* fx *is called the* **truth value of** *(or extent to which)* 'x belongs *to* g'.

Exercise 8: The domain of the map v must be a terminal object.

Exercise 9: (uniqueness of the truth value space) Between any two truth value objects in the same category there is exactly one isomorphism for which the true points correspond.

The truth values often include many elements besides truth and total falsity. Together they have a rich structure partially reflecting the particular nature of the cohesion and motion in which all objects of $\mathcal{C}$ participate. Somewhat as the mere existence of map spaces forces the products and coproducts to satisfy distributivity, so also the mere existence of Ω in $\mathcal{C}$ has profound effects all over $\mathcal{C}$. In Session 33 it is shown that Ω itself has a rich algebraic structure (which is sometimes known as logic in the narrow sense); this in turn forces the system of parts of any object X to have properties quite different from those of the analogous systems of subobjects in linear algebra or group theory.

For abstract structureless sets and several other categories, $\Omega = 1 + 1$, but for many categories of cohesion and variation (such as graphs and dynamical systems) the determination of the detailed structure of the truth-value object is an important step in understanding the whole category and its workings. (See the pictures in section 2 of Session 33.)

Exercise 10: By the general properties of exponentiation applied to the particular base Ω, any map $X \xrightarrow{f} Y$ induces a map $\Omega^Y \longrightarrow \Omega^X$. Show that, applied to the points of the space Ω^X and interpreted as in Definition 4, this induced map represents the inverse image operation on parts and the substitution operation on conditions.

The foregoing discussion gives a brief introduction to the algebra of parts. This algebra admits broad development, especially in the study of the behavior of the logical operators when the object in which the parts live is itself varied along a map, and in the studies (in functional analysis and general topology) of the parts of map-objects. The developed logical algebra serves as a useful auxiliary tool in the study of the core content of mathematics, which is the variation of quantities and spaces; indeed it was a particular form of that variation, known as 'sheaf', which led to the first discovery, by A. Grothendieck in 1960, of the class of categories known as toposes. Thus the Greek word *topos* signifying location or situation, was adopted to mean mode of cohesion or category (kind) of variation.

SESSION 32

Subobjects, logic, and truth

1. Subobjects

We are going to find a remarkable object that connects **subobjects**, **logic**, and **truth**. What should be meant by a 'subobject,' or 'part,' of an object? Suppose that we have an abstract set X such as

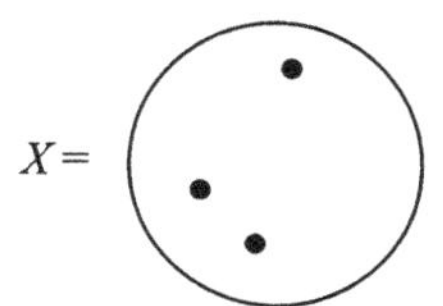

and we look at some of its elements, for example those indicated in the picture

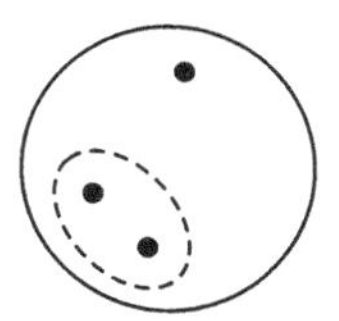

These constitute what can be called a *part* of the set X. This concept of 'part' has two ingredients. First, a part has a *shape*, which in this example is a set S with precisely two elements,

$S =$

but there is no meaning in saying that S itself is a part of X, because there are different parts of X which have the same shape. So, a second ingredient is necessary in order to determine a part of X: an 'inclusion' map which indicates the particular way in which the set S is inserted into X:

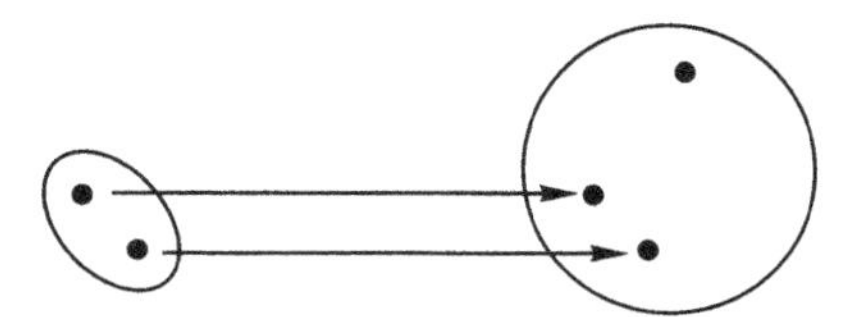

A part of X (of shape S) should therefore be a map from S to X. But there are many maps from S to X (in the example above there are precisely $3^2 = 9$) most of which we do not wish to call parts, for example

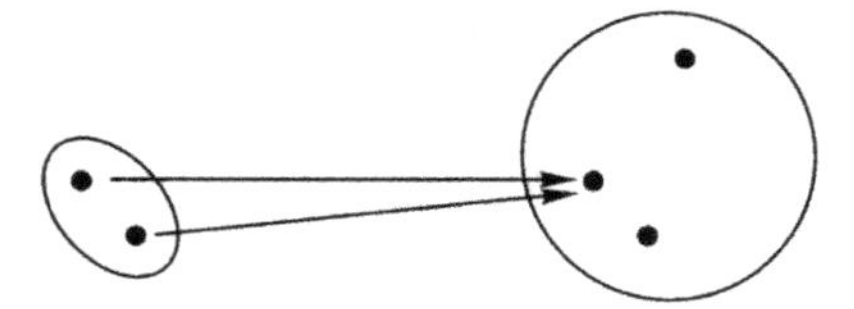

There is a part of X arising from this map, but it is not a part of shape S; it is a map from **1** to X, namely

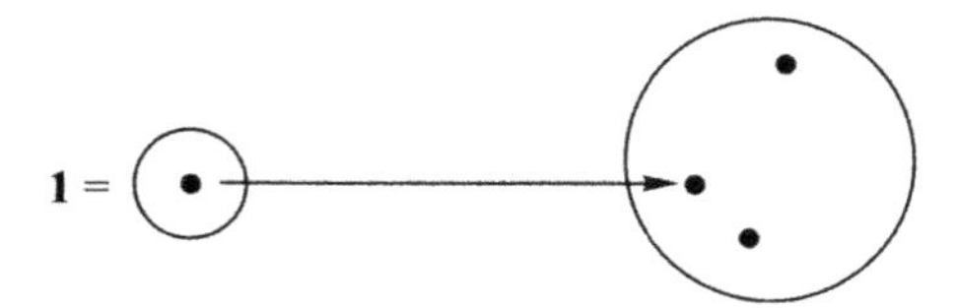

What should a map $S \longrightarrow X$ satisfy to be called an 'inclusion map?'

Definition: *In any category, a map* $S \xrightarrow{i} X$ *is an* **inclusion**, *or* **monomorphism**, *or* **monic map**, *if it satisfies:*

For each object T and each pair of maps s_1, s_2 from T to S
$is_1 = is_2$ implies $s_1 = s_2$

In many categories one doesn't need to use all 'test-objects' T. For example:

Exercise 1:
Show that in the category of sets, if $S \xrightarrow{i} X$ is such that for all *points* $\mathbf{1} \xrightarrow{s_1} S$ and $\mathbf{1} \xrightarrow{s_2} S$, $is_1 = is_2$ implies $s_1 = s_2$, then i is an inclusion. Briefly: if i preserves distinctness of points, then i is an inclusion. (Recall our old friend the 'contrapositive': '$is_1 = is_2$ implies $s_1 = s_2$' says the same thing as '$s_1 \neq s_2$ implies $is_1 \neq is_2$.')

Exercise 2:
(a) Show that in the category of graphs, if $S \xrightarrow{i} X$ satisfies both:
(i) for each pair $D \xrightarrow{d_1} S$ and $D \xrightarrow{d_2} S$ of *dots* of S,
$id_1 = id_2$ implies $d_1 = d_2$
and
(ii) for each pair $A \xrightarrow{a_1} S$ and $A \xrightarrow{a_2} S$ of *arrows* of S,
$ia_1 = ia_2$ implies $a_1 = a_2$,
then i is an inclusion.
(b) Find a simple example in the category of graphs of a map $S \xrightarrow{i} X$ which preserves distinctness of *points*, but is not an inclusion.

Other names for 'inclusion' or 'inclusion map' are: 'monic map' and 'non-singular map', or, especially in sets, 'injective map' and 'one-to-one map.' There is a special notation to indicate that a map $S \xrightarrow{i} X$ is an inclusion; instead of writing a plain arrow like $\longrightarrow$ one puts a little hook, $\subset$, on its tail, so that $S \overset{i}{\hookrightarrow} X$ indicates that i is an inclusion map.

According to our definition the map

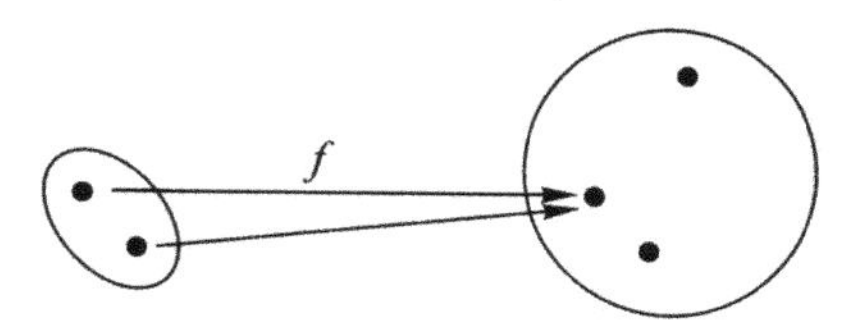

is not injective because there are two *different* maps from some set to

$S =$

which composed with f give the same result (in fact, in this particular case, it is even true that *all* the maps from *any* given set to S give the same result when composed with f).

A good example of an inclusion map in sets is the map which assigns to each student in the class, the chair occupied by that student. (For this to be a map, all students must be seated, and it's an inclusion map if there is no lap-sitting!) In this example, S can be taken to be the set of all the students in the class and X the set of all chairs in the classroom. The example illustrates that the same set S may underlie different parts of the set X, because on another day the students may sit in different places and thus determine a different part of the set of chairs. So, a part of X is not specified by just another set, but by another set together with an inclusion map from that set to X.

For an example in the category $\boldsymbol{\mathcal{S}}^{\downarrow\downarrow}$ of irreflexive graphs, consider the graphs

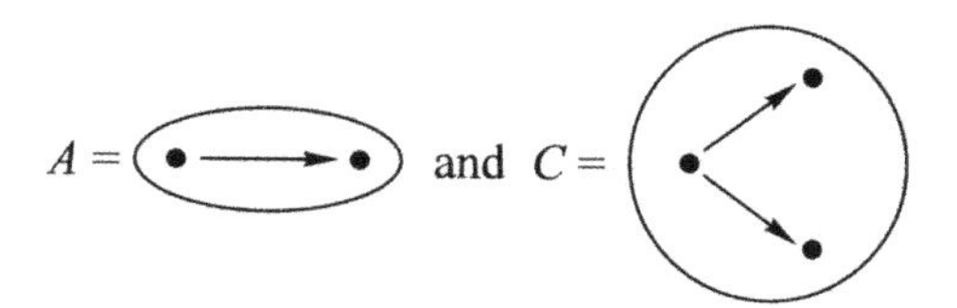

We can see two different parts or *subgraphs* of C with shape A:

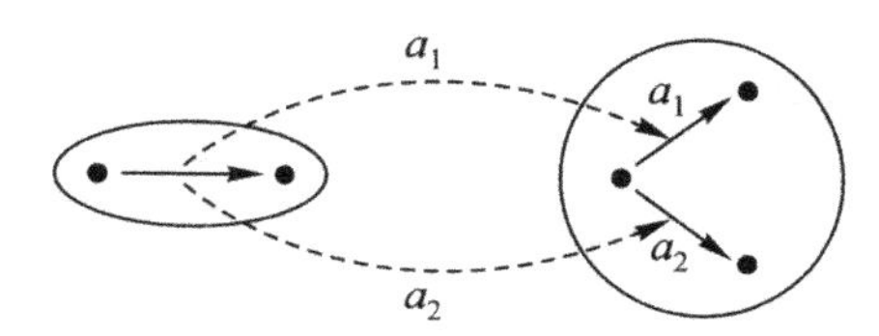

of course, a_1 and a_2 are only two of the several subgraphs that C has. Another subgraph is specified by

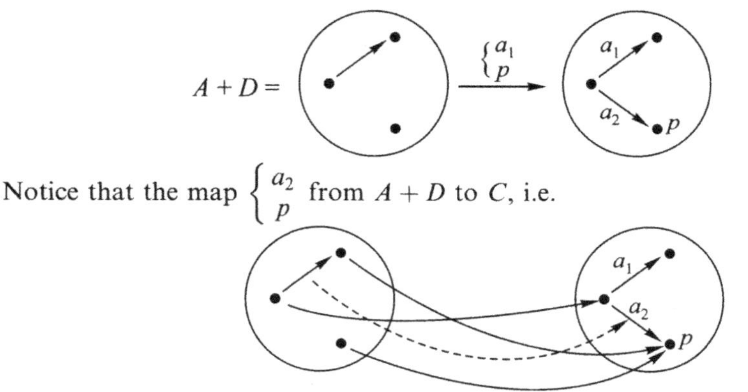

Notice that the map $\begin{cases} a_2 \\ p \end{cases}$ from $A + D$ to C, i.e.

does not give a subgraph because it is not injective.

2. Truth

Let me illustrate that if you have a part of X and you choose any particular element or figure in X, this element or figure may be already included in the part. How can this idea be expressed just in terms of maps and composition? Suppose that we have a figure $T \xrightarrow{x} X$ and we have a part $S \overset{i}{\hookrightarrow} X$. If this refers to the example of students and chairs mentioned above, x may just be one particular chair. Then to ask whether this chair x is included in that part of chairs determined by the seating of the students is just to ask whether the chair x is occupied, and this, in turn, is just to ask whether there is a map f to fill the diagram

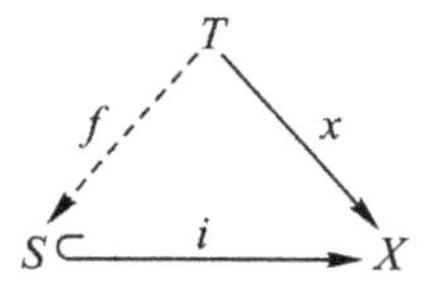

namely, a map f such that $i \circ f = x$. The injectivity of i implies that there can only be at most one map f that 'proves' that the chair x is occupied, which means included in the part S, i.

In the above example where the figure $T \xrightarrow{x} X$ is just a point of the set of chairs, the object T is terminal, but we must emphasize that T may be any object, since the same concept applies to arbitrary figures of arbitrary shape.

As an example in the category of graphs we can take the maps we had before; as the figure x we take

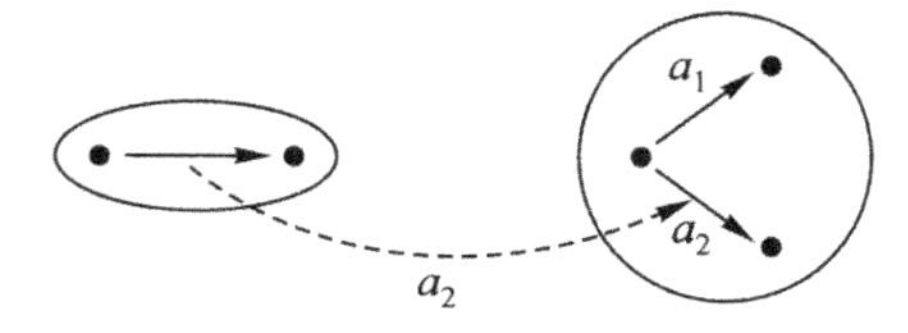

and as the part S, i we take

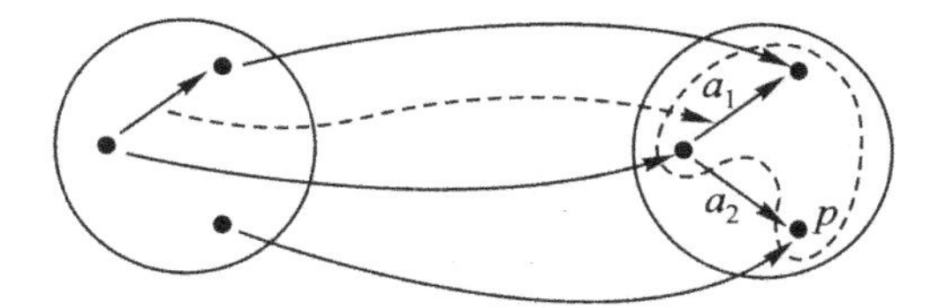

and now we ask: Is the arrow a_2 of X included in the part S, i? The answer to this is clearly 'no.' This is obvious from the picture

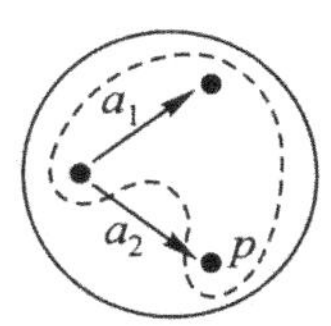

and one can also verify that none of the maps from A to $A + D$ composed with i give x. However, one can't avoid the feeling that this answer doesn't do the question complete justice, because although the figure a_2 is not in the indicated subgraph of X, both its source and its target are. So, there is some degree of truth in the statement that a_2 is included in the subgraph S, i; it is not completely false. This suggests that it is possible to define different degrees of truth appropriate for the category of graphs, so that we can do complete justice in answering such questions.

Let's go back to the category of sets and see what the situation is there. If we have a part of a set and a point of that set, and someone says that the point is included in the part, there are only two possible levels of truth of that statement: it is either true or false. Thus, in the category of sets the two-element set $\mathbf{2} = \{true, false\}$ has the following property (at least for points, x, but in fact for figures of any shape): If X is a set and $S \overset{i}{\hookrightarrow} X$ is any subset of X, there is exactly one function $\varphi_S : X \longrightarrow \mathbf{2}$ such that for all x, x is included in S,i if and only if $\varphi_S(x) = true$.

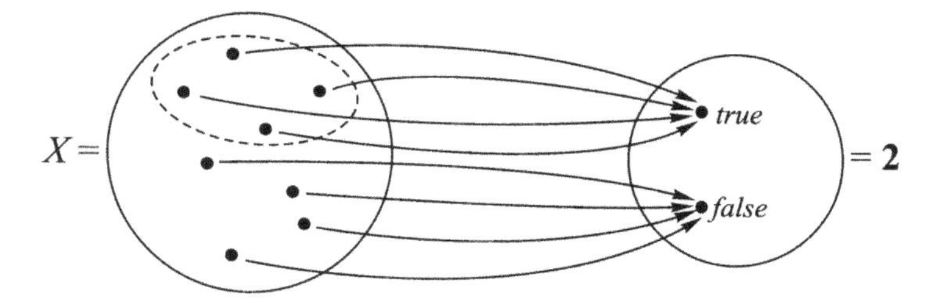

Thus once we have chosen a point $\mathbf{1} \xrightarrow{true} \mathbf{2}$, we have a one-to-one correspondence

$$\frac{\textit{parts of } X}{\textit{maps } X \longrightarrow \mathbf{2}}$$

In particular, this allows you to count the number of parts of X, which is equal to the number of maps from X to $\mathbf{2}$, and this, in turn, is equal to the number of points of $\mathbf{2}^X$.

Given a part of a set X, say $S \overset{i}{\hookrightarrow} X$, its corresponding map $X \xrightarrow{\varphi_s} \mathbf{2}$ is called the *characteristic* map of the part S, i, since at least φ_S characterizes the points of X that are included in the part S, i as those points x such that $\varphi_S(x) = true$. Actually, the map φ_S does much better than that since, as indicated above, this characterization is valid for all kinds of figures $T \xrightarrow{x} X$ and not only for points; the only difference is that when T is not the terminal set, we need a map $true_T$ from T to $\mathbf{2}$ rather than the map $true : \mathbf{1} \longrightarrow \mathbf{2}$. The map $true_T$ is nothing but the composite of the unique map $T \longrightarrow \mathbf{1}$ with $true : \mathbf{1} \longrightarrow \mathbf{2}$,

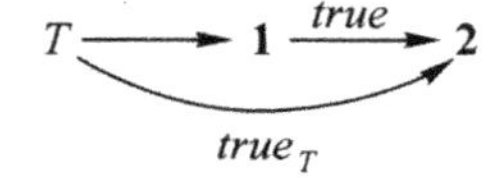

Summing up: The fundamental property of the characteristic map φ_S is that for *any* figure $T \xrightarrow{x} X$, x is included in the part S, i of X if and only if $\varphi_S(x) = true_T$.

3. The truth value object

Let's see now how to do something similar in the category of graphs. Giving a part of an object X in this category (i.e. giving a subgraph of a graph X) amounts to giving a part S_A of the set of arrows of X, and a part S_D of the set of dots of X such that the source and target of every arrow in S_A is a dot in S_D:

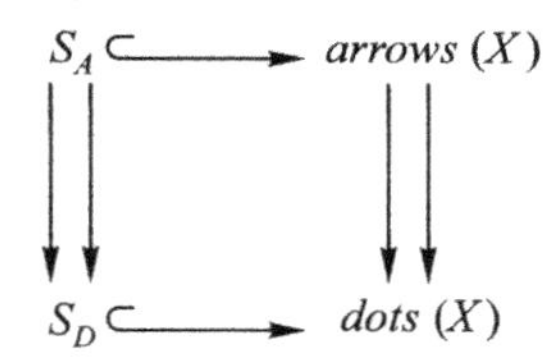

Now we need a graph that plays the same role for graphs that the set $\mathbf{2}$ played for sets. We want a graph Ω, together with a specified point $\mathbf{1} \xrightarrow{true} \Omega$, with the following property: The maps from any graph X to Ω are to correspond with the parts of X

$$\frac{S \hookrightarrow X}{X \xrightarrow{\varphi_S} \Omega}$$

in such a way that for each figure $T \xrightarrow{x} X$ of X, $\varphi_S x = true_T$ if and only if x is included in the part S. If such a pair Ω and $\mathbf{1} \xrightarrow{true} \Omega$ exists, it is uniquely characterized by the property above. It will be a surprising bonus that for each figure $T \xrightarrow{x} X$ the map $T \xrightarrow{\varphi_S x} \Omega$ tells us the 'level of truth' of the statement that x is included in the given part S.

Fortunately, such a pointed object exists in the category of graphs and it is the following $\mathbf{1} \xrightarrow{true} \Omega$:

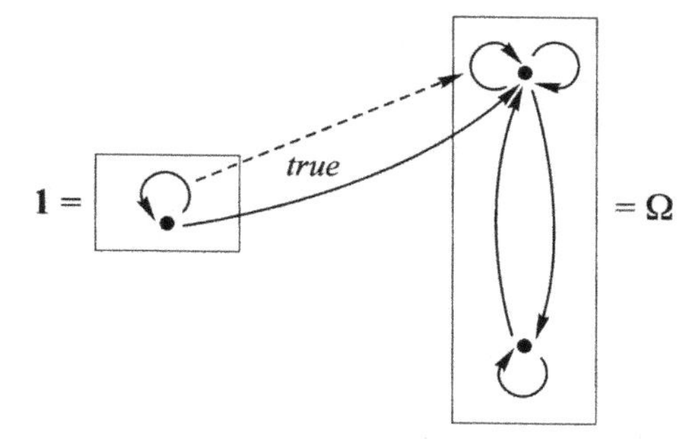

This graph Ω plays the role among graphs that the set **2** plays among sets. There are five arrows and two dots in Ω, which represent the various degrees of truth that a statement may have. (Here we consider statements of the form 'a certain figure is included in a certain subgraph.') These seven elements represent the seven possible relations in which an element of X (arrow or dot) may stand with respect to a given subgraph of X (there are five possibilities for an arrow and two for a dot). They are the following:

(a) For arrows:

1. The arrow is indeed included in the subgraph. Examples of this are the arrows x and y, with respect to the indicated subgraph of the following graph.

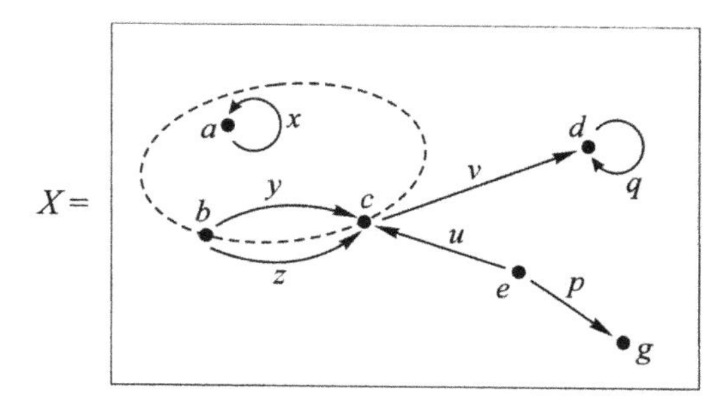

2. The arrow is not in the subgraph, but its source and its target are (e.g. the arrow z in the graph above).
3. The arrow is not in the subgraph and neither is its source, but its target is (e.g. the arrow u above).
4. The arrow is not in the subgraph and neither is its target, but its source is (e.g. the arrow v above).
5. The arrow is not in the subgraph and neither is its source nor its target (e.g. the arrows p and q above).

(b) For dots:

1. The dot is in the subgraph (e.g. dots a, b, and c above.)
2. The dot is not in the subgraph (e.g. dots d, e, and g in the graph pictured above).

Thus, in the example above, the characteristic map of the indicated part is the following:

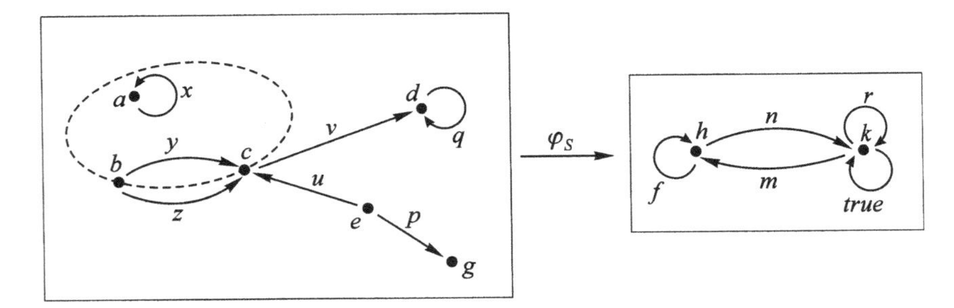

defined on arrows as $\varphi_S(x) = \varphi_S(y) = true$, $\varphi_S(z) = r$, $\varphi_S(u) = n$, $\varphi_S(v) = m$, $\varphi_S(p) = f$, and $\varphi_S(q) = f$ (and similarly on dots).

The category of dynamical systems also has a truth-value object Ω; it is surprising that it has an infinite number of elements or 'truth-values,' and yet it is not equal to the natural numbers with the *successor* endomap, it is rather opposite to it in the sense that the dynamics goes in the opposite direction. This object of truth values in the category of dynamical systems has the following picture:

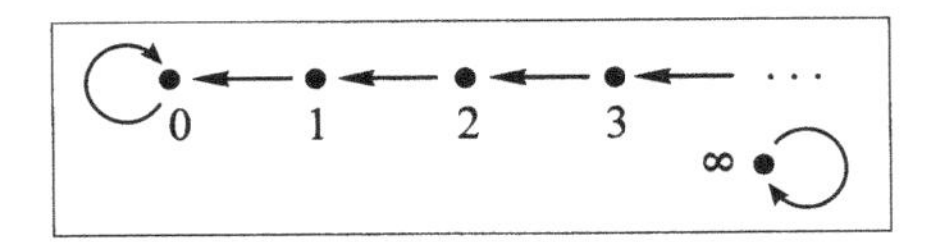

The explanation of this is that a subsystem is a part of a dynamical system that is closed under the dynamics and if you pick a state x and ask whether x is included in the subsystem, the answer may be 'no, but it will be included in one step,' or in two steps, etc. For example, consider the subdynamical system indicated below.

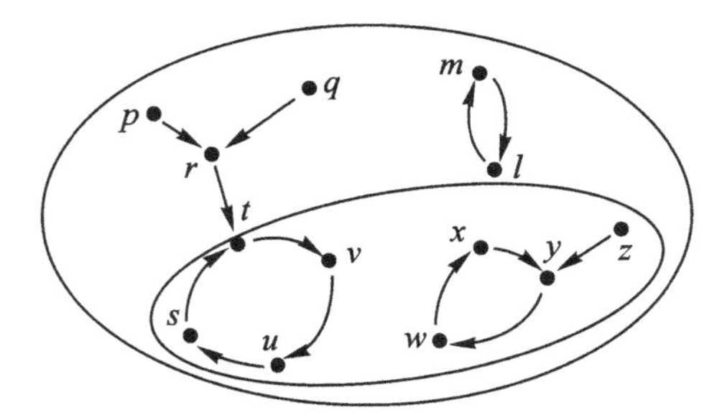

If we ask whether the state p is in the subsystem, the answer is Ω is 'No, but yes after two steps,' whereas the same question about m has the truth-value ∞, 'forever false'. Thus for every state in the big system, the statement that it is in the subsystem has a definite value in Ω; this value is *true* $= 0$ only for eight of the states.

DANILO: What about the case of an element leaving the subsystem?

We are speaking here about subdynamical systems, i.e. inclusion maps whose domains are also objects in the category of dynamical systems. Thus elements never 'leave'; the interesting feature is that they might 'enter' from the larger system. We can, of course, consider *subsets* of the underlying set of states of a given dynamical system.

There is, in fact, a larger *dynamical system* $\hat{\Omega}$ which we might call the space of 'chaotic truth-values' with the property that maps $X \xrightarrow{\varphi} \hat{\Omega}$ of dynamical systems correspond to these arbitrary subsets of X; only those φ which belong to $\Omega \hookrightarrow \hat{\Omega}$ correspond to actual subsystems. There are 'modal operators' $\hat{\Omega} \rightrightarrows \Omega$ which relate any subset A of any X to the smallest subsystem of X containing A and to the largest subsystem which is contained in A. As an exercise, can you make more specific what the states of $\hat{\Omega}$ must be and how they must evolve?

Exercise 3:
'Mathematical induction' is a term sometimes used for a particular kind of application of iteration, or recursion (see Exercise 3, page 334), namely to proving that a given part of $\mathbb{N}$ is actually the whole of $\mathbb{N}$. Show that the only subdynamical system of $N = (\mathbb{N}, \sigma)$ which contains the element 0 is N itself, by constructing via recursion a map which will be both a retraction and a section for the inclusion map of the given part.

A map space Ω^Y of the truth-value object has as its points the parts of Y, as discussed further in the next and final section. In many categories this enables us to obtain a further clarification of the construction of the coequalizer of an equivalence relation (see page 294) $X \rightrightarrows Y$:

Exercise 4:
If $X \rightrightarrows Y$ is a jointly monomorphic pair, then the subobject $X \longrightarrow Y \times Y$ of $Y \times Y$ has a characteristic map $Y \times Y \longrightarrow \Omega$. If $X \rightrightarrows Y$ is symmetric, then either of the two possible uses of the universal property of map spaces leads to the same map $Y \xrightarrow{\varphi_X} \Omega^Y$. If $X \rightrightarrows Y$ is also a transitive, reflexive graph, then its coequalizer gives a factorization $\varphi_X = i \circ q$, where a point of Ω^Y belongs to the part i iff it names a connected component (or 'equivalence class') of $X \rightrightarrows Y$.

SESSION 33

Parts of an object: Toposes

1. Parts and inclusions

In the previous session we used the idea of inclusion, which is the basis of truth and logic; now we will consider it in more detail. Logic (in the narrow sense) is primarily about subobjects; the important thing about subobjects is how they are related, and their most basic relationships are given by maps in a certain category.

If X is a given object of a category $\mathcal{C}$, then, as we have already explained, we can form another category $\mathcal{C}/X$: an object of $\mathcal{C}/X$ is a map of $\mathcal{C}$ with codomain X, and a map from an object $A = (A_0 \xrightarrow{\alpha} X)$ to an object $B = (B_0 \xrightarrow{\beta} X)$ is a map of $\mathcal{C}$ from A_0 to B_0 which β takes to α, i.e. a map $A_0 \xrightarrow{f} B_0$ such that $\beta f = \alpha$:

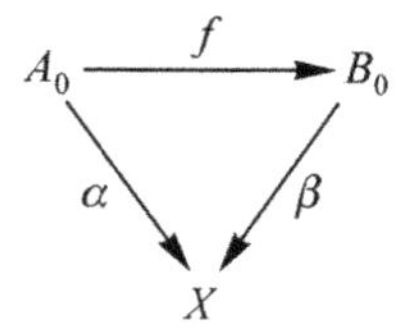

Of course, we obtain a category, because if we have another map

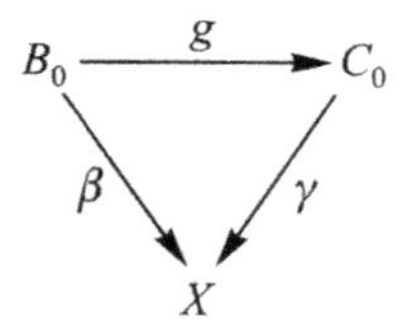

i.e. $\gamma g = \beta$, then gf is also a map in $\mathcal{C}/X$ since $\gamma(gf) = \alpha$:

$$\gamma(gf) = (\gamma g)f = \beta f = \alpha$$

We want to define a part of this category, denoted by $\mathcal{P}(X)$,

$$\mathcal{P}(X) \subseteq \mathcal{C}/X$$

which is called *the category of parts of* X. The *objects* of $\mathcal{P}(X)$ are all objects α of $\mathcal{C}/X$ which are *inclusion maps in* $\mathcal{C}$, i.e. the objects of $\mathcal{P}(X)$ are the parts or subobjects of X in $\mathcal{C}$. The *maps* of $\mathcal{P}(X)$ are all the maps between its objects in $\mathcal{C}/X$; but as was pointed out in the last session, given any two objects $A_0 \xrightarrow{\alpha} X$ and $B_0 \xrightarrow{\beta} X$ in $\mathcal{P}(X)$, there is *at most one* map $A_0 \xrightarrow{f} \beta_0$ in $\mathcal{C}$ such that $\beta f = \alpha$.

If the category $\mathcal{C}$ has a terminal object $\mathbf{1}$, then we can form the category $\mathcal{C}/\mathbf{1}$, but this turns out to be none other than $\mathcal{C}$, since it has one object for each object of $\mathcal{C}$ (a map $A_0 \longrightarrow \mathbf{1}$ contains no more information than just the object A_0 of $\mathcal{C}$) and its maps are precisely the maps of $\mathcal{C}$. Therefore the category of parts of $\mathbf{1}$, $\mathcal{P}(\mathbf{1})$, is a subcategory of $\mathcal{C}$, precisely the subcategory determined by those objects A_0 whose unique map $A_0 \longrightarrow \mathbf{1}$ is injective. Thus, while a subobject of a general object X involves both an object A_0 and a map $A_0 \xrightarrow{\alpha} X$, when $X = \mathbf{1}$ only A_0 need be specified, so that 'to be a part of $\mathbf{1}$' can be regarded as a *property* of the object A_0, rather than as an additional *structure* α.

As an example of the category of parts of a terminal object we can consider the parts of the terminal set in the category of sets. The objects of this category are all the sets whose map to the terminal set is injective. Can you give an example of such a set?

DANILO: The terminal set itself.

Yes. In fact, in any category all the maps whose domain is a terminal object are injective. And a map from a terminal object to a terminal object is even an isomorphism. Any other example?

FATIMA: The empty set.

Yes. The only map $\mathbf{0} \longrightarrow \mathbf{1}$ is also injective because any map with domain $\mathbf{0}$ is injective: for any set X there is at most one map $X \longrightarrow \mathbf{0}$ and therefore it is not possible to find two different maps $X \longrightarrow \mathbf{0}$ which composed with the map $\mathbf{0} \longrightarrow \mathbf{1}$ give the same result. Are there any other sets whose map to the terminal set is injective? No. Therefore the category of parts or subsets of the terminal set is very simple: it only has two non-isomorphic objects $\mathbf{0}$ and $\mathbf{1}$, and only one map besides the identities. It can be pictured as

$$\mathcal{P}(\mathbf{1}) = \boxed{\mathbf{0} \longrightarrow \mathbf{1}}$$

In this category it is usual to name the two objects $\mathbf{0}$ and $\mathbf{1}$ as '*false*' and '*true*' respectively, so that $\mathcal{P}(\mathbf{1})$ is also pictured as

$$\mathcal{P}(\mathbf{1}) = \boxed{\mathit{false} \longrightarrow \mathit{true}}$$

What about the category of graphs $\mathcal{C} = \mathcal{S}^{\downarrow\downarrow}$? What is the category of parts of the terminal object in this category? To answer this we must start by determining those graphs X such that the unique map $X \longrightarrow \mathbf{1}$ to the terminal graph is injective. For this it is useful to remember that a graph is two sets (a set of arrows and a set of dots) and two maps,

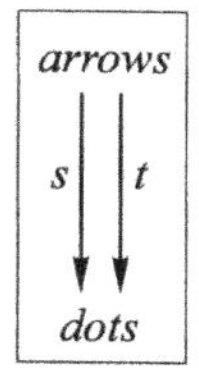

and that for the terminal graph both sets are singletons, so that we have to find the different possibilities for the sets *arrows* and *dots* for which the only map of graphs

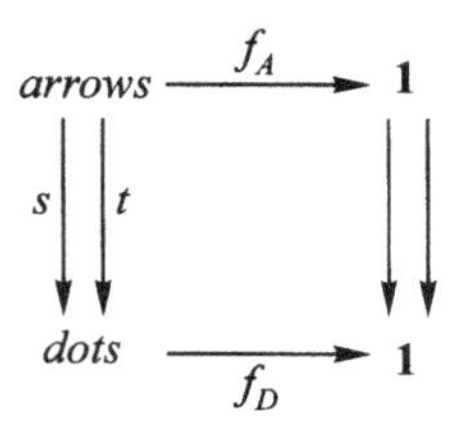

is injective. This means (exercise) that the two maps of sets f_A and f_D must be injective; each of the sets *arrows* and *dots* must either be empty or have one single element. Thus every subgraph of the terminal graph is isomorphic to one of these three graphs:

$$\boxed{\begin{array}{c}\mathbf{0}\\ \downarrow\downarrow\\ \mathbf{0}\end{array}} = \mathbf{0} = \bigcirc \qquad \boxed{\begin{array}{c}\mathbf{0}\\ \downarrow\downarrow\\ \mathbf{1}\end{array}} = D = (\bullet) \qquad \boxed{\begin{array}{c}\mathbf{1}\\ \downarrow\downarrow\\ \mathbf{1}\end{array}} = \mathbf{1} = (\bullet\circlearrowleft)$$

These three graphs and the maps between them form a category which can be pictured as

$$\mathcal{P}(\mathbf{1}) = \boxed{\mathbf{0} \longrightarrow D \longrightarrow \mathbf{1}}$$

The graphs **0, 1** are also called '*false*' and '*true*' respectively, so that we can put

$$\mathcal{P}(\mathbf{1}) = \boxed{\textit{false} \longrightarrow D \longrightarrow \textit{true}}$$

Here the graph D represents an intermediate 'truth-value' which can be interpreted as 'true for dots but false for arrows.'

The answers we got for 'parts of **1**' look familiar, because we have seen them before: the map $X \longrightarrow \mathbf{1}$ is injective if and only if X is idempotent.

As was pointed out at the beginning of this session, given any two objects $A \overset{\alpha}{\hookrightarrow} X$, and $B \overset{\beta}{\hookrightarrow} X$ in $\mathcal{P}(X)$ there is at most one map $A \overset{f}{\longrightarrow} B$ in $\mathcal{C}$ such that $\beta f = \alpha$. Thus, the category of parts of an object is very special. For any two of its objects there is at most one map from the first to the second. A category which has this property is called a *preorder*. Thus, *the category of subobjects of a given object in any category is a preorder*.

Therefore, to know the category of subobjects of a given object X, we need only know, for each pair of subobjects of X, whether there is or there is not a map from the first to the second. To indicate that there is a map (necessarily unique) from a subobject $A \overset{\alpha}{\hookrightarrow} X$ to a subobject $B \overset{\beta}{\hookrightarrow} X$ we often use the notation

$$A \subseteq_X B$$

(read: A is *included* in B over X); the 'A' is an abbreviation for 'the pair A, α', and similarly for B. The 'X' underneath helps remind us of that.

FATIMA: Can you explain the inclusion of one part into another with a diagram?

Yes. Suppose that some of the desks in the classroom have a chair behind them. If B is the set of chairs, we have the injective map 'is behind' from B to the set of desks, let's call it $B \xrightarrow{\beta} X$. We also have an injective map α from the set of A of students to the set X of desks – each student at one desk. Then the diagram of the two inclusions is as follows:

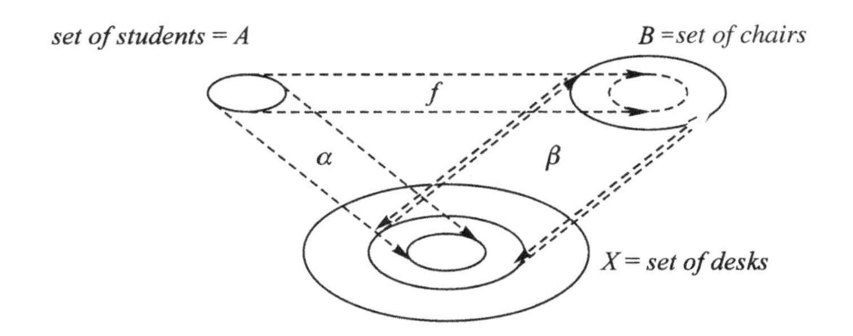

which shows that the desks occupied by students are included in the desks that have chairs. The reason or 'proof' for this inclusion is a map $A \xrightarrow{f} B$ (assigning to each student the chair behind their desk) such that $\beta f = \alpha$. This map is the (only) proof of the relation $A \subseteq_X B$.

DANILO: But if each chair is occupied by some student, the obvious map is from B to A, assigning to each chair its occupant.

Yes, there can be a map $B \xrightarrow{g} A$, but unless each chair is occupied $\alpha g \neq \beta$.

DANILO: So one must say '*if f* exists.'

That's right! There might not be any such f such that $\beta f = \alpha$, but there cannot be more than one since β is apart.

On the other hand, in some cases the map g may also be in the category $\mathcal{P}(X)$; i.e. it may be compatible with α and β ($\alpha g = \beta$). If so, it is also true that

$$B \subseteq_X A$$

Then, in fact, the maps f and g are inverses of each other, so that A and B are isomorphic objects; and more than that: $A \overset{\alpha}{\hookrightarrow} X$ and $B \overset{\beta}{\hookrightarrow} X$ are isomorphic objects in $\mathcal{P}(X)$. Thus, we have:

$$\text{If } A \subseteq_X B \quad \text{and} \quad B \subseteq_X A \quad \text{then} \quad A \cong_X B.$$

What does an isomorphism of subobjects mean? Suppose that on Friday and on Monday sets F and M of students occupied exactly the same chairs in the classroom. Then we have two different maps to the set of chairs, but they are isomorphic:

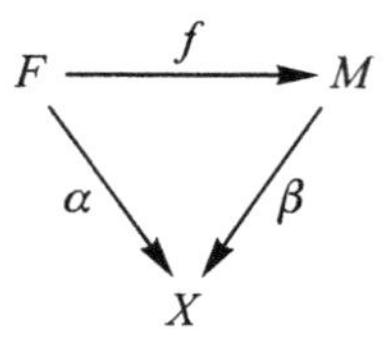

Since between any two isomorphic subobjects there is only one isomorphism, we treat them as the 'same subobject.'

The idea of an occupied chair can be expressed in the following way: suppose that we have a subobject $A \overset{\alpha}{\hookrightarrow} X$ and a figure $T \xrightarrow{x} X$ (which is not assumed to be injective). To say that x is in the subobject $A \overset{\alpha}{\hookrightarrow} X$ (written $x \in_X A$) means that there exists some $T \xrightarrow{a} A$ for which $\alpha a = x$. Now, since α is injective, there is at most one a that proves that $x \in_X A$. For example, if Danilo sits on this chair, then Danilo is the proof that this chair is occupied. According to the above definition, if we have $x \in A$ and $A \subseteq B$ (the X being understood) then we can conclude that $x \in B$, the proof of which is nothing but the composite of the maps $T \xrightarrow{a} A \xrightarrow{i} B$ proving respectively $x \in A$ and $A \subseteq B$.

The property above (if $x \in A$ and $A \subseteq B$ then $x \in B$) is sometimes taken as the *definition* of inclusion, because of the result of the exercise below.

Exercise 1:
Prove that if for all objects T and all maps $T \xrightarrow{x} X$ such that $x \in A$ it is true that $x \in B$, then necessarily $A \subseteq B$.
It suffices to consider $T = 1$ for sets, or
$T =$ the dot and $T =$ the arrow for irreflexive graphs.
The notation $x \in A$ is common when the figure x
has one or a few preferred shapes T.

2. Toposes and logic

It is clear from the above that one can discuss the category $\mathcal{P}(X)$ and the relations $\subseteq$, $\in$ in any category $\mathcal{C}$, but the 'logical' structure is much richer for those categories known as *toposes*.

Definition:
A category $\mathcal{C}$ is a **topos** *if and only if:*

1. *$\mathcal{C}$ has* **0**, **1**, $\times$, $+$, *and for every object X, $\mathcal{C}/X$ has products.*
2. *$\mathcal{C}$ has map objects Y^X, and*
3. *$\mathcal{C}$ has a 'truth-value object'* $\mathbf{1} \longrightarrow \Omega$ *(also called a 'subobject classifier').*

Most of the categories that we have studied are toposes: sets, irreflexive graphs, dynamical systems, reflexive graphs. (Pointed sets and bipointed sets are not toposes, since having map objects implies distributivity.)

We saw last session that the truth-value object in the category of sets is $\mathbf{2} = \{true, false\}$, while those in dynamical systems and irreflexive graphs are respectively

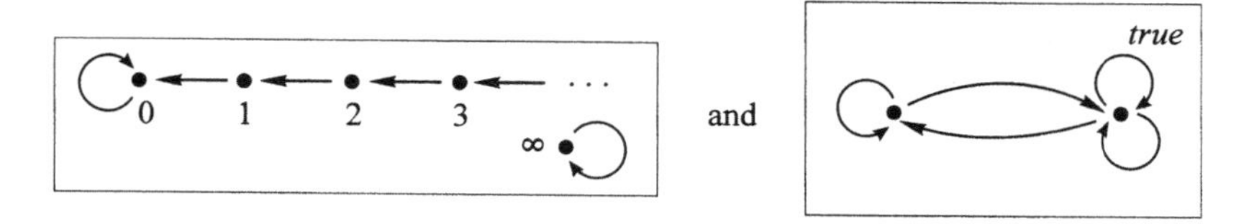

The defining property of a truth-value object or subobject classifier $\mathbf{1} \xrightarrow{true} \Omega$ was that for any object X the maps $X \longrightarrow \Omega$ are 'the same' as the subobjects of X. This idea is abbreviated symbolically as

$$\frac{X \longrightarrow \Omega}{? \hookrightarrow X}$$

This means that for each subobject, $A \overset{\alpha}{\hookrightarrow} X$ of X there is exactly one map

$$X \xrightarrow{\varphi_A} \Omega$$

having the property that for each figure $T \xrightarrow{x} X$, $\varphi_A x = true_T$ if and only if the figure x is included in the part $A \overset{\alpha}{\hookrightarrow} X$ of X.

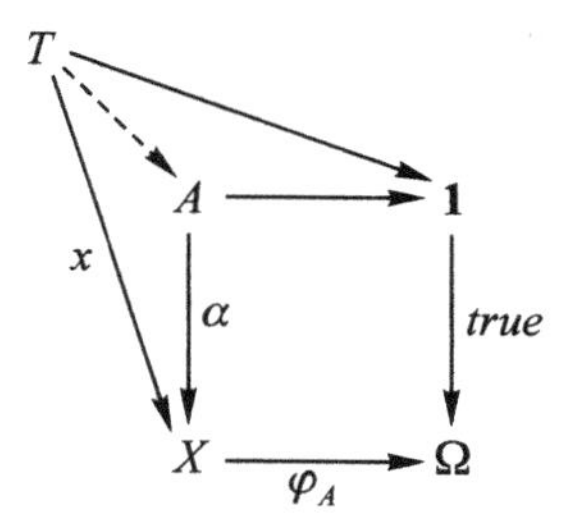

The consequence of the existence of such an object Ω is that everything one may say about subobjects of an object X can be translated into speaking about maps from X to Ω.

What is the relation between this and logic? We can form the product $\Omega \times \Omega$ and define the map $\mathbf{1} \xrightarrow{\langle true,\, true\rangle} \Omega \times \Omega$. This is injective because any map whose domain is terminal is injective; therefore this is actually a subobject and it has a classifying or characteristic map $\Omega \times \Omega \longrightarrow \Omega$. This classifying map is the logical operation '*and*,' denoted in various ways such as '&' and '$\wedge$.' The property of this operation is that for any $T \xrightarrow{a} \Omega \times \Omega$, say $a = \langle b, c\rangle$ where b and c are maps from T to Ω, the composite

$$T \xrightarrow{a} \Omega \times \Omega \xrightarrow{\wedge} \Omega \quad (b \wedge c)$$

(which is usually denoted $b \wedge c$ instead of $\wedge \circ \langle b, c\rangle$, just as we wtite $5+3$ instead of $+ \circ \langle 5, 3\rangle$) has the property that $b \wedge c = true_T$ if and only if $\langle b, c\rangle \in \langle true,\ true\rangle$, which means precisely: $b = true_T$ and $c = true_T$.

Now, because b is a map whose codomain is Ω, by the defining property of Ω it must be the classifying map of some subobject of T, $B \hookrightarrow T$. In the same way, c is the classifying map of some other subobject, $C \hookrightarrow T$, and the subobject classified by $b \wedge c$ is called the *intersection* of B and C.

Exercise 2:
Show that the intersection of two subobjects of T is, in fact, the product of these objects considered as objects of $\mathcal{P}(T)$.

Another logical operation is 'implication,' which is denoted '$\Rightarrow$.' This is also a map $\Omega \times \Omega \longrightarrow \Omega$, defined as the classifying map of the subobject $S \hookrightarrow \Omega \times \Omega$ determined by all those $\langle \alpha, \beta\rangle$ in $\Omega \times \Omega$ such that $\alpha \subseteq \beta$.

There is a third logical operation called '*or*' (disjunction) and denoted '$\vee$,' and there are relations among the operations $\wedge, \Rightarrow, \vee$, which are completely analogous to the relations among the categorical operations $\times$, map object, and $+$. Remember that these relations were

$$\frac{X \longrightarrow B_1 \times B_2}{X \longrightarrow B_1,\ X \longrightarrow B_2} \qquad \frac{X \longrightarrow Y^T}{T \times X \longrightarrow Y} \qquad \frac{B_1 + B_2 \longrightarrow X}{B_1 \longrightarrow X,\ B_2 \longrightarrow X}$$

The particular cases of these in the category $\mathcal{P}(X)$ of subobjects of X are the following 'rules of logic':

$$\frac{\xi \subseteq \beta_1 \wedge \beta_2}{\xi \subseteq \beta_1 \quad \text{and} \quad \xi \subseteq \beta_2} \qquad \frac{\xi \subseteq (\alpha \Rightarrow \eta)}{\xi \wedge \alpha \subseteq \eta} \qquad \frac{\beta_1 \vee \beta_2 \subseteq \xi}{\beta_1 \subseteq \xi \quad \text{and} \quad \beta_2 \subseteq \xi}$$

The middle rule is called the *modus ponens rule of inference*.

FATIMA: Shouldn't the last one say 'or' instead of 'and'?

No. In order that the disjunction '$\beta_1 \vee \beta_2$' be included in ξ it is necessary that both β_1 AND β_2 be included in ξ. This is another manifestation of the fact that products are more basic than sums. The conjunction 'and' is really a product, yet it is necessary in order to explain the disjunction 'or,' which is a sum.

A remarkable thing about the classifying map of a subobject is that although the subobject is determined by just the elements on which the classifying map takes value '*true*,' the classifying map also assigns many other values to the remaining elements. Thus, these other values are somehow determined by just those elements on which the map takes value '*true*.'

It is also possible to define an operation of negation ('not') by

$$not\ \varphi \stackrel{\text{def}}{=} [\varphi \Rightarrow false]$$

Then one can prove the equality

$$\varphi \wedge not\ \varphi \ = \ false$$

and the inclusion

$$\varphi \subseteq not\ not\ \varphi$$

In most categories this inclusion is not an equality. The universal property (rule of inference) for $\Rightarrow$ implies that for any subobject A of an object X, $not(A)$ is the subobject of X which is largest among all subobjects whose intersection with A is empty. Here is an example in graphs.

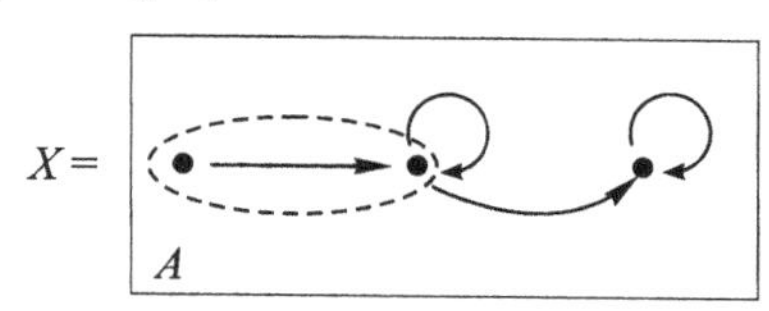

Exercise 3:
With the A pictured above, *not A* is the subgraph

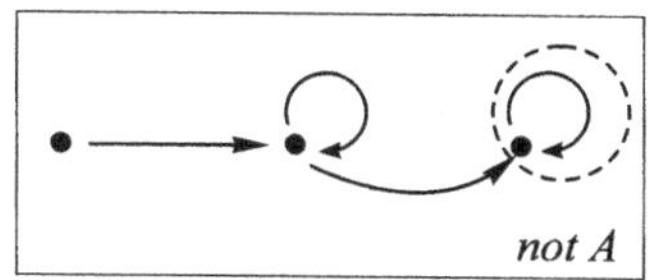

and *not not A* is

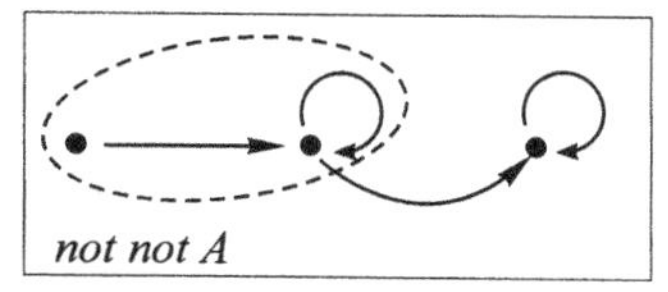

which is larger than A.

The logic in a topos such as this is said to be *non-Boolean*; the algebraist and logician G. Boole treated the special case in which *not not* $A = A$. Notice, however, that in this example

$$not\ not\ not\ A = not\ A$$

Exercise 4:
The '3 = 1 rule' (above) for '*not*' is correct in any topos.

We promised to show you a more general version of the proof by Cantor that $X < 2^X$ in sets (in the order given by monomorphisms); the conclusion will be that $X < \Omega^X$ in any non-trivial cartesian closed category with truth value object Ω.

Exercise 5:
The fact that $X \leq \Omega^X$ means that there exists a monomorphism. A standard example is the singleton map s, which is the exponential transpose of the characteristic map of the diagonal $X \longrightarrow X \times X$. Show that s is mono.

Exercise 6:
The inequality is strict, because if there were a mono $\Omega^X \longrightarrow X$, then by reasoning similar to that of Exercise 3, Session 29, it would follow that Ω has the fixed-point property. But the endomap *not* never has a fixed-point unless the whole category is trivial.
Hint: First show that the only possible fixed-point x would be x = false.

The fact encountered earlier in this section, that disjunction ('or') must be explained in terms of conjunction ('and') and not in terms of 'or', gave pause to many people meeting it for the first time (not just to Fatima). Indeed, there are often situations in daily life where the transformation from 'or' to 'and' can be puzzling if we try to explain it. Perhaps the following exercises will help to illuminate what is behind such a transformation. The algebra of parts (to which Article VI with Sessions 32 and 33 is a brief introduction) is useful for illuminating relations of actual spaces and transformations as they occur in mathematics. Conversely, the interplay of actual spaces and transformations can help to clarify the more abstract algebra of parts which is its reflection.

Exercise 7: Typically, each of the set C of customers who enter a restaurant for lunch on a certain day, has already decided whether to eat in OR take out. But approaching a restaurant one day, I happened to read the sign which said "Eat in AND take out". Surely, the owners do not want to force me to do both? If M denotes the selection of meals offered and hence M^C denotes the possibilities for the lunch transactions that day, use the exponential law

$$M^{C_1 + C_2} = M^{C_1} \times M^{C_2}$$

to explain why 'and' and 'or' are both correctly applied, but to different aspects. (Recall that 'and' is typically used in naming elements of product sets, see Session 1.)

Exercise 8: Of course, some customers do want to eat in and also to take out; show that by splitting C into three parts instead of two, this possibility is explained in the same way: the right hand side of the equation still involves categorical product.

To visualize these situations, it may be helpful to imagine each element of M^C as a possible record of the day's lunch transactions; each record is a list of 'eat in' sales AND a list of 'take out' sales. On the other hand, each customer occurs in one OR the other of these lists. (There is no significant change if we include three kinds of customer preferences as in Exercise 8; then the one record is really one list AND another list AND another list.)

ARTICLE VII

The Connected Components Functor

1. Connectedness versus discreteness

Besides map spaces and the truth space, another construction that is characterized by a 'higher universal mapping property' objectifies the counting of *connected components*. Reflexive graphs and discrete dynamical systems, though very different categories, support this 'same' construction. For example, we say that dots d and d' in a reflexive graph are **connected** if for some $n \geq 0$ there are

dots $d = d_0, d_1, \ldots, d_n = d'$ and
arrows $a_1, \ldots, a_n$ such that
for each i either the source of a_i is d_{i-1} and the target of a_i is d_i
or the source of a_i is d_i and the target of a_i is d_{i-1}.

The graph as a whole is **connected** if it has at least one dot and any two dots in it are connected; it is noteworthy that to prove a given graph to be connected may involve arbitrarily long chains of elementary connections a_i, even though the structural operators s, t, i are finite in number (Sessions 13 to 15).

By contrast, this aspect of steps without limit does not arise in the same way for dynamical systems, even though the dynamical systems themselves involve infinitely many structural operators α^n, effecting evolution of a system for n units of time. We say that the states x, y are **connected** if there are n, m such that $\alpha^n x = \alpha^m y$, and that the system is connected if it has at least one state and every two states are connected.

Exercise 1: Suppose that there exist n, m so that $\alpha^n x = \alpha^m y$. If moreover y is connected to a third state z, for example if $\alpha^k y = \alpha^p z$, show that x is connected to z.
Hint: The addition of natural numbers is commutative.

In these examples we can see intuitively that every reflexive graph or dynamical system is a coproduct (disjoint sum) of connected pieces, usually called *components*. The number of pieces can be precisely defined before we get into the difficulties of computing that number in a particular category. The idea of this definition involves the contrast between arbitrary objects and discrete objects. In many categories we can define a subcategory of 'discrete' objects. Here we give two examples, reflexive graphs and dynamical systems, and we will give more in section 3 of this article.

Definition 1: *A reflexive graph is* **discrete** *if it has no arrows other than the degenerate loop at each dot (roughly, we may say that it 'consists only of dots'). A dynamical system is* **discrete** *if all states are rest states.*

Clearly, if $X \longrightarrow S$ is a map to a discrete object S then any two dots (or states) that are connected in X must map to the same dot (or state) in S. Thus we are led to the following universal property, which makes sense for graphs, dynamical systems, and many other categories in which a subcategory of objects called discrete has been specified.

Definition 2: *A map* $X \longrightarrow \pi_0 X$ *with discrete codomain is* **universal** *if for any map* $X \longrightarrow S$ *with discrete codomain there is exactly one map* $\pi_0 X \longrightarrow S$ *making a commutative triangle. Such a* $\pi_0 X$ *(clearly unique up to a unique isomorphism) is often called the* **space of components** *of* X. *If* $1 \xrightarrow{i} \pi_0 X$ *is any point of it, then the inverse image* $X_i \hookrightarrow X$ *under the universal map is called the* **i-th connected component** *of* X.

Exercise 2: If universal components maps as in Definition 2 are specified for both X and Y and if $X \xrightarrow{f} Y$ is any map, then there is exactly one map $\pi_0 X \longrightarrow \pi_0 Y$ giving rise to a commutative square. These 'induced maps' compose.

Exercise 3: In reflexive graphs and dynamical systems, X is connected if and only if $\pi_0 X = 1$.

In our examples and many others of interest, the subcategory of discrete spaces is equivalent as a category in its own right to the category $\mathcal{S}$ of abstract sets. In these, a more detailed description of the π_0 construction can be given as follows: The category of discrete objects can be considered in its own right, with its inclusion $I : \mathcal{S} \longrightarrow \mathcal{X}$ into the category $\mathcal{X}$ of interest. Given X we consider FX an abstract set of tokens (or names) for the connected components of X, and we note that any abstract set W can be construed as a discrete space IW. Then, $\pi_0 X = IFX$. The processes I and F are both functors and the universal property relating them is a special case of adjointness as defined in Appendix II. In fact, F is left adjoint to the inclusion I. (Remark: Exercise 4 of Session 32 suggests a way to construct F.)

When are two figures $A \rightrightarrows X$ connected, i.e. become the same on composing with the natural $X \longrightarrow \Pi_0 X$? That question is difficult in general, but in two special cases we already gave the answer: nodes $1 \rightrightarrows X$ in a reflexive graph (or states in a dynamical system) are connected if and only if they satisfy the criterion we used as definition at the beginning of this article.

2. The points functor parallel to the components functor

The above description of π_0 as a composite process contrasts with a parallel process involving the right adjoint to same inclusion I of discrete spaces in all spaces of a given category $\mathcal{X}$.

Definition 3: *A map* $|X| \longrightarrow X$ *with discrete domain is universal if for every map* $S \longrightarrow X$ *with discrete domain, there is exactly one* $S \longrightarrow |X|$ *making a commutative triangle. This* $|X|$ *is the* **space of points** of X.

> **Exercise 4:** Since 1 is discrete, show that every point $1 \longrightarrow X$ belongs to $|X|$.

But $|X|$ may in some cases involve more than collecting the bare points of X, because it has the structure of an X-space, even if in a relatively trivial way.

> **Exercise 5:** Show that there is a process G giving an $\mathcal{S}$-object for every $\mathcal{X}$-object which via construal I becomes $|\ \ |$, i.e.
>
> $$|X| = IG(X).$$
>
> **Exercise 6:** There is a standard map
>
> $$|X| \longrightarrow \pi_0 X$$
>
> for any X, giving more information about X. Give examples of reflexive graphs X for which this standard map is an isomorphism, as well as other examples where $\pi_0 X = 1$ while $|X|$ is arbitrarily large. Show that in the category of dynamical systems (indeed, of actions of any given commutative monoid, see Definition 4) the standard map is always a monomorphism.

3. The topos of right actions of a monoid

A class of toposes defined over $\mathcal{S}$ for which the components and points functors can be studied, are the toposes of actions, or 'presheaves'. We will focus only on actions of a monoid (see Session 13). Monoids that happen to be small can be described in an equivalent way.

Definition 4: *A* **small monoid** *is a set* **M** *with a given associative multiplication* $M \times M \longrightarrow M$ *with unit* $1 \longrightarrow M$. *A* **representation** *of* M, *or* **right action** *of* M *on a set* X *is a given map* $X \times M \longrightarrow X$, *denoted by juxtaposition, satisfying*

$$x(ab) = (xa)b$$
$$x1 = x$$

where we have used juxtaposition ab *to denote the composition in* M *and* 1 *to denote the unit of the monoid.*

For example, a right action of the additive monoid of natural numbers is just a dynamical system; a right action of the three-element monoid, in which the two non-identity elements satisfy the four equations $s_i s_j = s_i$, is nothing but a reflexive graph (with dots defined to be the common fixed elements under the two idempotent 'source and target' actors s_0, s_1).

Exercise 7: For any monoid M, define 'maps of right M-actions on sets', generalizing the notions of map for reflexive graphs and for dynamical systems. The resulting category is denoted $\mathcal{S}^{M^{op}}$.

Exercise 8: Show that among right M-actions on sets there is a special one $\underline{M}$ whose states are just as many as the elements of M and for which the states of any X correspond to the maps $\underline{M} \xrightarrow{x} X$ that preserve the action as in Exercise 7. In particular, the maps $\underline{M} \xrightarrow{a} \underline{M}$ correspond to left multiplications by the elements of M in such a way that the condition to be a map $X \xrightarrow{f} Y$ becomes a special case of the associativity of the compositon in $\mathcal{S}^{M^{op}}$.

$$f(xa) = (fx)a$$

The content of Exercise 8 is due to Cayley (in the case where all elements of M are invertible) and was generalized by Yoneda in the 1950's to presheaves, i.e. right actions of a small category with more than one object. A very simple case of that generalization involves the irreflexive graphs as actions of a two-object category, but more general cases justify the figures-and-incidence interpretation of general geometric structures as in Appendix I.

Exercise 9: For any monoid M, the truth-value object Ω in the category of right actions has as many elements as M has right ideals, that is, subsets V of M for which ma is in V whenever m is in V and a is in M.

Exercise 10: The map space Y^X of two right actions X, Y of M has as many states as there are maps (in the category of actions) $\underline{M} \times X \longrightarrow Y$. These reduce to mere maps $X \longrightarrow Y$ of the underlying sets only in case every element of M is invertible. (See Exercise 6, Session 34).

SESSION 34

Group theory and the number of types of connected objects

The connected components functor F on a category $\mathcal{C}$ can reveal much about $\mathcal{C}$. For example, how many connected objects does $\mathcal{C}$ have?

Definition 1: C *is called* **connected** *if it has exactly one component, i.e.* $FC = 1$ *(where* F *is the left adjoint to the inclusion* I *of the category of discrete objects). This is equivalent to* $\pi_0 C = 1$, *where* $\pi_0 = IF$.

> **Exercise 1:** The terminal object is connected.
> Hint: use the fullness of I.
> C is connected if and only if it has exactly two maps to $1 + 1$.

Most of the categories we have studied have infinitely many non-isomorphic connected objects. For example, for any given non-empty set S, there is a standard reflexive graph X with $FX = 1$ but with S dots, given by the projections and the diagonal map

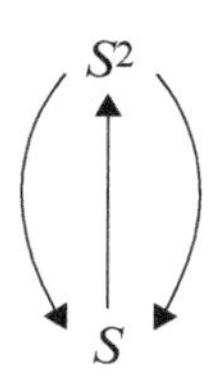

It has S^2 arrows: for each pair p, q of dots, there is a connecting arrow from p to q. But even a graph X with fewer arrows can have S dots and $FX = 1$, because p might be connected to q only by a string of arrows.

> **Exercise 2:** In the category $\mathcal{S}^{\downarrow}$ whose objects are maps of sets (even finite sets) there are infinitely many non-isomorphic objects with one component (and also infinitely many with exactly one point).

Roughly speaking, if all structural actors are invertible, there is a major simplification, important in group theory.

Definition 2: *A* **group** *is a monoid in which every element has an inverse.*
While the theory of monoids and their actions already has much content, the group case has special features which arise mainly from these three:

(1) Each congruence relation on a group arises from a 'normal subgroup' (Exercise 3 below);
(2) for any two representations, the underlying set of the map space is the map set of the underlying sets (Exercise 6 below);
(3) given elements x and y in a connected representation, there is a group element g with $xg = y$ (Exercise 11 below).

Feature (3) of group theory implies that there are only a small number of connected representations, as we will prove in this session.

Exercise 3: Let $G \xrightarrow{F} H$ *be* a homomorphism of groups, i.e. $F(xy) = F(x)F(y)$. Study the pairs a_1, a_2 for which $F(a_1) = F(a_2)$ in the following way: Define Ker(F) to be the subgroup of all maps a in G for which $F(a)$ is the identity element of H. Then
(a) Ker(F) is a subgroup, and moreover for any t in G

a is in Ker(F) if and only if $t^{-1}at$ is in Ker(F).

(b) $F(a_1) = F(a_2)$ if and only if $a_1a_2^{-1}$ is in Ker(F).
(c) Any 'normal' subgroup K of G, i.e. a subgroup satisfying the condition in (a), arises as $K = \text{Ker}(F)$ for some functor to some group H.

Our main goal is the theorem after Exercise 11. A hint of its content is already in Exercise 4, which does not mention actions.

Exercise 4: If

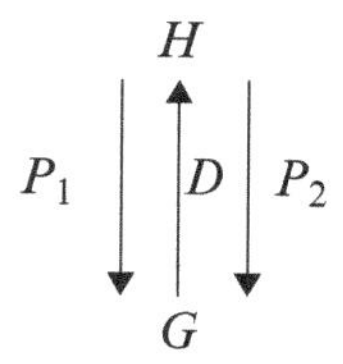

are three homomorphisms of monoids for which $P_1D = P_2D = 1_G$, the identity homomorphism, consider the (obviously reflexive) relation R on G given by

$$a_1Ra_2 \text{ if and only if } P_1(h) = a_1 \text{ and } P_2(h) = a_2 \text{ for some } h.$$

If H and G are groups, then R is also symmetric and transitive. Indeed, there is a group G/R and a homomorphism $G \xrightarrow{Q} G/R$ such that $QP_1 = QP_2$ and any homomorphism from G coequalizing P_1, P_2 factors uniquely as $(\;) \circ Q$.

The fact that R is automatically transitive expresses itself in the category of representations of a group G.

Exercise 5: An action of a group G is often called a representation of G by permutations, because if X has a given action of G then for each element g of G, the map $X \xrightarrow{\phi} X$ defined by $\phi(x) = xg$ is an isomorphism of sets. What is its inverse?

The category of actions of a group is actually a topos, but this takes place in a very special way compared to the topos of actions of a category (or even a monoid) which is not a group; Exercises 6 and 7 spell this out.

Exercise 6: Given two actions X, Y of a group G, consider the set Y^X of all abstract maps (i.e. not necessarily preserving the actions) from X to Y, equipped with the action

$$(f{\cdot}a)(x) = (f(x\,a^{-1}))a$$

Show that this is actually the determination of the exponential object or map space in the category of actions.

Exercise 7: For any group G and any sub-action X_1 of a G-action X, there is another sub-action X_2 disjoint from X_1 such that $X_1 + X_2$ is the whole X. In other words, there is a unique map of actions $X \longrightarrow 2$ such that X_1 is the inverse image of one of the points; take X_2 to be the inverse image of the other point.

Exercise 7 implies that the logic of the topos of actions of a group is Boolean in the sense of Session 33. Actions of a groupoid also have the property that every subobject has a complement.

Exercise 8: Every element of G determines a map of G representations $\underline{G} \longrightarrow \underline{G}$. Conversely, every map $\underline{G} \longrightarrow \underline{G}$ preserving the action is effected by left multiplication by a unique element of $\underline{G}$. (The special action $\underline{G}$ from Exercise 8, Article VII is often called the **regular representation** of G.)

Exercise 9: For each state x in a representation X of G, there is exactly one map of representations $\underline{G} \longrightarrow X$ whose value at 1 is x. For simplicity we identify states with $\underline{G}$-shaped figures.

Exercise 10: 'Points' (1-shaped figures) $1 \longrightarrow X$ in the category of G-representations are just equilibria (fixed points of the action).

In fact, the 'Cayley – Yoneda' results, i.e. Exercises 8, 9, and 10, are true for monoids and categories; they do not depend on the invertibility of the actors from G. By contrast the next result requires the invertibility.

Exercise 11: If G is a group with regular representation $\underline{G}$ and if X is a *connected* object in the topos of representations, then for any state x in X, the map $\underline{G} \xrightarrow{x} X$ is *surjective.*

Hint: Consider the image, in the sense of Article VI, of x (obviously a sub-representation $X_1 \hookrightarrow X$. The complementary subset X_2 is also a sub-representation. (This is the crucial fact: using the invertibility, for y in X_2 none of the yb are in X_1.) Thus if X_2 is not zero, there is a surjective map of G-representations $X \longrightarrow 2 = 1 + 1$, showing that X is not connected.

Theorem: *The number of non-isomorphic connected representations of a group* G *is at most the number of subgroups of* G *(and so is finite if* G *is).*

Proof: Any connected representation X of G is determined up to isomorphism by the stabilizer $G(x)$ of any chosen one of the states x and X. Here $G(x)$ is the subgroup of all those a for which $xa = x$.

Exercise 12: If two connected representations X, Y are isomorphic (by f) and if x, y are given states of X, Y, then the stabilizer subgroups $G(x)$, $G(y)$ are 'conjugate', that is, there exists g in G such that $gG(x) = G(y)g$.

Hint: y is surjective, so choose g for which $yg = fx$. Then for any a such that $xa = x$, it follows that $b = gag^{-1}$ is such that $yb = y$. Conversely, if b is any element of G that stabilizes y, then $g^{-1}bg$ stabilizes x.

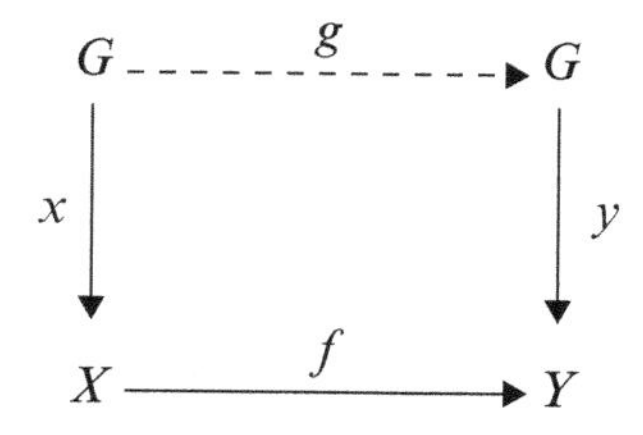

SESSION 35

Constants, codiscrete objects, and many connected objects

1. Constants and codiscrete objects

We saw in Session 34 that among the right actions of a finite group there are only a finite number of non-isomorphic connected objects. Conversely, for a finite monoid M that is not a group, one can always find an infinite number of non-isomorphic finite connected right M-actions. We will show this here only for a special class of monoids.

Definition 1: *A* **constant** *of a monoid* M *is an element* c *such that* cm = c *for all elements* m *of* M.

Exercise 1: Maps in the topos of right M-actions, from 1 to $\underline{M}$, correspond to constants of M.

Exercise 2: If a group M has a constant, then M has only one element.

Exercise 3: If c is a constant in a monoid M and if m is any element of M, then mc is also a constant.

Let C be the set of all constants of a monoid M and let S be any set. Then the map set

$$J(S) = S^C$$

has the natural structure of a right M-set, namely

$$(y \cdot m)(c) = y(mc)$$

defines a new $C \xrightarrow{y \cdot m} S$ for each given $C \xrightarrow{y} S$ (see Exercise 3). This construction J takes sets to right M-actions in a way opposite to the discrete inclusion I. If C is not empty, the values $J(S)$ are called **codiscrete**.

Exercise 4: Let M be a monoid with at least one constant. For any right M-action X, each M-map $X \longrightarrow J(S)$ is induced by exactly one set map $G(X) \longrightarrow S$, where $G(X)$ is the set of fixed points of the action. In particular, the identity map on $J(S)$ corresponds to a map $GJ(S) \longrightarrow S$, which is invertible; $GJ(S) \approx S$. The only maps from a codiscrete action to a discrete action are constant.

Because of the universal mapping property established in Exercise 4 we say that the codiscrete inclusion J is right-adjoint to the fixed-points functor G. Indeed, we have a string of four adjoints

$$F \dashv I \dashv G \dashv J$$

relating the M-actions to the abstract sets. While $FI \approx GI \approx GJ$ are all equivalent to the identity functor on $\mathcal{S}$, the composite FJ assigns, to every set S, the set of (tokens for) connected components of the codiscrete action on S^C. In the next section we show that $FJS = 1$ for $S \neq 0$; but for that we will need at least two constants in M.

Exercise 5: If M has exactly one constant, then $J = I$, i.e. the codiscrete and discrete subcategories of actions coincide.

2. Monoids with at least two constants

This is a big generalization of the example of reflexive graphs which, however, by our coarse measurements acts rather like that example.

Proposition: *If a monoid* M *has at least two constants, then for every nonempty set* S, *the codiscrete action* J(S) *is connected.*

Proof: Among the maps $C \longrightarrow S$ are the constant maps $C \xrightarrow{\overline{s}} S$, one for each s in S. For any $C \xrightarrow{x} S$ and c in C, $x{\cdot}c = \overline{x(c)}$. Let c_0 and c_1 be distinct in C. Then for any s_0 and s_1 in S, we can choose $C \xrightarrow{z} S$ with $z(c_0) = s_0$ and $z(c_1) = s_1$. To connect the pair $C \overset{x}{\underset{y}{\rightrightarrows}} S$, let $s_0 = x(c_0)$ and $s_1 = y(c_1)$. Then the chosen z satisfies $x{\cdot}c_0 = z{\cdot}c_0$ and $z{\cdot}c_1 = y{\cdot}c_1$, completing the proof.

In fact, there are even more connected M-actions. It can be shown that if M has at least two constants, then the truth-value space Ω in the category of right M-actions is connected. Indeed, for any right M-action X, the map space Ω^X is connected. Hence any X is the domain of a subobject of a connected object, for example $X \longrightarrow \Omega^X$ by the 'singleton' inclusion.

APPENDICES

Toward Further Studies

The goal of this book has been to show how the notion of composition of maps leads to the most natural account of the fundamental notions of mathematics, from multiplication, addition, and exponentiation, through the basic notions of logic and of connectivity.

Your further work with mathematics may apply it to physics, computer science or to other fields. In each of these, illuminating guides to the formulation and solution of problems often come from explicit recognition of structures occurring in commutative algebra, functional analysis, algebraic topology, etcetera. Clarifying unification of these branches has been developed during the last 60 years, using the categorical methods that you have begun to learn. To begin to deepen your knowledge of categories, here are four appendices which, although too brief for learning the subjects thoroughly, outline some important connections, in formulations which you will recognize in your subsequent encounters with mathematical topics.

Appendix I A general description of the geometry of figures in a space and the algebra of functions on a space, together with their basic functorial behavior.

Appendix II The description of Adjoint Functors and how they are exemplified in the categories of directed graphs and dynamical systems.

Appendix III A very brief history of the emergence of the theory of categories from within various mathematical subjects.

Appendix IV An annotated bibliography to guide you through elementary texts, monographs, and historical sources.

May you enjoy the fruits of your perseverance.

Bill and Steve

APPENDIX I

Geometry of figures and algebra of functions

1. Functors

The notion of category embraces at least three related sorts of mathematical environments:

(a) an abstract theory $\mathcal{A}$ of some notion (such as 'preorder' or 'group' or 'dynamical evolution');
(b) a background $\mathcal{B}$ (such as smooth spaces) in which that notion might be realized;
(c) the concrete totality $\mathcal{C}$ of realizations of $\mathcal{A}$ in $\mathcal{B}$.

The example (a) reflects the observation that the substitutions in a theory $\mathcal{A}$ can be composed, hence are the maps in a category. A realization R of $\mathcal{A}$ in $\mathcal{B}$ is then a *functor* from $\mathcal{A}$ to $\mathcal{B}$: a functor is a transformation that turns objects and maps in $\mathcal{A}$ into objects and maps in $\mathcal{B}$ in such a way as to satisfy the equations

$$R(\varphi\psi) = R(\varphi)R(\psi)$$
$$R(1_A) = 1_{RA}$$

that express compatibility of R with composition of maps in $\mathcal{A}$ and $\mathcal{B}$. The many examples of this kind demanded the recognition of the 'category of categories': an object is a category and a map is a functor. The category of categories has map objects, called 'functor categories'. That is, we have for each $\mathcal{B}$ and $\mathcal{A}$ a category $\mathcal{B}^{\mathcal{A}}$ and a bijective correspondence

$$\frac{\text{functors } \mathcal{X} \longrightarrow \mathcal{B}^{\mathcal{A}}}{\text{functors } \mathcal{A} \times \mathcal{X} \longrightarrow \mathcal{B}}$$

Specializing $\mathcal{X}$ to be $\mathbf{1}$ shows that an object of $\mathcal{B}^{\mathcal{A}}$ is a functor $\mathcal{A} \longrightarrow \mathcal{B}$; specializing to $\mathbf{2}$ (the two-object, three-map preorder) shows that a map in $\mathcal{B}^{\mathcal{A}}$ is a functor $\mathcal{A} \times \mathbf{2} \longrightarrow \mathcal{B}$; in particular, a map γ in $\mathcal{B}^{\mathcal{A}}$ from $R_0 : \mathcal{A} \longrightarrow \mathcal{B}$ to $R_1 : \mathcal{A} \longrightarrow \mathcal{B}$ is called a 'natural transformation': it assigns to each object A in $\mathcal{A}$ a map $R_0 A \xrightarrow{\gamma_A} R_1 A$ in $\mathcal{B}$, in a way compatible with each map α in $\mathcal{A}$, in the sense that if $\alpha: A' \longrightarrow A$, then the following diagram pictures a representation of $\mathbf{2} \times \mathbf{2}$ in $\mathcal{B}^{\mathcal{A}}$, i.e. is a commutative square in $\mathcal{B}^{\mathcal{A}}$:

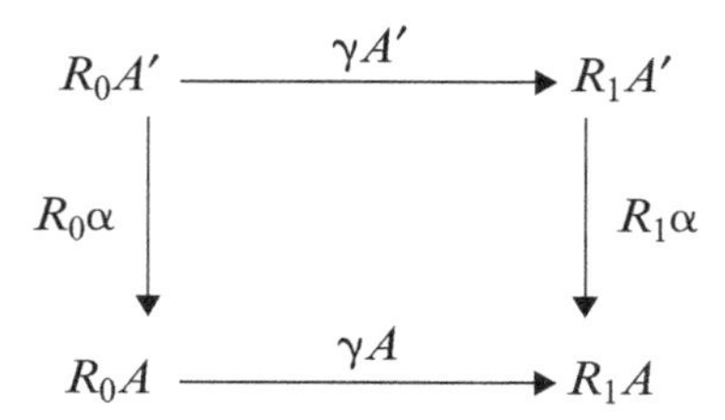

These functor categories give rise to the interpretation (c) of the realizations of $\mathcal{A}$ in $\mathcal{B}$ as a category $\mathcal{C} = \mathcal{B}^{\mathcal{A}}$.

This use of **1**, **2**, **2** × **2**, is based on the insight that these very simple categories are 'basic figure shapes' for the analysis of general categories; it is an example of a general method for analyzing the 'inside' of objects in any category, as explained below.

2. Geometry of figures and algebra of functions as categories themselves

Given an object X in a category $\mathcal{C}$, we will construct two categories $\mathcal{F}ig(X)$ and $\mathcal{F}cn(X)$; the picture illustrates $\mathcal{F}ig(X)$ and $\mathcal{F}cn(Y)$ for two objects X and Y.

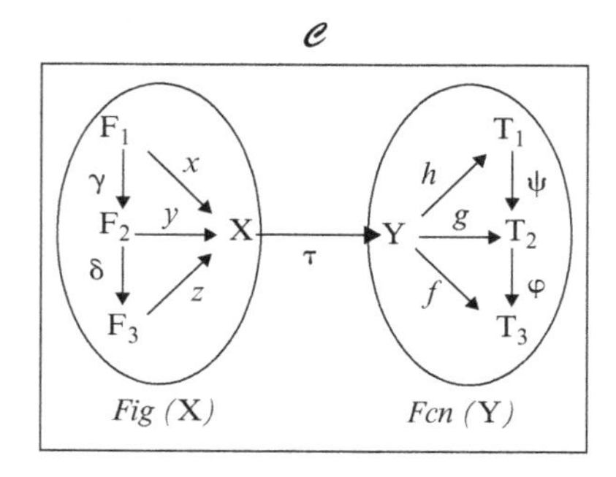

In a category of spaces, maps with codomain X may be considered as *figures* in X, and the domain of a figure may also be called its shape. A figure x may be said to be in another figure y in X, in a specific way, if a map γ is given, such that $x = y\gamma$; these γ are the maps of the *category of figures in* X; maps of figures and their compositions enable geometers to fully express the 'incidence relations' between figures that give X its geometrical structure. In particular, y is a *part* of X if for all x there is at most one way to be in y, i.e. at most one γ demonstrating the incidence. The 'x is in y' relation, like the notation used by Dedekind and Banach, subsumes both 'membership' (the case where y is a part and x is of some specified shape) and 'inclusion' (the case where both are parts).

Of equal importance with the geometry of figures is the algebra of *functions*: maps with domain Y may be considered as intensively variable quantities on Y; the codomain of such a function is often called its 'type'. A function f may be said to be *determined by* g if we can find a map φ which operates algebraically to implement the determination i.e. $f = \varphi g$; in that case one often says simply that f is a function of g. The maps φ are the maps of a category whose objects are functions on Y. This category often has categorical products and hence each instance of the action of a map $T \times T \longrightarrow T$ (for example addition) as a binary operation on functions is a map in this category.

The transformation of X into Y by a map τ induces transformations on the function algebras and the figure geometries. The induced transformation τ^* on function algebras is called contravariant because it goes back from the functions on Y to the functions on X, as shown in the figure above; it is called a homomorphism because it preserves all the algebraic operations on functions. On the other hand, the transformation $\tau_!$ on figure geometries is covariant and is called smooth because it preserves the incidence relations without tearing them. Even if x is a part, a transform τx need not be (for example, a triangle may degenerate into a line segment or even to a point, or a convergent sequence may happen to be constant). Figures that are not parts are often called 'singular' figures. Both of these induced transformations are simply compositions with τ, and the elementary properties just mentioned follow follow from the associativity of composition in the ambient category $\mathcal{C}$.

The smooth maps and homomorphisms induced by a transformation τ of spaces can be considered as functors as follows. The smooth map $\tau_!$ of geometries

$$\mathcal{F}ig(X) \xrightarrow{\tau_!} \mathcal{F}ig(Y)$$

preserves the shapes of the figures and the homomorphism τ^* of algebras

$$\mathcal{F}un(Y) \xrightarrow{\tau^*} \mathcal{F}un(X)$$

preserves the types of functions. Composite transformations induce composite functors: $(\sigma\tau)^* = \tau^*\sigma^*$. These generalities about associativities take on very particular significance if we choose a subcategory $\mathcal{F} \longrightarrow \mathcal{C}$, often with only three or four objects, to serve as shapes $F_1, F_2, F_3 \ldots$ of figures. Likewise we can choose a subcategory $\mathcal{T} \longrightarrow \mathcal{C}$ to serve as types $T_1, T_2, T_3 \ldots$ of functions. The fact that the induced homomorphisms (respectively smooth maps) preserve types (respectively shapes) is then explicitly expressed by the commutativity of the diagrams of functors.

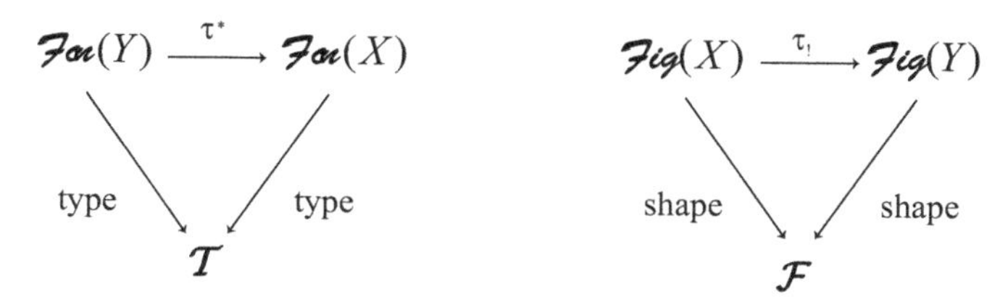

By judicious choice of the subcategory $\mathcal{F}$ of shapes we often achieve the 'adequacy' described by Isbell: we are justified in identifying a space X with its geometry of $\mathcal{F}$-figures provided that every functor defined just for its $\mathcal{F}$-figures and preserving shapes is induced by exactly one $\mathcal{C}$-map τ from the space X to Y. Dually, via the choice of a subcategory $\mathcal{T}$ (of objects like the space of real numbers or of truth values) we can represent $\mathcal{C}$ in the opposite of a category of algebras.

APPENDIX II

Adjoint functors with examples from graphs and dynamical systems

Much mathematical struggle (e.g. 'solving equations') aims to partially invert a given transformation. In particular, in the case of a functor Φ from one category to another, Kan (1958) noticed that there is sometimes a uniquely determined functor in the opposite direction that, while not actually inverting Φ, is the 'best approximate inverse' in either a left- or a right-handed sense. The given functor is typically so obvious that one might not have mentioned it, whereas its resulting adjoint functor is a construction bristling with content that moves mathematics forward.

The uniqueness theorem for adjoints permits taking chosen cases of their existence as axioms. This unification guides the advance of homotopy theory, homological algebra and axiomatic set theory, as well as logic, informatics, and dynamics.

Roughly speaking these reverse functors may adjoin more action, as in the free iteration $\Phi_!(X)$ of initial data on X, or the chaotic observation $\Phi_*(X)$ of quantities in X, where Φ strips some of the 'activity' from objects Y in its domain. When Φ is instead the full inclusion of constant attributes into variable ones, then the reverse functors effect an 'averaging', as in the existential quantification $\Phi_!(X)$ and the universal quantification $\Phi_*(X)$ of a predicate X. Exponentiation of spaces satisfies the exponential law that $\Phi_* = (\ )^L$ is right adjoint to $\Phi = (\ ) \times L$. Similarly, implication $L \Rightarrow (\ )$ of predicates is right adjoint to $(\ )\ \&L$.

For example, consider the process *points* that, for every reflexive graph, extracts its set of dots and forgets its arrows. By noting that this is a functor from the category of graphs to the category of abstract sets, we have the possibility to investigate whether it has adjoints:

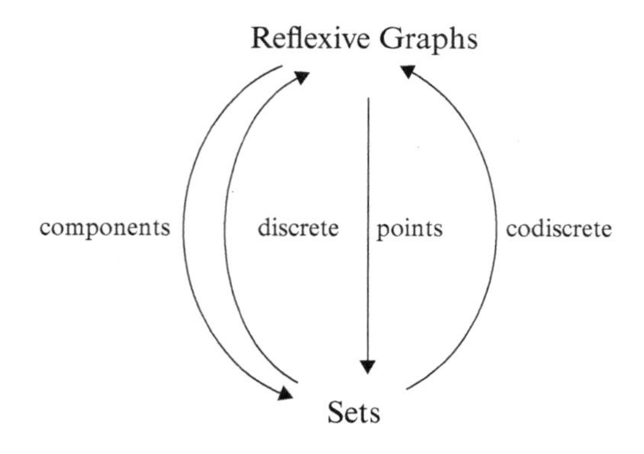

It turns out that the left adjoint to *points* constructs, for every given set S, the graph that has S as dots, but no arrows except for the trivial loop at each dot; this is often called a 'discrete' graph. The right adjoint instead constructs the graph having S as dots and exactly one connecting arrow between every ordered pair of dots, sometimes called a 'codiscrete' graph. The united opposites, discrete and codiscrete, can play a significant role in the organized investigation of the more general graphs of interest. But from the adjointness perspective, the discrete functor itself is not really as trivial as it may seem, because it has itself a further left adjoint, which is uniquely determined as the construction of the set of connected components of any graph, a non-trivial ingredient in the investigation of the form of a graph. The universal mapping property of adjointness in this case just means that a graph map from any graph X to a discrete graph is determined by a map on the set $F(X)$ of names for the connected components of X.

It is a general theorem of Kan that left adjoint functors preserve sums (coproducts) and that right adjoint functors preserve products. The laws of exponents and the resulting distributivity are examples of such preservation (see Article V). But in particular cases we can investigate whether, for example, a left adjoint functor F of interest might also preserve products in the sense that the natural map

$$F(X\times Y)\longrightarrow F(X)\times F(Y)$$

(induced by functoriality) is an isomorphism in the codomain category.

In the case of the components functor F, the above special feature of product preservation holds for it on the category of reflexive graphs, as well as on many other categories of a geometric character. Hurewicz made use of the product-preserving property of the connected components functor, in order to define a new category with the same objects, but with its map-sets $[X, Y]$ defined as the components of the map spaces by

$$[X, Y] = F(Y^X).$$

The study of the higher connectivity of a space Y involves not only the question of whether Y itself is connected, but also the connectedness of various map spaces Y^A and the relations $F(Y^B)\longrightarrow F(Y^A)$ between their sets of connected components that are induced by maps $A\longrightarrow B$. For example, the circle A, which is the boundary of a disk D, shares with the disk the property of being connected, i.e. $F(A) = F(D) = 1$; but whereas D^A is also connected, by contrast $F(A^A)$ has infinitely many elements, as studied by H. Poincaré and L. E. J. Brouwer before 1920. That first step in higher connectivity was extended, using similar properties of the contrast between a ball B and its boundary sphere A, to analyze higher connectivity of general spaces Y via the reduced figures $[A, Y]$ of spherical shape.

By contrast, for the category of irreflexive graphs there is also a components functor F, but it does not preserve products; the example of the non-trivial arrow X shows that (we saw in Session 23 that $F(X^2) = 3$, whereas $F(X)^2 = 1$). This irreflexive components functor F is also a left adjoint. Its right adjoint notion of 'discrete' assigns to each set S the graph with S dots and one loop at each. The right adjoint of the right adjoint in this case is not the 'dots' functor, but rather extracts from each irreflexive graph the set of *loops*.

Because *loops* has no right adjoint, this adjoint sequence stops with three. Instead of extracting *loops*, we could extract the set of *dots* or the set of *arrows*, and find that each of those two functors has a right adjoint and a left adjoint, but that those have no further adjoints.

In order to verify the above assertions, one needs to know the precise conditions that characterize adjoint functors, so we begin with the definitions.

Suppose F and G are functors, with

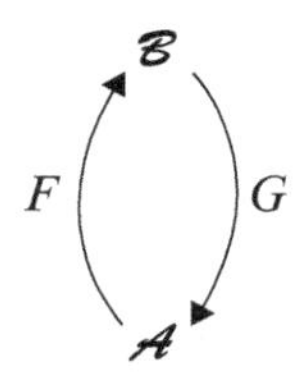

Definitions: An *adjunction* for the pair F, G is a natural correspondence

$$\frac{FA \longrightarrow B}{A \longrightarrow GB}$$

We say the adjunction makes F *left adjoint* to G, or equivalently, makes G *right adjoint* to F.

'Natural' above means 'compatible with maps'. That is, if $FA \xrightarrow{f} B$ corresponds to $A \xrightarrow{g} GB$, then

(1) for each $A' \xrightarrow{a} A$, the composite $FA' \xrightarrow{Fa} FA \xrightarrow{f} B$ corresponds to $A' \xrightarrow{a} A \xrightarrow{g} GB$, and
(2) for each $B \xrightarrow{b} B'$ the composite $FA \xrightarrow{f} B \xrightarrow{b} B'$ corresponds to $A \xrightarrow{g} G\,B \xrightarrow{Gb} GB'$.

It follows that such a natural correspondence is given by a universal mapping property: Given A, there is an adjunction unit $\eta_A : A \longrightarrow GFA$ such that any function $f : A \longrightarrow GB$ (thus whose type is given as a value of G) depends on η_A, and uniquely so. That is, in

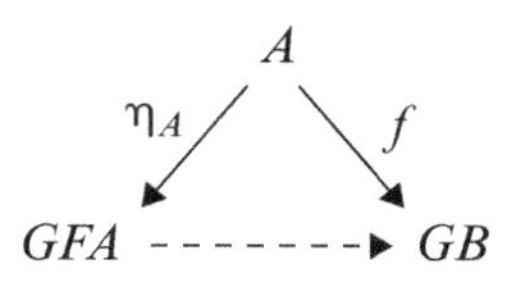

there is only one t for which $G(t)$ implements the dependence in the sense that $f = G(t)\eta_A$. This t is then the correspondent above the bar in the schematic picture. Then the family $\eta_A : A \longrightarrow GFA$ of adjunction units actually constitutes a natural transformation η from the identity functor on $\mathcal{A}$ to GF.

Dually, there is an adjunction co-unit $\varepsilon_B : FGB \longrightarrow B$ such that every figure b, whose shape is given as a value of F, is in ε_B, and in a unique way. That is, in

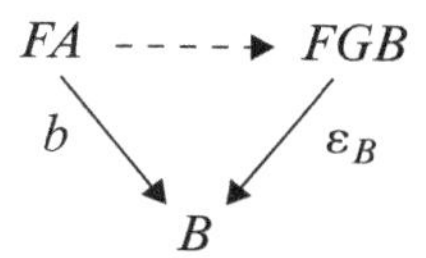

there is just one $s : A \longrightarrow GB$ such that $F(s)$ demonstrates the incidence in the sense that $b = \varepsilon_B F(s)$. Again, the adjunction co-unit is a natural transformation, but from the composite FG to the identity functor on the category $\mathcal{B}$. Essentially the same proof that shows the uniqueness of terminal objects also shows that 'the' left or right adjoint of a given functor is unique up to a unique invertible natural transformation. In fact, there is a category $\mathcal{M}ap(F, B)$ in which ε_B is terminal.

We sometimes denote the left adjoint of a functor Φ, if any, by $\Phi_!$, and the right adjoint of Φ by Φ_*, so that we have the natural correspondences

$$\frac{\Phi_! X \longrightarrow Y}{X \longrightarrow \Phi Y} \qquad\qquad \frac{Y \longrightarrow \Phi_* X}{\Phi Y \longrightarrow X}$$

Sometimes we use the symbol Φ^* for the functor Φ itself, emphasizing that the three functors together describe a single complex relation between the two categories.

For example, a second relation between reflexive graphs and sets is based on the functor 'arrows'. The reader can verify that this functor has both left and right adjoints; the right adjoint assigns to every set S a specific reflexive graph with S dots and S^3 arrows, sometimes called a co-free reflexive graph.

Instead of graphs versus sets, consider the relation between dynamical systems and sets, given by the functor Φ that remembers the state space, but forgets the action. The left-adjoint $\Phi_!$ of this functor gives rise to the whole story of natural numbers and recursion, because the natural one-to-one correspondence in this case is that between functions defined by recursion and the recursion data that defines them. On the other hand, its right adjoint Φ_* produces the chaotic dynamical systems (see secton 4 of Article V).

The other basic adjointness relating dynamical systems and their background sets of states involves the 'trivial' dynamics $\Psi_!$ in which no state moves; that $\Psi_!$ is the left adjoint to the important functor Ψ which extracts the set $\Psi(X)$ of fixed points (or equilibrium states) from each dynamical system X. It is easy to verify that this functor Ψ has no right adjoint, essentially because the fundamental shape of dynamical figures, the natural numbers, has no fixed points. But consider a somewhat richer category $\mathcal{A}$ of 'augmented dynamical systems', where there is not only a dynamical action for each time n, but moreover, an action for time $= \infty$, which corresponds to the ultimate equilibrium or 'destiny' that a state may have; this additional ideal time satisfies the equation $n + \infty = \infty$. The dynamics of every object in this category satisfies that equation as a further law. Then one can show that the right adjoint action does exist.

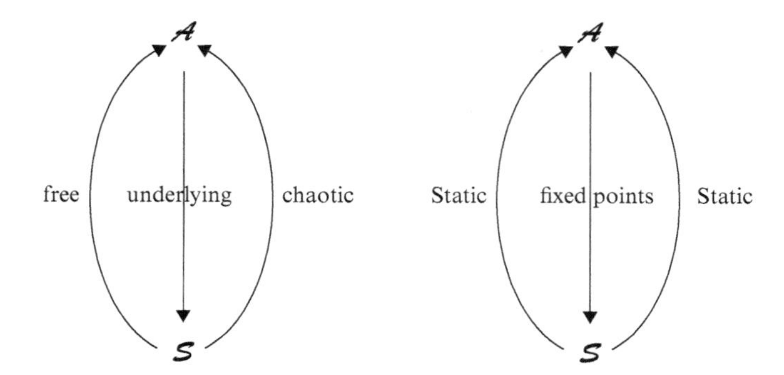

The 'underlying' functor has a left adjoint, assigning the free dynamical system $N_\infty \times (\ )$, and a right adjoint assigning the chaotic dynamical system $(\ )^{N_\infty}$. The 'fixed point' functor has as its left adjoint the inclusion of discrete (or static) systems; its right adjoint happens to be, exceptionally, the same functor *static*.

Certain augmented dynamical systems are much used under the name of *Newton's method*. The mathematical struggle of solving equations, in one-dimensional, finite-dimensional, and even infinite-dimensional calculus treats a given smooth function f from X to to Y, and a given possible output y. One wishes to find an x for which $f(x) = y$; that is, to partially 'invert' f, even though there is no known section for f. Newton's method for this problem uses the new map

$$\varphi(x) = x + (f'(x))^{-1}(y - f(x))$$

where f' is the derivative of f in the sense of calculus. Often one can find a 'small' space A in X such that the formula φ defines an endomap of A and such that there is just one point x_∞ in A for which $f(x_\infty) = y$. Since clearly x_∞ is a fixed point of φ, we see that A, φ, x_∞ defines an action of N_∞ i.e. an augmented dynamical system. The purpose of the whole construction derives from the fact that often the sequence a, $\varphi(a)$, $\varphi(\varphi(a))$, . . . 'converges' to the solution x_∞ from any starting point a in A. This is often the most practical way to find approximations to x_∞. (Unlike the simple case of a quadratic polynomial f, usually no explicit formula for x_∞ is known.) To define what it means for A to be 'small' and for the sequence above to 'converge' to x_∞, requires an environment $\mathcal{B}$ somewhat richer than abstract sets (such $\mathcal{B}$ are studied in topology and analysis).

The study of reflexive graphs and generalized graphs, of dynamical systems and augmented dynamical systems, and many other important examples can start from a given (finite or small) category $\boldsymbol{E}$, a chosen object D in $\boldsymbol{E}$, and a background topos $\mathcal{B}$ (for example finite or small abstract sets). Then if $\mathcal{C}$ is the topos of actions of $\boldsymbol{E}$ on spaces in B there are six functors connecting $\mathcal{B}$ and $\mathcal{C}$, all determined by adjointness from the functors D^* = 'underlying space of states of D', and P^* = 'trivial action'. All these are useful for analyzing and comparing the actions of interest.

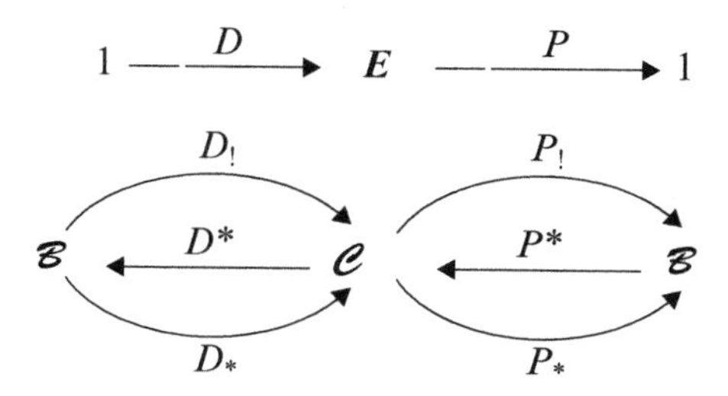

Properties of the small category $\boldsymbol{E}$ and the chosen object D determine answers to questions like:

(a) do the six collapse to fewer (as in the case of augmented dynamical systems $P_* = P_!$)?

(b) do any of the six have additional adjoints (such as the codiscrete functor $P^!$ for reflexive graphs)?

(c) does some left adjoint (such as $P_!$ = the connected components functor for reflexive graphs) preserve products?

(d) what do the functors do to the truth value space? (For example, consider the truth-value graph (Ω): for the fixed-point set we have $P_*(\Omega) = 2$, but for its underlying set we have $D^*(\Omega) = 5$.

In turn these answers (even if simple) will assist in the analysis and measurement of the more intricate $\boldsymbol{E}$-actions of scientific interest.

APPENDIX III

The emergence of category theory within mathematics

The unification of mathematics is an important strategy for learning, developing, and using mathematics. This unification proceeds from much detailed work that is punctuated by occasional qualitative leaps of summation. The 1945 publication by Samuel Eilenberg and Saunders Mac Lane of their theory of categories, functors, and natural transformations, was such a qualitative leap. It was also an indispensable prerequisite for a further leap, the 1958 publication by Daniel Kan of the theory of adjoint functors. The application of algebra to geometry had forced Eilenberg and Mac Lane to create their general theory; geometric methods developed by Alexander Grothendieck on the basis of that general theory were used 50 years later in the Andrew Wiles proof of Fermat's Last Theorem and in many other parts of algebra.

In the 1940s, the application which had given rise to the Eilenberg and Mac Lane summation, namely the study of qualitative forms of space in algebraic topology, began to be worked out by Eilenberg & Steenrod and others, and this development still continues in this century.

In the 1950s Mac Lane categorically characterized linear algebra; Yoneda showed that maps in any category can be represented as natural transformations; and Grothendieck made profound applications to the continuously-variable linear algebra which arises in complex analysis.

Rapidly, these and other advances permitted the organic incorporation, into one new theory, of the ideas of earlier giants among whom are the following:

Hermann Grassmann (1840s) verified the Leibniz conjecture that geometry is a form of algebra, showing that the geometric figures themselves are algebraic entities, because they are subject to definite operations (such as taking the midpoint of the segment connecting any two given points) and their relevant properties are determined by equations involving such operations.

Richard Dedekind (1870s) mathematically characterized the operations of recursion and induction using infinite intersections of families of sets, and cautiously considered sets of rational approximations as points in a model of the continuum. These advances made much use of *preorders*, now defined as categories in which there is at most one map connecting any given ordered pair of objects. He envisaged number theory and algebraic geometry as one subject; to achieve that required not only a theory $\mathcal{A}$ that could be interpreted both covariantly and contravariantly, but also required the abstract sets that had been made explicit by his friend Georg Cantor, to serve as an initial background $\mathcal{B}$ in which such $\mathcal{A}$ could be realized. His work inspired the work of Emmy Noether.

Felix Klein (1870s) classified geometrical objects using *groupoids*, now defined as categories in which all maps are isomorphisms.

Vito Volterra (1880s) recognized that the *elements* in a space X are figures $A \longrightarrow X$ of various given shapes A (not only punctiform $1 \longrightarrow X$) including smooth curved lines $L \longrightarrow X$. He emphasized that, along with functions $X \longrightarrow T$ on a space X, mathematics is concerned with functions $X^A \longrightarrow T$ on the space of A-shape figures in X; such functions were later called *functionals*. For the study of smooth functionals (in the calculus of variations) he analyzed figures in X^A, for example lines, as $(A \times L)$-shaped figures in X, i.e. $A \times L \longrightarrow X$.

Emmy Noether (1920s) advanced the application of general algebraic methods in geometry and physics.

Witold Hurewicz (1930s & 1940s) captured the higher connectivity of X using the connectivity of spaces X^A and demanded that a reasonable category of spaces should contain (along with X and A) a space X^A satisfying the *exponential law*

$$(X^A)^L \cong X^{A \times L}$$

that objectifies the above analysis.

George Mackey (1940s and 1950s) devised the concept of Mackey convergence in bornological vector spaces and isolated the essence of the duality functor in functional analysis. He was a leader in applying 20th century mathematics to 20th century physics.

José Sebastião e Silva (1940s and 1950s) recognized the need for a very general theory of homomorphisms in connection with his pathbreaking work that was among the inspirations for Grothendieck's work in functional analysis. The Nazi occupation of Rome during a crucial period delayed contact between Silva and his contemporaries Mackey, Eilenberg, and Mac Lane.

****

The groupoids of Klein and the preorders of Dedekind remain useful, but many important categories (such as the categories of directed graphs and of dynamical systems) contain non-invertible maps as well as parallel maps; in fact their maps include with equal status

- figures in a space,
- functions on a space,
- projections from product spaces,
- transformations of one space into another
- inclusions of subspaces,
- functionals.

All these kinds of maps are not merely accumulated into one large set; rather in their category we recognize their totality organized as one system sorted by precise *domain and codomain* relations, which are symbolized by arrow diagrams.

In the 1950's and 1960's, the use of categorical instruments by Grothendieck, Isbell, Kan, Yoneda, and others began to make explicit and generally applicable the insights of their predecessors and of themselves. The geometrical roles of figures, of map spaces, and of functor categories began to reveal a rational guide to the construction of needed concepts; thus the awkward 19th century attempt, by Frege, to subsume concepts as properties was superceded. At the same time new insights into logic itself were obtained:

(1) Fregean quantifiers are special Kan adjoints to the more basic substitution functionals;

(2) the models of logical theories are parameterized by classifier toposes;

(3) the rigidly hierarchical 'sets', while serving as a useful calculational tool in certain foundational investigations, can be replaced as a framework for mathematical practice by the relationship between Cantorian abstract sets and cohesively variable spaces.

At the beginning of the 21st century, the various fields of mathematics such as algebraic geometry, functional analysis, combinatorics, etcetera, have the external appearance of esoteric specialties. However, the categorical instruments and insights already obtained give promise of still further simplifications and unifications which will form an important component of the effort to make the mathematical sciences useable to people who will need them.

APPENDIX IV

Annotated Bibliography

Undergraduate texts:

J. Bell, *A Primer of Infinitesimal Analysis*, Cambridge University Press 1998.
Using some basic category theory, the traditional use of nilpotent infinitesimals in the classical analysis and geometry of Euler, Lie, and Cartan is placed on a rigorous footing and applied to numerous traditional calculations of central importance in elementary geometry and engineering. Like *Conceptual Mathematics*, this book could serve as a detailed guide to the construction of a course at the beginning level.

M. La Palme Reyes, G. E. Reyes, H. Zolfaghari *Generic Figures and their Glueings*, Polimetrica, 2004.
Several examples of 'presheaf' toposes, with particular properties of map spaces and truth value spaces, are discussed in detail, continuing the elaboration of our present Part V and Appendix II beyond the simple graphs and dynamical systems, with 'windows' into various mathematical applications.

R. Lavendhomme, *Basic Concepts of Synthetic Differential Geometry*, Kluwer Academic Publishers, 1996.
This was the first text that introduced differential geometry synthetically, using the categorical foundation of that subject as developed by Kock and others.

F.W. Lawvere, R. Rosebrugh, *Sets for Mathematics,* Cambridge University Press, 2003.
This text explains on a higher level many of the topics from Conceptual Mathematics, with more advanced examples from set theory. Axioms for the category of sets serve as a foundation for the mathematical uses of set theory as a background for topological, algebraic, and analytical structures, their manipulation, and their description by means of logic. The appendices briefly explain many of the topics from mathematical fields which gave rise to category theory and to which it is applied.

Advanced texts:

S. Mac Lane, *Categories for the Working Mathematician*, Springer Verlag New York, 1971.

This classic text uses examples such as rings, modules, and topological spaces, to illustrate the uses of categories, and develops in detail such topics as Kan extensions and homology in abelian categories.

S. Mac Lane, I. Moerdijk, *Sheaves in Geometry and Logic, a first introduction to Topos Theory*, Springer Verlag New York, 1992.

This text explains clearly the powerful method of presheaf categories, the logical significance of classifying toposes, and some of the more refined theorems of the theory of toposes.

P. Johnstone, *Sketches of an Elephant: a Topos Compendium*, Oxford University Press, 2002.

These rich and comprehensive volumes include much recent research not easily available elsewhere, and contain detailed structure theorems for toposes as well as their application to geometry and model theory.

A. Kock, *Synthetic Differential Geometry*, second edition, London Mathematical Society Lecture Note Series 333, Cambridge University Press, 2006.

The traditional study of multi-dimensional calculus as used in geometry and mechanics required the existence and basic properties of general map spaces, as well as the objective representation of infinitesimals, but these were lacking in the available formalization. The categorical developments expounded in this book finally provided such a foundation. Some of the many developments since the first (1981) edition are indicated here. Both editions present two approaches, an 'abstract' axiomatic presentation expressing intuitive properties for direct use, and a construction of models for the axioms using topos theory.

Basic early articles:

S. Eilenberg, S. Mac Lane, General Theory of Natural Equivalences, Transactions of the American Mathematical Society, **58**, 1945, 231–294

Eilenberg initially expected that this paper, though important, would be the last one necessary in its field, but that proved to be wrong. It strikingly foresees some of the later developments, and is still interesting to read.

A. Grothendieck, Sur quelques points d'algèbre homologique, Tohoku Math. Journal **9**, 1957, 119–221.

This pivotal work applied a branch of linear algebra, known as homological algebra, to the sheaves of holomorphic functions and differential forms that arise in complex analysis, and thus prepared the basis for contemporary algebraic geometry.

D. Kan, Adjoint Functors, Transactions of the AMS, **87**, 1958, 294–329.
Kan's continuing work on the combinatorial foundations of algebraic topology led him to this work which already contains uniqueness theorems and many typical applications. Both the general continuity properties and the particular examples now known as Kan extensions are explicitly clarified. The fact that map spaces are uniquely defined in the categories that were later called 'presheaf toposes' (such as the categories of graphs and dynamical systems) was established in this work 50 years ago.

F.W. Lawvere, Functorial Semantics of Algebraic Theories, Proceedings of the National Academy of Sciences USA, **50**, 1963, 869–872. Full version available on-line in the Reprints section of the journal Theory and Applications of Categories (www.tac.mta.ca).
Here it was shown that not only the concrete aspect, but also the abstract theoretical aspect of an algebraic notion, such as 'Lie algebra' or 'lattice', can be objectified as a category. The categories arising in universal algebra could therefore be characterized in their particularity. A great number of methods of construction of algebras were shown to be instances of a refined form of Kan extension, and a method for extracting the algebraic aspect (similar to cohomology operations) of very general constructions was expressed by a large-scale adjointness.

F.W. Lawvere, Adjointness in Foundations, Dialectica **23**, 1969, 281–296. Available on-line in the Reprints section of the journal Theory and Applications of Categories (www.tac.mta.ca).
All the axioms of intuitionistic higher-order number theory, as well as the semantical relation between theory and example, were shown to be instances of adjointness.

Index

Definitions appear in **bold** type

abbildung 245
absolute value 34, 140, 188
accessibility 138
  as positive property 173
action **218f**, 303
adjoint functors **372ff**
adjunction **374**
algebra of functions **370**
analysis of sound wave 106
arithmetic of objects 327
arrow in graph **141**
  generic arrow **215**
associative law
  for composition **17**, 21
  sum, product (objects) 220ff, 281ff
  versus commutative law 25
automaton (= dynamo syst.) **137**, 303
automorphism **55**
  in sets = permutation 57, 138, 155, 180
  category of **138**

balls, spheres 120ff
Banach's fixed point theorem 121
base point 216, 295
Bell, J. 381
belongs to **335**
binary operation 64, **218**, **302**
bird watcher 105
bookkeeping **21**, 35
Boole, G. 355
Brouwer 120, 309
  fixed point theorems 120ff
  retraction theorems 122

cancellation laws 43, 44, 59
Cantor, Georg 106,291, 304ff
  Cantor-Bernstein theorem 106
  diagonal theorem 304, 316

cartesian closed category 315, 322
cartesian coordinates 42, 87
category **17**, 21
  of dynamical systems **137**
  = of endomaps **136**
  of graphs **141**, 156
  of parts **344, 348**
  of parts of X 348
  of permutations **57**
  of pointed sets **216**, 223, 295ff
  of smooth spaces 135
  of topological spaces 135
category of categories **369**
category of figures **370**
category of parts of a terminal object 349
center of mass 324
Chad's formula 75, 117
characteristic map **344**
chaotic **317**
  truth values 347
Chinese restaurant 76
choice problem **45**, 71
  examples 46, 71
  section as choice of representatives 51, 100, 117
clan 163
codiscrete **366**
cofigure (dual of figure) 272
cograph **294**
cohesive 120, 135
commuting diagram 50, **201**
components functor 358, 359
composition of maps 16, 114
  and combining concepts 129
computer science 103, 185
concept, coconcept 276, 284
connected **358**, **362**
configuration 318

congressional representatives 51, 95
congruent figures 67
conic section 42
connected component **359**
constant map **71**
constant monoid **366**
constructivist idealization 306ff
continuous map 6, 120
contraction map 121
contrapositive **124**
convergence to equilibrium 138
coproduct (= sum)
coordinate pair 42, 87
  system 86
counterexample 115
cross-section 93, 105
crystallography 180
cycle 140, 176, 183, 187, 271
  multiplication of 244

Dedekind, Richard 378
determination
  examples **45**, 47
  or extension problem 45, 68
Descartes, R. 41, 322
diagonal argument, Cantor's 304, 316
diagram
  commuting **201**
  external **15**
  internal, of a set **13**
    of a map of sets **14**, 15
  of shape G 149ff, **200**ff
differential calculus 324
directed graph (*see* graph)
discrete **359**
disk 8, 121, 236
disjunction **354**
distinguished point 295
distributive category **223**, **276**ff, 295
distributive law **223**
  general **278**
division problems **43**
domain of a map **14**
dot, in graph **141**
  naked **215**
dual 215, 284
dynamical system,
  discrete **137**, 161, 303
    equivariant map of 182
    objectifying properties of 175

Eilenberg, Samuel 3, 382
Einstein, Albert 309
electrical engineering 227
empty set, maps to and from 30
  as initial set 254
endomap **15**
  automorphisms **55**
    category of 138f
  category of 136ff
  idempotent 99ff, 118
  internal diagram of 15
  involution **139**
epimorphism **53**, 59
equality of maps, test for
  in dyn. syst. 215, 246f
  in graphs 215, 250f
  in sets 23, 115
equalizer **292**, 337
equilibrium state 214
equinumerous = isomorphic 41
equivariant (map of dyn. syst.) 182
Euclid's category 67
Euclidean algorithm 102
evaluation, as composition 19
  map **313**
Exemplifying
  = sampling, parameterizing 83
exponential object = map obj. 313,
    320ff
exponents, laws of 324ff
extension problem 45
  = determination problem
external diagram (*see* diagram)
eye of the storm 130

factoring 102
faithful **318**
family 82
  of maps 303
family tree 162
fiber, fibering 82
Fibonacci (Leonardo of Pisa) 318
figure **83**
  incidence of 344ff
  shape of **83**
  of shape **1** (*see* point)
finite sets, category of 13
fixed point 117, 137
  and diagonal theorem 303
  as point of dyn. syst. 214, 232f

formulas and rules of proof 306
full (insertion) 138, 146
fraction symbol 83, 102
free category 200, 203
function (= map) 14, 22
  space (= map object) 313
functor 167
functor categories **369**

Galileo 3, 47, 106, 120, 199, 216, 236, 257, 308, 322
gender 162, 181
genealogy 162f
generator 183, 247
Gödel, Kurt 306ff
  numbering 307
graph
  as diagram shape 149,200
  irreflexive **141**, 189, 196ff
  reflexive 145, 192
graph of a map **293**
Grassman, Hermann 378
gravity 309
Grothendieck, A. 352, 383
group **362**

hair 183, 187
Hamilton, William Rowan 309
helix 240
homeomorphic 67
Hooke, Robert 129
Hurewicz, Witold 379

idempotent endomap **54**, 108, 117f, 187
  category of 138
  from a retract 54, 100
  number of (in sets) 20, 35
  splitting of **102**
idempotent object **289**
identity
  laws **17**, 21, 166, 225
  map 15, 21
  matrix **279**
image of a map **335**
implication **354**
incidence relations 245, 249ff, 258
inclusion map 122, **340**, 344
  = injective map
incompleteness theorem 106, 306ff
inequality 99
infinite sets 55, 106, 108
initial object **215**, 216, 254, 280
  in other categories 216, 280
  in sets 30, 216
  uniqueness 215
injection maps for sum **222**, 266ff
injective map **52**, 59, 146ff, 267, **340**
integers 140, 187
internal diagram 14
intersection 353f
  of subobjects **354**
inverse of a map **40**
  uniqueness 42,62
inverse image **336**
invertible map (isomorphism) **40**
  endomap (automorphism) 55, 138, 155
involution **118**, 139, 187
irreflexive graph (*see* graph)
is in **335**
isomorphic **40**
isomorphism **40**, 61ff
  as coordinate system 86ff
  Descartes' example 42, 87
  reflexive, symmetric, transitive 41
iteration 179

Jacobi, Karl 309
Johnstone, P. 382

Kan, D. 383
Klein, Felix 180, 379
knowledge 84
Kock, A. 382

labeling (= sorting)
  laws of categories 21
  of exponentiation 324ff
  (*see also* identity, associative, commutative, distributive)
La Palme, M. 381
Lavendhomme, R. 381
Lawvere, F.W. 381, 383
left adjoint **374**
Leibniz, Gottfried Wilhelm 129
linear category 279ff
locally in 336
logic 339ff, 344ff
  and truth 339ff
  rules of 354f

logical operations 180, 353f
logicians 306ff
loop, as point 232f

Mac Lane, Saunders 3, 382
Mackey, George 379
map, of sets 14, 22
  in category 17, 21
mapification of concepts 127
map object (exponential) 313ff, 320ff
  definition vs product 330
  and diagonal argument 316
  in graphs 331ff
  and laws of exponents 314f, 324ff
  points of 323
  in sets 331
  transformation (for motion) 323f
maps, number of
  in dynamical syst. 182
  in sets 33
map space for M-actions 361
map space in $S^{M^{op}}$ 361
mathematical universe
  category as 3, 17
matrilineal 181
matrix
  identity matrix **279**
  multiplication **279**ff
modal operators 347
modeling, simulation of a theory 182
modus ponens rule of inference 354
Moerdijk, I. 382
momentum 318
monic map (= monomorphism) **340**
monoid **166**ff
monomorphism **52**, 59, 340
  test for in: sets 340; graphs 340
motion 3, 216,236, 320
  of bodies in space 323ff
  periodic 106
  state of 318
  uniform 120
  of wind (or fluid) 130ff
multigraph, directed irreflexive
    {= graph)
multiplication of objects (= product) of
  courses 7
  cycles 244
  disc and segment 8
  dynamical systems 239ff
  plane and line 4
  sentences 8
multiplication of matrices (*see* matrix)

naming (as map)
  in dynamical systems 176ff
  in sets 83
navigation, terrestial and celestial 309
negation **355**
negative of object 287
negative properties 173, 176
Newton, Isaac 199, 309
Newton's method 376
Noether, Emmy 379
non-distributive categories 295f
non-singular map (= monomorphism) **340**
numbers, natural
  analog, in graphs 267
    in distributive category 327
  isomorphism classes of sets 39ff
  monoid of 167ff
  to represent states of dyne syst. 177f
numbers, rational 83, 102

objectification
  of concepts as objects, maps 127
  in dynamical systems 175ff
  in the subjective 181
objective
  contained in subjective 84, 181ff
  in philosophy 84
observable **317**
  chaotic **317**
one-to-one map 340
operation, unary, binary, etc. **302**
operator **14**
origin or base point **295**

paradox 306
parameterizing **83**
  of maps 303f, 313
  of maps, weakly **306**
parity (even vs odd) 66, 174
partitioning **82**
parts of an object 339ff
  category of **344**ff
permutation, set-automorphism 56ff
  category of 57, **138**ff
philosophical algebra 129
philosophy **84**

Pick's formula 47
plot 86
pointed sets (cat. of) **216**, 223, 295ff
point (= map from terminal object)
  distinguished **295**
  in general **214**
  in graphs, dynamical systems 214
  of map object 314, 323
  in part 343
  of product 217, 258
  in sets **19**
  of sum 222
points functor **359**
polygonal figure (Euclid's cat.) 67
positive properties 170ff
preorder **350**
presentation, of dyne syst. 182ff
  of graph 253
preserve distinctness 106
  see injective
probe, figure as, in dyne system 180
product of objects **216**, 236ff
  projections 217
  uniqueness of 217, 255ff, 263ff
  points of 217, 258
  (*see* also multiplication)
projection maps (*see* product)

quadratic polynomial 292
QED (quod erat demonstrandum)
quiz 108, 116

rational numbers 83, 102
reality 84
reciprocal versus inverse 61
reduced fraction 102
regular representation 364
relations (in presentation) 183
retraction (for map) **49**, 59, 108, 117
  as case of determination 49, 59, 73
  and injectivity 52, 59
  is epimorphism 59, 248
  number in sets (Danilo) 106, 117
  (*see* section, retract, idempotent)
Retract **99**
  as comparison 100f
  and idempotent 100ff
Reyes, G. & M. 381
right adjoint **374**
Rosebrugh, R. 381
Russell, Bertrand 306ff

sampling **82**
Sebastião e Silva, José 379
section (for a map) **49**, 72ff
  as case of choice 50, 72
  of a composite 54
  and epimorphism 53, 59
  is monomorphism 52, 59
  number in sets (Chad) 75, 94, 117
  and stacking, sorting 74
  (*see* also retraction, idempotent)
separating **215**
  (*see* also equality of maps)
shadow
  as map 4, 236
  vs sharper image 136f
shape
  (graph) domain of diagram 149, **200**ff
  (object) domain of figure **83**
shoes and socks rule
  for inverse of composite 55
singleton set
  and constant map 71
  as domain of point **19**
  as terminal object 29, 225
  (*see* terminal object, point)
singular figure 245
small monoid **360**
smooth categories 120, 135, 323ff
sorting **81**, 103, 104
  gender as 162
  in graphs 270f
sorts (as codomain of map) 81ff
source, target **141**, 150, 156, 189, **251**
space
  motion in 4ff, 323f
  as product 4ff
  travel 199
spheres and balls 120ff
splitting of idempotent **102**, 106, 117
stable conditions 335
  belongs to **335**
  exact **337**
  is in **335**
  locally in **335**
state (in dynamical system) 137
  and configuration 318
  naming of 177ff

structure
  in abstract sets 136
  types of 149ff
structure-preserving map 136, 152ff, 175f
subcategories 138, 143
subgraph 341ff (*see* subobject)
subjective contained in objective 84, 86
  in dynamical systems 180ff
subobject 339ff
subobject classifier **337**
  (*see* truth value object)
substituting 336
successor map on natural numbers
  as dynamical system 177ff, **247**
  vs truth value object 346
sum of objects **222**f, 265ff
  distributive law **222**, 275ff, 315
  as dual of product 260
  injections **222**
  uniqueness 266
supermarket 71
surjective for maps from T 51, 59

target (*see* source)
Tarski, Alfred 306ff
terminal object **213**, 225ff
  in dynamical systems 214, 226f
    in sets 213, 214
  in graphs 214, 227f
  point as map from **214**
  uniqueness of 213
time (as object) 4, 217, 323
topological spaces 120, 135
topology 67
topos 338, **352**
transformation of map objects 323
  lambda-calculus 319
true 337
truth 306ff, 342ff
  level of 342ff
  truth value object 306ff, **344**ff
  in dynamical systems 346
    chaotic 347
  in graphs 344f
  in sets 343f
truth value object **337**
truth value object for M-Actions **361**
truth value object in graphs 344ff
Turing, Alan 309
type of structure **149**ff

unary operation **302**
underlying configuration 318
uniqueness of
  initial object 215
  inverse 42, 54, 62
  product 239, **263**
  sum 266
  terminal object 213

velocity 324
vertices (in Pick's formula) 47
violin string 106
Volterra, Vito 379
wishful thinking 328

zero maps **279**
Zolfaghari, H. 381